SAP PRESS e-books

Print or e-book, Kindle or iPad, workplace or airplane: Choose where and how to read your SAP PRESS books! You can now get all our titles as e-books, too:

- By download and online access
- For all popular devices
- And, of course, DRM-free

Convinced? Then go to www.sap-press.com and get your e-book today.

Margin Analysis with SAP S/4HANA®

SAP PRESS is a joint initiative of SAP and Rheinwerk Publishing. The know-how offered by SAP specialists combined with the expertise of Rheinwerk Publishing offers the reader expert books in the field. SAP PRESS features first-hand information and expert advice, and provides useful skills for professional decision-making.

SAP PRESS offers a variety of books on technical and business-related topics for the SAP user. For further information, please visit our website: www.sap-press.com.

Stoil Jotev
Configuring SAP S/4HANA Finance (3rd Edition)
2025, 744 pages, hardcover and e-book
www.sap-press.com/5920

Tritschler, Walz, Rupp, Mucka
Financial Accounting with SAP S/4HANA: Business User Guide (2nd Edition)
2023, 571 pages, hardcover and e-book
www.sap-press.com/5698

Mehta, Aijaz, Parikh, Chattopadhyay
SAP S/4HANA Finance: An Introduction (2nd Edition)
2023, 394 pages, hardcover and e-book
www.sap-press.com/5606

Janet Salmon, Stefan Walz
Controlling with SAP S/4HANA: Business User Guide (2nd Edition)
2025, 697 pages, hardcover and e-book
www.sap-press.com/6063

Stefanos Pougkas
SAP S/4HANA Financial Accounting Certification Guide: Application Associate Exam (2nd Edition)
2021, 449 pages, hardcover and e-book
www.sap-press.com/5310

Kathrin Schmalzing

Margin Analysis with SAP S/4HANA®

Profitability Analysis for SAP®

Editor Rachel Gibson
Acquisitions Editor Emily Nicholls
Copyeditor Julie McNamee
Cover Design Graham Geary
Photo Credit Shutterstock: 1018123756/© Anna Kucherova
Layout Design Vera Brauner
Production Kelly O'Callaghan
Typesetting III-satz, Germany
Printed and bound in Canada, on paper from sustainable sources

ISBN 978-1-4932-2711-2
3rd edition 2025

© 2025 by:
Rheinwerk Publishing, Inc.
2 Heritage Drive, Suite 305
Quincy, MA 02171
USA
info@rheinwerk-publishing.com
+1.781.228.5070

Represented in the E.U. by:
Rheinwerk Verlag GmbH
Rheinwerkallee 4
53227 Bonn
Germany
service@rheinwerk-verlag.de
+49 (0) 228 42150-0

Library of Congress Cataloging-in-Publication Control Number: 2025030591

All rights reserved. Neither this publication nor any part of it may be copied or reproduced in any form or by any means or translated into another language, without the prior consent of Rheinwerk Publishing.

Rheinwerk Publishing makes no warranties or representations with respect to the content hereof and specifically disclaims any implied warranties of merchantability or fitness for any particular purpose. Rheinwerk Publishing assumes no responsibility for any errors that may appear in this publication.

"Rheinwerk Publishing" and the Rheinwerk Publishing logo are registered trademarks of Rheinwerk Verlag GmbH, Bonn, Germany. SAP PRESS is an imprint of Rheinwerk Verlag GmbH and Rheinwerk Publishing, Inc.

All screenshots and graphics reproduced in this book are subject to copyright © SAP SE, Dietmar-Hopp-Allee 16, 69190 Walldorf, Germany.

SAP, ABAP, ASAP, Concur Hipmunk, Duet, Duet Enterprise, ExpenseIt, SAP ActiveAttention, SAP Adaptive Server Enterprise, SAP Advantage Database Server, SAP ArchiveLink, SAP Ariba, SAP Business ByDesign, SAP Business Explorer (SAP BEx), SAP BusinessObjects, SAP BusinessObjects Explorer, SAP BusinessObjects Web Intelligence, SAP Business One, SAP Business Workflow, SAP BW/4HANA, SAP C/4HANA, SAP Concur, SAP Crystal Reports, SAP EarlyWatch, SAP Fieldglass, SAP Fiori, SAP Global Trade Services (SAP GTS), SAP GoingLive, SAP HANA, SAP Jam, SAP Leonardo, SAP Lumira, SAP MaxDB, SAP NetWeaver, SAP PartnerEdge, SAPPHIRE NOW, SAP PowerBuilder, SAP PowerDesigner, SAP R/2, SAP R/3, SAP Replication Server, SAP Roambi, SAP S/4HANA, SAP S/4HANA Cloud, SAP SQL Anywhere, SAP Strategic Enterprise Management (SAP SEM), SAP SuccessFactors, SAP Vora, TripIt, and Qualtrics are registered or unregistered trademarks of SAP SE, Walldorf, Germany.

All other products mentioned in this book are registered or unregistered trademarks of their respective companies.

No part of this book may be used or reproduced in any manner for the purpose of training artificial intelligence technologies or systems. In accordance with Article 4(3) of the Digital Single Market Directive 2019/790, Rheinwerk Publishing, Inc. expressly reserves this work from text and data mining.

Contents at a Glance

1 Introduction to Profitability Analysis 19
2 Configuring the Operating Concern and Basic Settings for Profitability Analysis 41
3 Configuring Characteristics 71
4 Value Flows of Margin Analysis 127
5 Costing-Based Profitability Analysis 209
6 Planning 285
7 Reporting 295

Contents

Introduction 13

1 Introduction to Profitability Analysis 19

1.1 Purpose of Profitability Analysis 19
1.2 Profitability Concepts 21
  1.2.1 Profitability Segments 21
  1.2.2 Characteristics 24
  1.2.3 Value Fields 26
1.3 Types of Profitability Analysis 27
  1.3.1 Margin Analysis (Account-Based Profitability Analysis) 27
  1.3.2 Costing-Based Profitability Analysis 30
  1.3.3 Comparison of Profitability Analysis Types 34
1.4 Technical Setup 36
  1.4.1 Technical Setup of Costing-Based Profitability Analysis ... 36
  1.4.2 Technical Setup of Margin Analysis 37
1.5 Summary 38

2 Configuring the Operating Concern and Basic Settings for Profitability Analysis 41

2.1 Creating the Operating Concern 41
  2.1.1 Assign a Controlling Area to the Operating Concern 41
  2.1.2 Create Operating Concern 42
  2.1.3 Maintain Further Settings for the Operating Concern 43
  2.1.4 Define Operating Concern Currency 45
  2.1.5 Assign Value Fields to Operating Concern 48
2.2 Activating Profitability Analysis 52
2.3 Currencies in the Operating Concern 54
2.4 Creating an Extension Ledger for Profitability Analysis 56
2.5 Creating Number Ranges for Profitability Analysis 59
2.6 Assigning Versions to the Operating Concern 62

7

2.7	Setting the Operating Concern	65
2.8	Transporting the Operating Concern	66
2.9	Summary	69

3   Configuring Characteristics   71

3.1	Characteristics		73
	3.1.1	Creating, Changing, and Updating Characteristics	73
	3.1.2	Creating a Characteristic from an SAP Table	76
	3.1.3	Creating Characteristics with Your Own Value Maintenance	80
	3.1.4	Creating Characteristics Without Value Maintenance	83
	3.1.5	Creating Characteristics with Reference to Existing Values	86
	3.1.6	Assigning Characteristics to the Operating Concern	88
	3.1.7	Maintaining Characteristic Values	91
3.2	Characteristics Derivations		92
	3.2.1	Creating Characteristic Derivations with Derivation Rules	92
	3.2.2	Creating Characteristic Derivations with Table Lookup	98
	3.2.3	Creating Characteristic Derivations with Moves	103
	3.2.4	Creating Characteristic Derivations to Clear Fields	109
	3.2.5	Creating Characteristic Derivations with Enhancements	112
	3.2.6	Sequence of Characteristic Derivation Rules	113
	3.2.7	Derivation Analysis	116
	3.2.8	Realignment of Characteristics	119
3.3	Summary		126

4   Value Flows of Margin Analysis   127

4.1	Predictive Accounting		128
	4.1.1	Predictive Accounting: Incoming Sales Order	128
	4.1.2	Predictive Accounting Commitment Management	138
4.2	Billing Data Transfer		141
	4.2.1	Price Determination in the Sales Order	141
	4.2.2	Create Sales Order	142

	4.2.3	Account Determination	145
	4.2.4	Activate Transfer of Statistical Sales Conditions	147
	4.2.5	Create Billing Document	147
	4.2.6	Review Accounting Documents	148
4.3		**Costs of Goods Manufactured in Margin Analysis**	150
4.4		**Splitting the Cost of Goods Sold**	151
4.5		**Variance Calculation**	159
	4.5.1	Production Order	161
	4.5.2	Define Variance Categories	164
	4.5.3	Create General Ledger Accounts for Variance Split	168
	4.5.4	Check Production Order	172
4.6		**Order Settlement**	178
4.7		**Overhead Allocation**	187
	4.7.1	Advantages of Universal Allocations	187
	4.7.2	Types of Cost Centers	188
	4.7.3	Create an Overhead Allocation Cycle	189
	4.7.4	Execute the Overhead Allocation Cycle	196
4.8		**Direct Account Assignment**	198
4.9		**Top-Down Distribution**	202
4.10		**Summary**	208

5   Costing-Based Profitability Analysis   209

5.1		**Value Fields**	212
5.2		**Quantity Fields**	216
5.3		**Assigning Value Fields and Quantity Fields to the Operating Concern**	218
5.4		**Transferring Incoming Sales Orders**	218
5.5		**Transferring Quantity Fields**	223
5.6		**Billing Data Transfer**	230
	5.6.1	Transferring Cost of Goods Sold	230
	5.6.2	Assigning Value Fields to Pricing Conditions	232
	5.6.3	Assigning Quantity Fields to Quantities	235
	5.6.4	Reviewing Profitability Document	238
	5.6.5	Resetting Value and Quantity Fields for Credit and Debit Memos	240

5.7	**Costs of Goods Manufactured**	246
5.8	**Cost of Goods Sold Transfer and Split**	249
	5.8.1 Defining Valuation Strategies	253
	5.8.2 Setting Up Valuations Using Material Cost Estimates	256
5.9	**Variance Calculation**	269
	5.9.1 Cost Center Types	269
	5.9.2 Variance Calculation and Settlement of Process Order	272
5.10	**Cost Center Assessment**	276
5.11	**Direct Account Assignment**	277
5.12	**Summary**	283

6 Planning — 285

6.1	**Planning in Margin Analysis**	286
6.2	**Sales Planning**	289
6.3	**Cost of Goods Sold Planning**	293
6.4	**Overhead Cost Planning**	294
6.5	**Summary**	294

7 Reporting — 295

7.1	**Profitability Analysis Reporting Basics**	295
7.2	**Predictive Accounting**	298
	7.2.1 Displaying Predictive Documents	298
	7.2.2 Displaying the Profit and Loss and Balance Sheet Report	299
	7.2.3 Monitoring Predictive Accounting	303
	7.2.4 Incoming Sales Order Report	305
7.3	**Showing Line Items for Margin Analysis**	312
7.4	**Profitability Analysis Reporting**	317
	7.4.1 Gross Margin	317
	7.4.2 Sales Accounting Overview	319
	7.4.3 Product Profitability with Production Variances	320
	7.4.4 Profit and Loss Statements	321

		7.4.5	Market Segments Actual	323
		7.4.6	Report Writer	324
7.5	Other Reporting Possibilities with Development			324
		7.5.1	ABAP Reports	324
		7.5.2	Core Data Services Views	325
		7.5.3	SAP Datasphere	326
		7.5.4	SAP Business Technology Platform	327
7.6	Summary			328

Appendices 329

A	Changes to the Data Model	331
B	Additional Sources of Information	333
C	The Author	335

Index	337

Introduction

With SAP S/4HANA Finance, numerous functions in SAP S/4HANA financial planning and analysis (FP&A, also known as SAP ERP Controlling) have been changed and modernized. *Profitability analysis* (also known as CO-PA) is one of them. You're probably familiar with the types of profitability analysis that were available in SAP ERP: account-based CO-PA and costing-based CO-PA. SAP revised the long-term strategy for profitability analysis. Costing-based CO-PA isn't being further developed and is only available in SAP S/4HANA (not SAP S/4HANA Cloud). Account-based CO-PA is only available in SAP ERP. The equivalent of account-based CO-PA in SAP S/4HANA is called margin analysis and provides many more functionalities than account-based CO-PA. Margin analysis is being further developed with every release, and we strongly recommend activating margin analysis in SAP S/4HANA. This book helps you understand and use the new functions to improve your analytics capabilities and make faster and better-informed business decisions.

Because nearly every process generates a financial accounting document and/or profitability analysis document, it's important that you understand the value flows in order to configure profitability analysis properly. This book first describes the processes in detail and then explains the configuration settings for profitability analysis to make sure you get the desired output and can analyze the process flows correctly.

SAP S/4HANA FP&A is modernizing the finance and analytics functionalities. We'll introduce you to the new functionalities that help you improve reporting with real-time information and predictive insights to leverage the full potential of SAP S/4HANA FP&A for your profitability analysis solution.

Target Group

This book is intended for key users who are familiar with CO in SAP ERP and/or the SAP S/4HANA FP&A module in the SAP system and know how to set up the configuration for account-based and/or costing-based profitability analysis. The available margin analysis functionalities are linked to your SAP S/4HANA release, whereas the costing-based profitability analysis functionalities aren't necessarily linked to any release.

Objective and Content

Margin analysis and costing-based profitability analysis

This book describes the new functions, latest innovations, and differences between margin analysis and costing-based profitability analysis. We divided this book into different parts with detailed end-to-end descriptions of process flows followed by the functions and innovations of SAP S/4HANA FP&A for all areas of profitability analysis.

> **System Requirements**
>
> The functions of costing-based profitability analysis haven't changed in SAP S/4HANA Finance. Numerous core functions aren't linked to releases, but the margin analysis functions are linked to SAP S/4HANA. The screenshots in this book were created on the latest SAP S/4HANA 2023 system.

Structure of the book

This book is divided into seven chapters. **Chapter 1** provides an overview of SAP S/4HANA and profitability analysis in general. **Chapter 2** and **Chapter 3** give you a detailed description of the organizational settings and master data that are essential for setting up costing-based profitability analysis and margin analysis. **Chapter 4** is dedicated to the functions of margin analysis, whereas **Chapter 5** is dedicated to the functions of costing-based profitability analysis. We start the chapter with an end-to-end process overview before describing the configuration settings. **Chapter 6** and **Chapter 7** give you an overview of the planning and reporting functionalities, respectively. The following is a detailed description of the individual chapters of the book:

- **Chapter 1** introduces profitability analysis in SAP S/4HANA.
- **Chapter 2** explains the organizational settings and basic settings required for the activation of profitability analysis in SAP S/4HANA. You have to implement the basic settings before you can start configuring the value flows.
- **Chapter 3** discusses the creation and derivation of characteristics, which are identical for costing-based profitability analysis and margin analysis.
- **Chapter 4** is dedicated to the value flow and configuration settings of margin analysis. The chapter also describes the end-to-end processes and how they are transferred to profitability analysis, as well as the necessary configuration to display the processes correctly in margin analysis.
- **Chapter 5** is set up similarly to Chapter 4 with a focus on the end-to-end processes and how they are transferred to costing-based profitability analysis. We describe the configuration settings in detail followed by an example in which we test the configuration settings.

- **Chapter 6** introduces SAP's planning strategy within SAP S/4HANA, which mainly affects account-based profitability analysis. Furthermore, it describes the traditional planning methods for costing-based profitability analysis.
- **Chapter 7** focuses on reporting functionalities. It introduces traditional reports for costing-based profitability analysis as well as new reporting options with SAP S/4HANA for margin analysis.

The highlighted boxes in this book contain information that is good to know and useful, but is outside the context. To help you immediately identify the type of information contained in the boxes, we have assigned symbols to each box:

Information boxes

> **Note**
> Boxes marked with this symbol contain information about additional topics or important content that you should note.

> **Tip**
> This icon refers to specifics that you should consider. It also warns about frequent errors or problems that can occur.

> **Example**
> Examples, which are highlighted with this icon, refer to real-life scenarios and illustrate the functions described.

Universal Parallel Accounting and the Future of Margin Analysis

SAP is introducing a new functionality in SAP Finance: Universal Parallel Accounting (UPA). It represents a significant leap forward in the evolution of SAP S/4HANA Finance. It's designed to address the growing complexity faced by global organizations in meeting multi-GAAP, multi-currency, and multi-fiscal year reporting requirements. By integrating ledger-specific financial processes into a unified framework, UPA enables organizations to streamline financial operations, reduce reconciliation efforts, and enhance the transparency and accuracy of reporting.

The Purpose and Motivation Behind UPA

Global enterprises operate under diverse financial regulations, valuation methods, and fiscal structures. Traditional SAP systems have long required extensive adjustments and reconciliations to meet these requirements. UPA addresses this by enabling parallel accounting views across ledgers, each with its own chart of accounts, currencies, and fiscal year variants. This allows companies to do the following:

- Achieve audit-ready financial statements across multiple accounting principles
- Reduce manual adjustments and closing efforts
- Support multidimensional reporting for legal, managerial, and group purposes simultaneously

Key Features of UPA

UPA is deeply integrated into the Universal Journal (table ACDOCA), ensuring consistent and harmonized postings across all financial processes. Key innovations include these:

- Ledger-specific allocations and settlements in overhead controlling
- Ledger-based standard cost estimates and intercompany profit eliminations in product costing
- Real-time posting and valuation in multiple ledgers during production accounting
- Enhanced asset accounting (AA) with support for different fiscal year variants
- Continued support for event-based revenue recognition and integration with group reporting through preparatory ledgers

Business Scenarios Enabled by UPA

UPA enables a variety of advanced business cases, such as the following:

- Parallel valuations for both local and group GAAP
- Support for alternative fiscal year variants, which is essential for countries such as India or Japan
- Group valuation views with real-time intercompany elimination of profits
- Seamless multicurrency reporting, including functional and group currencies beyond local requirements

Implications for Core Finance Processes

UPA introduces simultaneous valuation across multiple accounting principles, impacting core finance processes by requiring changes to how transactions are recorded, valued, and reported. The following describes how UPA is impacting the various areas in SAP Finance:

- **Controlling (CO):**
  - Allocations and settlements now post per ledger, enabling ledger-specific profitability insights.
  - Old transactions are removed, and SAP Fiori apps become the interface of choice.
- **Product costing:**
  - Both unconsolidated and group valuation views are supported, with ledger-specific cost estimates.
  - Intercompany transfers reflect eliminated cost of goods sold (COGS) and revenues for accurate group margins.
- **Production accounting:**
  - Real-time calculations of input, overhead, work in process (WIP), and variances for each ledger remove the need for month-end batch runs.
  - Event-based production accounting is key to unlocking this functionality.
- **Asset accounting (AA):**
  - Simplified data structures and ledger-level postings allow different depreciation rules and fiscal calendars to coexist.
  - Quantity-based postings are now journal entries, not part of the asset master.
- **Group reporting:**
  - A preparatory ledger captures group reporting dimensions at the time of journal entry posting, eliminating the need for separate consolidation adjustments.
- **Margin analysis:**
  - UPA fully integrates margin analysis (the successor to account-based CO-PA) into the Universal Journal (table ACDOCA). All profitability characteristics (e.g., product, customer, region) are embedded directly in the journal. As a result, no separate operating concern is required or used.
  - UPA simplifies the architecture and eliminates data replication or reconciliation between SAP S/4HANA Finance and margin analysis in SAP S/4HANA . UPA is more than a technical enhancement—it's a

transformative shift in how finance operates within SAP S/4HANA. By aligning accounting processes with business complexity in real time and per ledger, UPA enables more accurate, timely, and compliant financial insights. Organizations that adopt UPA now will be better positioned for the future of finance, where real-time, multi-view reporting is the standard rather than the exception. Note that you still need the authorization from SAP to implement Universal Parallel Accounting, as it doesn't fulfill all process requirements yet.

Acknowledgments

My thanks go to my husband Paul; to my siblings, Constanze and Dorothee Schmalzing, for their motivational pearls of wisdom and support while writing this book; and to my lovely nieces and nephews.

Kathrin Schmalzing

Chapter 1
Introduction to Profitability Analysis

Margin analysis in SAP S/4HANA has been integrated into the Universal Journal, whereas no changes have been made to costing-based profitability analysis. In this chapter, you'll get an overview of the latest changes and all the information you need to understand the two types of profitability analysis available in SAP S/4HANA.

Profitability analysis is a subcomponent of controlling with SAP S/4HANA. Profitability analysis in SAP S/4HANA is used to display the internal view of accounting that supports managers in their decision-making. The data in profitability analysis is enriched with numerous characteristics derived from the value flow of predecessor documents and therefore can be displayed and analyzed from various angles.

Types of profitability analysis

In SAP ERP, there are two types of profitability analysis (CO-PA): account-based and costing-based. In SAP S/4HANA, on the other hand, you'll find costing-based profitability analysis and margin analysis. This chapter describes the differences between costing-based profitability analysis and margin analysis from both functional and technical viewpoints.

Costing-based profitability analysis isn't further developed in SAP S/4HANA, and with the migration to SAP S/4HANA, SAP recommends activating margin analysis. In this chapter, you'll learn which type of profitability analysis is your go-to for the future.

1.1 Purpose of Profitability Analysis

Processing and analyzing data comprise one of the most important tasks in a company. Data is required to know how a company is doing and to make both short-term and long-term decisions. There are two recipients for data who have different interests, which is why there are two data different views as well:

Profitability views

- **External view**
  Defined by commercial law and accounting principles, the external view gives an overview of the company's assets (e.g., cash, inventory, accounts

1   Introduction to Profitability Analysis

receivables), liabilities (e.g., accounts payable, loans payable, accruals), and the owner's equity.

Recipients include executive management, owners, banks, and investors.

- **Internal view**

    Defined by management requirements for the evaluation of business areas and profit centers, the internal view assigns costs to the responsible cost objects and identifies the most profitable products and services, as well as areas where overspending occurs.

    Recipients include executive management, business area owners, and product owners.

Traditional data models in enterprise resource planning (ERP) systems generally focus on the collection of data, aggregations, and reporting for the external view of profitability.

Requirements for profitability analysis

Today, the need for information in real time as well as flexibility in profitability continues to increase. Enterprises have different requirements now as they face a more competitive and innovative market. Profitability is measured at the lowest possible level to improve decision-making processes, to react quickly to changes in the market, and to even be a step ahead. Those new requirements are reasons for the recent developments and changes in the profitability analysis. Thanks to real-time analytics, this instant market information gives you better and faster insights and allows for better decision-making.

What Is Profitability Analysis?

The profitability analysis functionality is designed to meet the requirements of internal profitability analysis processes. Profitability analysis enriches actual and plan data with additional characteristics that can be freely defined and allow for detailed data analyses. The reporting in profitability analysis is very flexible and designed based on your specific needs. Profitability analysis in SAP S/4HANA has been revolutionized with margin analysis so that it gives you a snapshot of your management reporting not only in the past and present but also in the future, in addition to improving your analytical capabilities.

Structure for profitability analysis

SAP ERP provided CO-PA without any structures and master data, such as standard reports. You could thus structure and configure profitability analysis according to your specific needs and requirements. It almost resembled a SAP Business Warehouse (SAP BW) InfoCube with its multidimensional analysis of data. With margin analysis in SAP S/4HANA, you still

20

have a lot of freedom in designing and structuring your profitability analysis, but there are more preconfigured and predesigned functionalities you can take advantage of.

You rarely create postings in profitability analysis directly. Instead, profitability analysis receives postings from previous components and enriches them with characteristics. Therefore, it's very important to know where your data is coming from and how the processes are defined to make sure you receive the right data to analyze your data correctly.

As mentioned, there are two types of profitability analysis in SAP S/4HANA: costing-based profitability analysis and margin analysis. The following sections discuss these two types in detail. The functionalities of margin analysis are evolving significantly with almost every new release of SAP S/4HANA. Section 1.3.2 provides an overview of the basic differences as well as a recommendation for using profitability analysis in SAP S/4HANA Finance.

Postings in profitability analysis

Costing-based profitability analysis is still available in SAP S/4HANA but won't be further developed. In earlier releases, costing-based profitability analysis still had some functionalities that weren't available in margin analysis, but this has changed with SAP S/4HANA 2023.

1.2 Profitability Concepts

In this section, we'll explain the different profitability concepts, such as profitability segments, characteristics, and value fields, that lay the foundation for profitability analysis.

1.2.1 Profitability Segments

A *profitability segment* is created when a document is transferred to profitability analysis. There are no master records for profitability segments. The profitability segment is a combination of various characteristics. These can be SAP standard characteristics (customer, material, etc.) or custom-specific characteristics (ship-to country, continent, etc.). All of those characteristics are assigned to an operating concern, which is the organizational unit for profitability analysis. You'll learn more about the operating concern in Chapter 2.

Definition of profitability segment

Figure 1.1 shows an example profitability segment, which is multidimensional and resembles an SAP BW InfoCube. You can analyze your data in profitability analysis flexibly by displaying the data in various dimensions.

1   Introduction to Profitability Analysis

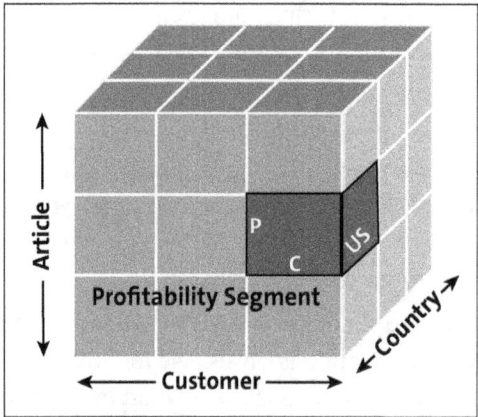

Figure 1.1 Profitability Segment in Profitability Analysis

Example of profitability segment

Amounts such as sales and costs are assigned to a profitability segment on the lowest level of the characteristic combination. For example, in Figure 1.2, you can see a sales order with two line items. After creating a goods issue for this sales order, you'll create a billing invoice for the same. Both at goods issue and billing invoice creation, a financial document will be created and posted into financial accounting. For margin analysis, you'll create a profitability segment at the same time the financial document is created. For costing-based profitability analysis, the profitability segment is only created when the billing invoice is created. If you activated predictive accounting for margin analysis and incoming sales orders in costing-based profitability analysis, a profitability segment will also be created when the sales order is created.

Sales Order: 9834
Customer: Profit 1000

Item Number	Material	Quantity in Pieces	Price in USD
10	109910	10	450.00
20	106640	5	550.00

Figure 1.2 Sales Order with Two Line Items

1.2 Profitability Concepts

How does this profitability segment look when the financial document is created? We mentioned earlier that the profitability segment is created based on the lowest level of characteristics that can be derived. Let's assume that in our example, the characteristics displayed in Table 1.1 are assigned to our operating concern and therefore are available for the profitability analysis. In Table 1.1, you can see two columns: **Fixed** and **Custom**. The fixed characteristics are assigned to every operating concern by default and therefore are available in every form of profitability analysis. The custom characteristics are specifically assigned to the operating concern.

Characteristics in profitability segments

Characteristic	Fixed	Custom
Sales order number	X	
Customer	X	
Ship-to country		X
Item number	X	
Material	X	
Material group		X
Unit of measure	X	
Currency	X	

Table 1.1 Characteristics Derived from the Sales Order

Next, how will the profitability segments look when the billing invoice is created? In Table 1.2, you can see the two profitability segments that will be created according to the characteristics we assumed to be assigned to the operating concern. (In addition to the characteristics we chose for our example, all organizational elements will be derived as fixed characteristics as well.)

Structure of profitability segments

Profit-ability Segment	Sales Order Number	Customer	Ship-To Country	Item Number	Material	Material Group	Unit of Measure
001	9834	Profit 1000	US	10	109910	025	PCS
002	9834	Profit 1000	US	20	106640	153	PCS

Table 1.2 Profitability Segments Created with Transfer of Billing Invoice

1 Introduction to Profitability Analysis

Amounts in profitability segment
If you're wondering where the amounts are in the profitability segment, the amounts get assigned to the profitability segment but aren't part of the profitability segment. The profitability segment itself is classified as a combination of characteristics.

1.2.2 Characteristics

Master data in profitability analysis
In the previous section, we talked about characteristics and how their values make up the profitability segment. Characteristics also make up the master data of an operating concern. When you define your reporting structure, you should think about the characteristics that you want to use for the analysis of your data to reveal information on the core of your business. You should ask yourself what the strategic goals for your business are and how profitability analysis might help you attain these goals. For example, if you're a shoe manufacturer, you'd consider which characteristics would be interesting for you to see in reporting to make decisions on how to develop your product portfolio or on how to measure success, such as the following:

- Shoe size
- Type of shoe (sneaker, sandal, etc.)
- Women's/men's/kids'
- Collection: Spring/Fall

Types of characteristics
After you've identified the characteristics, you need to assign those to your operating concern. There are two different types of characteristics in SAP S/4HANA:

- **Fixed fields**
  Fixed fields are all the fields around an organizational structure, such as company code, segment, and cost center, as well as most common master data, such as material master and cost object. Within the fixed fields, you'll also find technical fields, which are all currency and quantity fields, as well as time-dependent fields (month, day, year), and amounts for values.

- **User-defined characteristics**
  All other characteristics are user-defined characteristics. You're free to assign up to 60 characteristics of your choice to the operating concern. You can assign characteristics of certain data tables (e.g., the customer master), or you can create custom-specific characteristics.

You'll learn more about all types of characteristics and how to create them in Chapter 3.

Structure of characteristics
In Figure 1.3, you see the details of the **Characteristic BZIRK**, which is divided into five different segments:

1.2 Profitability Concepts

- **Texts**
  In this segment, you see the **Description** (**Sales District**) as well as **Short text** (**District**) of the characteristic.
- **ABAP Dictionary**
  This segment gives you information about the technical name as well as the length and data type of the characteristic. **BZIRK** has the data type **CHAR**, which means it can be any characteristic (not limited to numbers) and has a length of **6**.
- **Origin**
  In this segment, you can see the table in which the characteristic is stored. **BZIRK** is stored in data table KNVV (customer master data sales).
- **Further Properties**
  This segment gives the **Status** of the characteristic (**BZIRK** has status **A** for **active**). All characteristics that you assign to your operating concern must be saved in **Active** status; otherwise, you won't be able to assign the characteristic to the operating concern.
- **Validation**
  In this segment, you can see if there is a check table for the values of the characteristic. **BZIRK** has **Check table T171**, in which you'll find the names for the sales districts (**BZIRK**) that have been created.

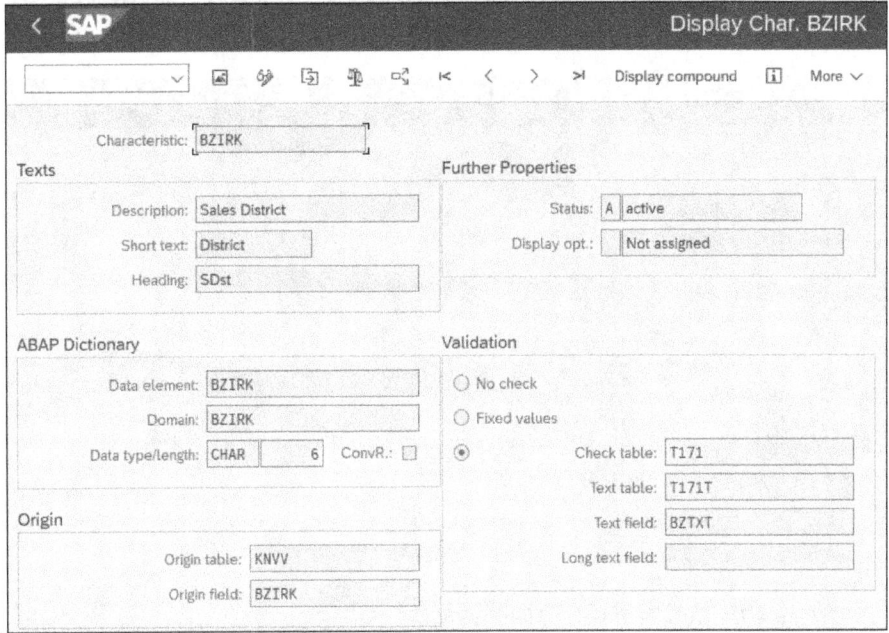

Figure 1.3 Details of Characteristic "BZIRK"

In Chapter 3, you'll learn much more about the creation and derivation of characteristics in profitability analysis.

1 Introduction to Profitability Analysis

1.2.3 Value Fields

Master data in costing-based profitability analysis

Value fields are only required in costing-based profitability analysis. Instead of accounts, costs and revenue are displayed on value fields. You can save both amounts and quantities on a value field. Therefore, value fields are also part of the master data in costing-based profitability analysis. Value fields don't match accounts 1:1; instead, they group costs and revenues according to their functional area to display a contribution margin report in cost of sales accounting. In configuration, you have to define how each value field receives data. You can analyze your revenues and costs on the value fields based on the characteristics.

Structure of value fields

In Figure 1.4, you can see the details of **Value Field VVMAT**. The screen is divided into three segments:

- **Texts**
  This segment shows the **Short text** and **Description** (here, **Material Cost**) of the value field.

- **Aggregation**
  In this segment, you'll see how the values are saved on the value field. **VVMAT** summarizes all values by fiscal year.

- **Other attributes**
  In this segment, you'll see if the value field is used for amounts or quantities. **Amount** is selected for **VVMAT**.

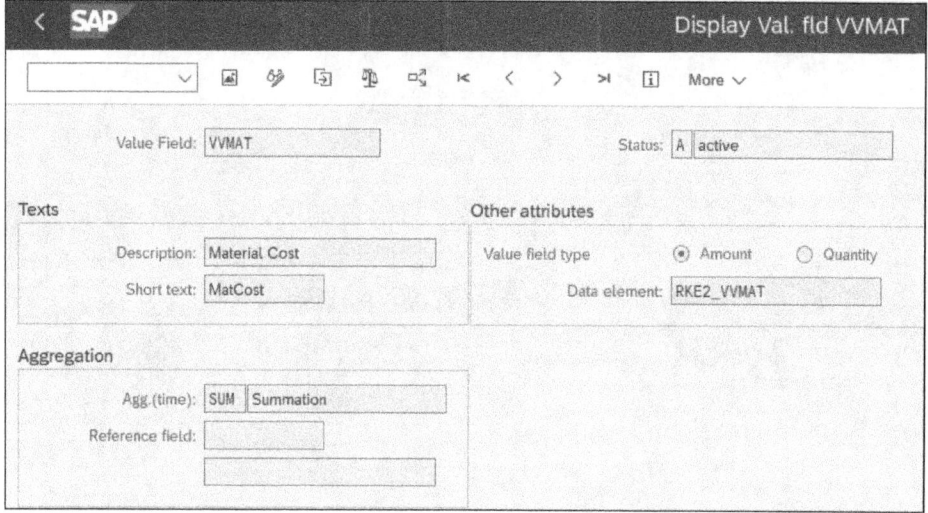

Figure 1.4 Details of Value Field "VVMAT"

You'll learn how to create amount and quantity value fields in Chapter 4.

1.3 Types of Profitability Analysis

In this section, we'll explain the difference between the types of profitability analysis. As mentioned earlier, SAP ERP has costing-based and accounting-based CO-PA. Both of these forms are available in SAP S/4HANA, but account-based CO-PA has been renamed as margin analysis. If you're implementing SAP S/4HANA Cloud, you can only activate margin analysis; costing-based profitability analysis isn't available.

1.3.1 Margin Analysis (Account-Based Profitability Analysis)

Account-based CO-PA hasn't been used that much in SAP ERP, as it was lacking a lot of functionality compared to costing-based CO-PA. Some users ran account-based CO-PA in parallel to costing-based CO-PA as it promised reconciliation of CO-PA with the general ledger (G/L), which wasn't guaranteed. Account-based CO-PA was based on cost elements in SAP ERP. In SAP S/4HANA, cost elements became integrated in G/L accounts. When creating a G/L account, you choose what type of G/L account you want to create. There are two G/L account types for cost elements:

- Primary costs or revenues
- Secondary costs

Merging of G/L account and cost element

In Figure 1.5, you can see the **G/L Account Type** field in the **General** tab of **G/L Account 00800000 Sales revenues-dom.**

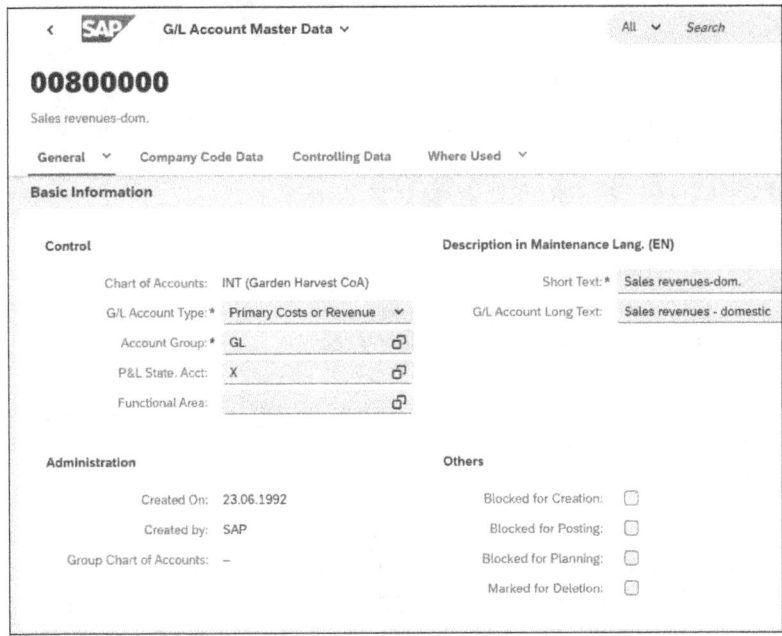

Figure 1.5 Displaying the G/L Account Type in the G/L Account

1   Introduction to Profitability Analysis

Choosing the G/L account type
When either **Primary Costs or Revenue** or **Secondary Costs** is selected from the **G/L Account Type** field, and the company code is assigned to a controlling area, then you'll see the **Settings in Controlling Area** section in the **Controlling Data** tab, as shown in Figure 1.6. The fields available in the section are the same fields that you maintained when you were creating a cost element:

- **Cost Element Category**
  The cost element category defines the usage of the cost element. For both primary cost elements and secondary cost elements, there are different cost element categories. The G/L account in Figure 1.6 is assigned to cost element category **11** for **Revenues**, which means this G/L account is transferred directly to profitability analysis when transferring a billing invoice. Postings with a cost element of cost element category 11 are only statistical and can't be settled to another recipient.

- **Unit of Measurement**
  The **Unit of Measurement** must be maintained when the indicator to record the quantity is activated. This field determines in which unit of measure quantity can be recorded on the cost element.

- **Record Quantity**
  If this indicator is set, you can record quantities on the cost element during posting.

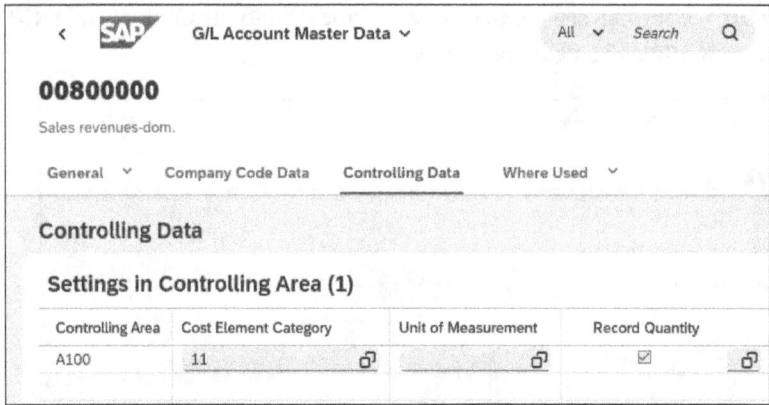

Figure 1.6 Displaying Settings in Controlling Area

Sample structure of account-based profitability analysis
Along with the cost elements, margin analysis also became integrated in the G/L. You can read more about the margin analysis data model in Section 1.4.2.

Integration of margin analysis in the G/L
Margin analysis is fully integrated in the Universal Journal. It's therefore reconciled by design. The internal and external views in accounting are harmonized and allow a comprehensive reporting with drilldown functionality in P&L statements as well as balance sheets to the lowest level of characteristics.

Margin analysis can be structured in both cost of sales accounting and periodic accounting. Figure 1.7 shows a sample structure of traditional account-based profitability analysis or margin analysis. In the left half of the figure, you can see the cost elements that are replaced by G/L accounts in SAP S/4HANA Finance. The values correspond to the posted values on the G/L account.

Account-Based Profitability Analysis		
P&L Accounts		**Value**
40000	Sales	USD 1,000,000
50000	Sales Deductions	USD 100,000
	Net Sales	USD 900,000
60000	COGS	USD 690,000
61000	Price Differences	USD 10,000
62000	Research and Development	USD 10,000
63000	Marketing	USD 50,000
64000	Sales	USD 40,000
	Operating Profit	USD 100,000

Figure 1.7 Sample Structure for Margin Analysis

Much less configuration needs to be done in margin analysis than in costing-based profitability analysis. After margin analysis is activated, most transactions automatically get a profitability segment assigned. There are some functionalities that allow you to split costs in margin analysis or predict outcomes. All these functionalities and more will be explained in Chapter 4. The following lists what you need to remember from this section about margin analysis:

- Cost elements have been integrated into G/L accounts.
- Margin analysis will get additional functionality in future releases.
- After margin analysis is activated, values are transferred to margin analysis without much configuration.
- We recommend activating margin analysis when implementing or transferring to SAP S/4HANA.

Configuration in margin analysis

1 Introduction to Profitability Analysis

1.3.2 Costing-Based Profitability Analysis

Costing-based profitability analysis in SAP S/4HANA

Costing-based CO-PA is used the most in the SAP ERP environment. No changes have been made to the data model, setup, and usage of costing-based profitability analysis in SAP S/4HANA. SAP won't invest in its further development, but it's still supported for usage and implementation in SAP S/4HANA (although not in SAP S/4HANA Cloud).

In the next section, you'll learn about the features and functionality of costing-based profitability analysis.

Setting up profitability analysis

To set up costing-based profitability analysis, you have to define characteristics and value fields. As explained in Section 1.2.2, characteristics include company code, customer, article, and so on. Value fields, as explained in Section 1.2.3, are groupings of costs and quantities, such as sales, material costs, and sales quantity. The structure in profitability analysis reporting is defined by value fields. Figure 1.8 shows an example reporting structure in costing-based profitability analysis based on cost of sales accounting (cost can't be structured in period accounting). You can see in the left column, the technical name for a value field followed by the description of the value field, for example, **VVREV** (**Sales**).

Structure of profitability analysis reporting

Costing-Based Profitability Analysis		
Value Field		Amount
VVREV	Sales	USD 1,000,000
VVQDI	Quantity Discount	USD 20,000
VVREB	Rebates	USD 80,000
	Net Sales	USD 900,000
VVMCT	Var Material Cost	USD 400,000
VVPCT	Var Production Cost	USD 190,000
VVPVR	Subcontracting Cost	USD 10,000
	Gross Margin I	USD 300,000
VVMAT	Overhead Material	USD 50,000
VVDIF	Overhead Production	USD 50,000
VVRED	Research and Development	USD 10,000
VVSMR	Sales, Admin, and Marketing	USD 90,000
	Gross Margin II	USD 100,000

Figure 1.8 Sample Structure of Costing-Based Profitability Analysis

The structure in costing-based profitability analysis follows a profit and loss (P&L) statement in cost of sales accounting or a contribution margin report. As costing-based profitability analysis doesn't use G/L accounts, you must use value fields to define the structure for your reporting. There shouldn't be a 1:1 relationship between G/L accounts and value fields, although it's technically possible. A value field symbolizes a group of costs, and you can drill down and display the cost on a profitability segment level. When creating a report in costing-based profitability analysis, you can also work with formula fields, for example, **Net Sales** in Figure 1.9.

Definition of a value field

When setting up your reporting structure and defining the value fields, you also must consider how these value fields are filled with amounts. Let's look at a few examples:

Setting up a reporting structure

- **Discounts**
  How are discounts in the SAP ERP Sales and Distribution module (SD) configured? Are they displayed with a separate sales condition? If so, you can display them on a separate value field.

- **Variable and fixed costs**
  Do you split your cost in variable and fixed costs? If so, you can transfer them on different value fields in costing-based profitability analysis.

When setting up your value fields, you can obtain details of the amounts by deriving characteristics. You shouldn't set up your value fields on a very detailed level. Let's look at a few examples:

- **Sales**
  If you want to separate domestic sales and international sales, you can separate the amounts on a value field by deriving the country of the goods recipient. We explain in Chapter 3 how to create this characteristics derivation.

- **Marketing costs**
  If you want to separate marketing costs by business area, you can derive the business area from the cost center and separate the amounts by business area on the marketing value field. You'll learn how to create a characteristics derivation for the business area in Chapter 3.

When defining your reporting structure and setting up your value fields, keep in mind that you can assign a maximum of 120 value fields to your operating concern.

Profitability analysis is delivered without any structures, so costing-based profitability analysis requires customer-specific configuration. Both characteristics and value fields are the master data of costing-based profitability analysis. Characteristics are the basis for the analysis of your operating

Customer-specific configuration

1 Introduction to Profitability Analysis

profit. When defining your characteristics, think of how you want to analyze your data. For example, if the material group is of interest for your analysis, then define the material group as a characteristic. The definition of the value fields greatly depends on the structure of your contribution margin accounting.

> **[»] Defining Value Fields**
>
> You shouldn't define your value fields on a very detailed level because you can analyze/split the values of the value fields with characteristics. Keep in mind that you can only assign a maximum of 200 value fields to your operating concern.

Supplying profitability analysis with data — After the definition of your characteristics and value fields master data, you need to define how profitability analysis is supplied with data. You rarely create documents in profitability analysis directly. Costing-based profitability analysis receives data from various upstream modules in SAP S/4HANA, such as sales and distribution (SD), materials management (MM), production planning (PP), and so on. The interfaces to profitability analysis in SAP S/4HANA determine to which value field the values from the previous components are posted. The characteristics can be derived from predecessor documents such as sales orders or purchase orders based on the settings in configuration when a profitability segment is created. Figure 1.9 illustrates the value flows in costing-based profitability analysis.

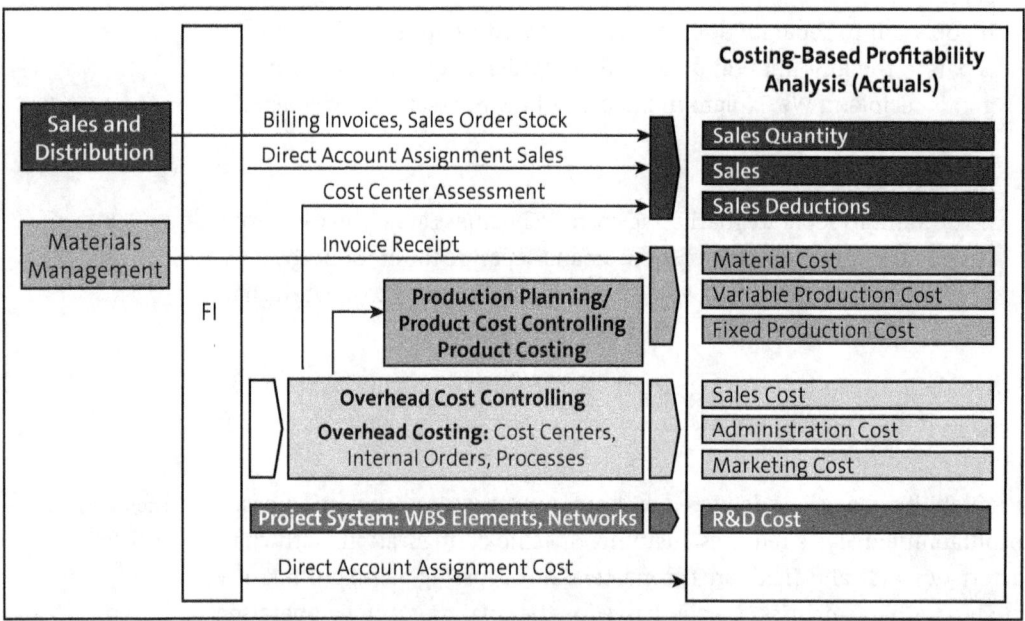

Figure 1.9 Value Flow in Costing-Based Profitability Analysis

Let's look closer at how the values are transferred in costing-based profitability analysis (the assumption is that the configuration for the data transfer into costing-based profitability analysis is in place):

> **Note**
> In Chapter 5, we'll explain the configuration settings step-by-step.

- **Incoming sales order**
  When creating a sales order, the sales quantity and the future revenue can be transferred to profitability analysis. The posting can take place on the same value fields as the actual sales postings because the postings occur with a different record type.

- **Billing invoice**
  At the time a billing invoice is created, a profitability segment for the actual sales and cost of goods sold (COGS) is created. Costing-based profitability analysis also allows the transfer of statistical SD conditions such as cash discounts. Make sure that you don't mix statistical costs with actual costs on a value field, or the reconciliation of costing-based CO-PA with the G/L will be nearly impossible.

  The SD standard process is that a customer order is created, the delivery is created, and the goods issue is posted. When the invoice is created and transferred to financial accounting, sales are posted, and sales and costs of goods sold are transferred to profitability analysis. Sales postings and goods issue postings in financial accounting are two separate postings that can happen at a totally different point in time, even in different months. However, costing-based profitability analysis receives the posting when the sales are posted. That being said, you can see that this can cause issues. At month end, goods in transit (i.e., where the goods issue has been posted but the invoice hasn't been issued yet) must be analyzed and accrued in financial accounting.

  Goods in transit in costing-based profitability analysis

- **Direct account assignment sales**
  If you don't use the SD module for your order-to-cash processes, you can transfer postings to the specific G/L accounts directly into costing-based profitability analysis with a direct account assignment. The profitability segment can be created manually when the posting is created.

- **Logistics invoice receipt**
  All invoices or goods receipt that aren't inventory postings can get transferred directly to costing-based profitability analysis with the direct account assignment.

1   Introduction to Profitability Analysis

- Product costing
  The COGS is recorded at the standard cost of the product when the goods issue is posted. The COGS is transferred with the billing invoice to costing-based profitability analysis. Variances between plan and actual cost resulting from overconsumption, underconsumption, or differing prices are transferred with the settlement of the production orders to costing-based profitability analysis.

- Overhead costing
  All overhead costs are transferred to costing-based profitability analysis with the settlement of the cost objects (e.g., internal orders, work breakdown structure [WBS] elements, or cost center assessment cycles) they are posted on.

Reconciling with financial accounting
One of the major challenges when reconciling costing-based profitability analysis with financial accounting is due to the structure of costing-based profitability analysis with value fields. There is no link to the G/L account of the original posting in financial accounting in costing-based profitability analysis.

Costing-based profitability analysis summary
In Chapter 5, you'll learn much more about costing-based profitability analysis and how to correctly configure the interfaces of the predecessor modules with costing-based profitability analysis. We'll also show you some SAP standard tools that help you reconcile costing-based profitability analysis with financial accounting. The following lists what you need to remember from this section about costing-based profitability analysis:

- Only cost of sales accounting is supported.
- Value fields and characteristics determine its structure.
- SAP S/4HANA doesn't add any new functions to costing-based profitability analysis.
- With SAP S/4HANA, performance isn't an issue anymore.

1.3.3   Comparison of Profitability Analysis Types

Different profitability analysis functions
This section introduces the changes to traditional account-based and costing-based CO-PA in SAP ERP as well as margin analysis in SAP S/4HANA Finance.

Table 1.3 provides an overview of the current and new profitability analysis functions.

Traditional CO-PA (Account-Based and Costing-Based)			SAP S/4HANA Finance
Area	Costing-based CO-PA	Account-based CO-PA	Margin analysis
Cost collector	Value fields	Cost elements	G/L accounts
Reconciliation with SAP ERP Financials	Not ensured	Yes	Yes
Incoming sales order	Yes	No	Yes
Statistical conditions	Yes	No	Yes
Valuation	Valuation with different costing variants possible	Valuation only with operational valuation in Financial Accounting (FI)	Valuation with different costing variants possible
Mapping of COGS	Mapping of COGS in cost items (e.g., material, employees, overhead) according to the cost component layout in product cost accounting	Single G/L account for COGS mapping	Mapping of COGS in cost items (e.g., material, employees, overhead) according to the cost component layout in product cost accounting
Production variances	Mapping of variance categories	Mapping of production variances on a P&L G/L account	Mapping of production variances with various accounts for each variance category
Quantity fields	Several quantity fields supported	One quantity field to map the sales quantity	Up to three quantity fields supported
Unit of measure	Several units of measure supported	One unit of measure supported	Up to three units of measure supported
Commitments in profitability analysis	No	No	Yes
Derivation of characteristics without profitability segment	No	No	Yes
Universal allocation	No	No	Yes

Table 1.3 Comparison of Profitability Analysis in SAP ERP and Margin Analysis in SAP S/4HANA Finance

1 Introduction to Profitability Analysis

Future of margin analysis

The comparison between the different forms of profitability analysis shows that margin analysis already has been enriched with new functionality to deliver 360-degree insight into financial performance. The objective for the future is to completely separate margin analysis and costing-based profitability analysis and deliver a margin analysis without operating concern based on the field catalog of the Universal Journal. In Chapter 4 and Chapter 5, we'll explain the configuration and usage of all the functionalities mentioned in this chapter.

1.4 Technical Setup

The technical setup describes the data structure behind the two types of profitability analysis. The insight into the technical setup might help you better understand the functionalities and reconciliation challenges for costing-based profitability analysis we mentioned earlier.

1.4.1 Technical Setup of Costing-Based Profitability Analysis

Table structure of costing-based profitability analysis

With SAP S/4HANA Finance, the current data model for costing-based profitability analysis stays the same as in SAP ERP. This section provides an overview of the tables in which the documents of costing-based profitability analysis are stored. The costing-based profitability analysis table structure hasn't been modified in SAP S/4HANA Finance. The costing-based profitability analysis documents aren't included in the Universal Journal, so the reconciliation of financial accounting with costing-based profitability analysis is a huge challenge.

Table 1.4 lists the individual table names and their description. The system generates these tables in the background when the operating concern for costing-based profitability analysis is created. The xxxx in the table are replaced with the technical name of the operating concern.

Table Name	Description
CE0xxxx	Logical Line-Item Structure
CE1xxxx	Actual Line-Item Table
hCE2xxxx	Plan Line-Item Table
CE3xxxx	Segment Level
CE4xxxx	Segment Table
CE4xxxx_KENC	Realignments

Table 1.4 Structure Overview in Costing-Based Profitability Analysis

1.4 Technical Setup

Table Name	Description
CE4xxxx_ACCT	Account Assignments
CE4xxxx_FLAG	Posted Characteristics
CE5xxxx	Logical Segment Level
CE7xxxx	Internal Help Structure for Assessments
CE8xxxx	Internal Help Structure for Assessments

Table 1.4 Structure Overview in Costing-Based Profitability Analysis (Cont.)

SAP doesn't provide any tool for transferring costing-based profitability analysis historical data into SAP S/4HANA. If you want to migrate to SAP S/4HANA Finance during the fiscal year and have already activated costing-based profitability analysis but want to use account-based profitability analysis in the future, costing-based profitability analysis should remain activated until the end of the year for reconciliation purposes and for analyses of historical data. You can then deactivate costing-based profitability analysis and continue to use account-based profitability analysis only.

Migration from SAP ERP to SAP S/4HANA

1.4.2 Technical Setup of Margin Analysis

In SAP S/4HANA, many finance and controlling tables have been consolidated into the Universal Journal (table ACDOCA). Figure 1.10 illustrates the consolidation of the tables from SAP ERP FI, CO-PA, CO, Asset Accounting (AA), and material ledger in the Universal Journal.

Consolidating tables into the Universal Journal

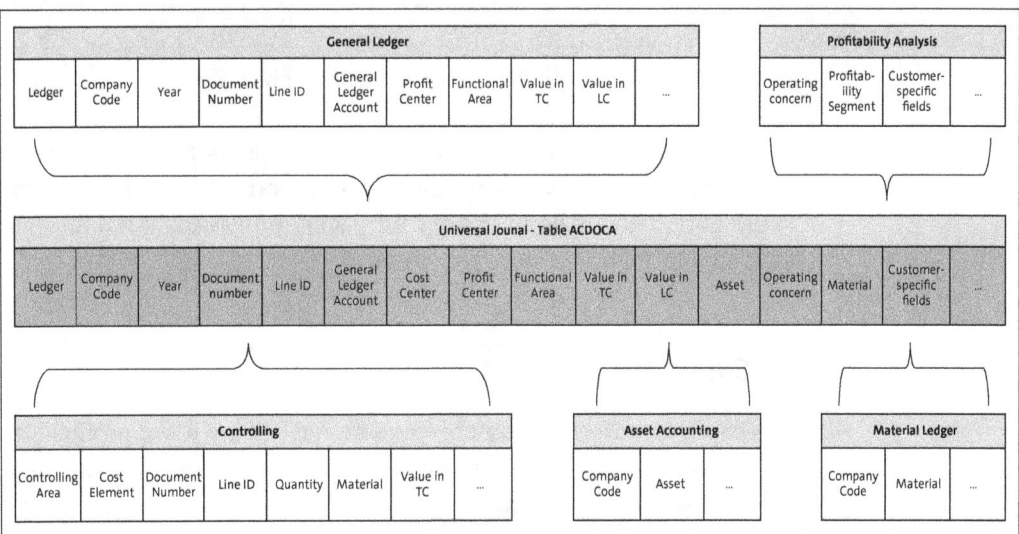

Figure 1.10 Universal Journal in SAP S/4HANA

1   Introduction to Profitability Analysis

All characteristics of the line-item table are transferred to the Universal Journal. Table BKPF for the document header (Document Header for Accounting) is still available in SAP S/4HANA. All other tables that are replaced by the Universal Journal are kept as views to ensure that the standard reports and standard transactions function properly.

Table structure in margin analysis In SAP S/4HANA, the account-based profitability analysis data and the margin analysis data are stored in the Universal Journal (table ACDOCA) in real time. The integration of margin analysis into the Universal Journal table provides many advantages, such as adding further characteristics to the Universal Journal, storing data in real time, and providing simplified reconciliation with financial accounting.

In Figure 1.11, you can see an extract of the Universal Journal (table ACDOCA) showing some actual line items for profitability segments (**Profit. segment** column). You can see that various characteristics are derived in the line item such as **Customer**, **Plant**, and so on.

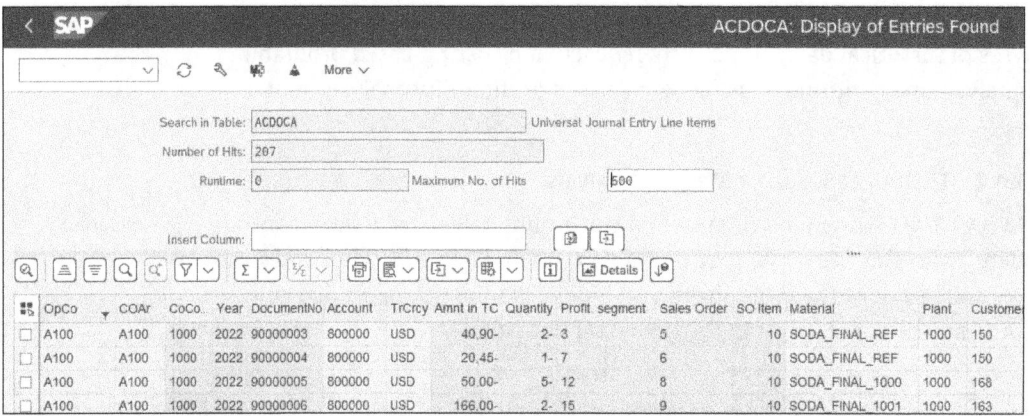

Figure 1.11 Display of Profitability Segment in the Universal Journal

The integration of margin analysis in the Universal Journal improves the processes by making accurate transaction data for the value chain available in one source. The single source of truth also improves analytical capabilities as all information is stored in one table.

1.5   Summary

This chapter described the differences between costing-based profitability analysis and margin analysis. It also summarized the differences and innovations in SAP S/4HANA Finance.

If you've used costing-based profitability analysis so far, we recommend activating margin analysis because this is the only form of profitability analysis SAP will enhance in the future and provide additional functions for.

If you currently use costing-based profitability analysis in SAP S/4HANA and want to continue to use it, you may additionally activate margin analysis. This doesn't result in larger data volumes because the documents in margin analysis are integrated into the Universal Journal (table ACDOCA). If you haven't used profitability analysis yet, you should also activate margin analysis in SAP S/4HANA Finance as it gives you a lot of additional functionality and will allow you to get a 360-view of your financial data.

The next chapter discusses the required basic settings for activating the organizational structure of profitability analysis in SAP S/4HANA.

Chapter 2
Configuring the Operating Concern and Basic Settings for Profitability Analysis

The basic settings lay the foundation for the use of profitability analysis. The central basic setting is the operating concern in which you define the general structure of profitability analysis.

The basic settings of profitability analysis form the basis for using it and define its general data structure. This chapter discusses the basic settings required to activate profitability analysis in SAP S/4HANA. It describes the individual configuration steps for creating the operating concern, number ranges, versions, and other basic settings that are required to use profitability analysis. We'll point out if there are differences in the settings for costing-based profitability analysis and margin analysis.

2.1 Creating the Operating Concern

A prerequisite for using profitability analysis is to create the *operating concern*. The operating concern is the highest organizational element in the SAP system—only the client is above the operating concern. Most of the configuration for profitability analysis is therefore cross-client, meaning that you must be very careful when working on the operating concern as you might block other people from executing their tasks.

Creating the operating concern

2.1.1 Assign a Controlling Area to the Operating Concern

To activate the operating concern, you assign your controlling area to it. You can assign one or more controlling areas to the operating concern, as shown in Figure 2.1. All controlling areas that are assigned to the operating concern are required to use the same fiscal year variant. The controlling areas that are assigned to the operating concern aren't required to have the same chart of accounts assigned. They can have different charts of accounts assigned and still be assigned to the same operating concern under the assumption that they are using the same fiscal year variant.

Assigning a controlling area to the operating concern

2   Configuring the Operating Concern and Basic Settings for Profitability Analysis

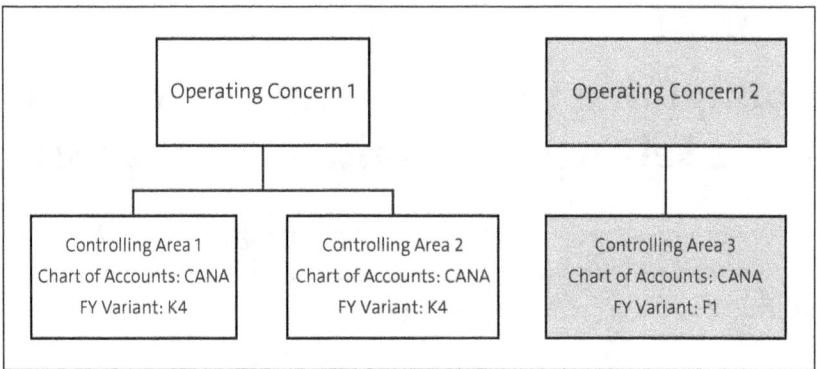

Figure 2.1 Organizational Structure of the Operating Concern

How many operating concerns?
SAP recommends using one operating concern and assigning all relevant controlling areas to this operating concern to allow for global reporting because the SAP standard doesn't provide for the creation of reports across multiple operating concerns. For more information relevant to choosing whether you should use one or several operating concerns, see SAP Note 1077293 (FAQ: Cross-Company Code Cost Accounting).

Looking to the future, we know that SAP is planning to remove the need for an operating concern for margin analysis. As of release 2023, you still need to create an operating concern to activate margin analysis. For costing-based profitability analysis, you'll always have to create an operating concern because SAP isn't planning to make any further improvements or developments related to costing-based profitability analysis.

2.1.2   Create Operating Concern

Creating the operating concern using Transaction KEA0
The operating concern is created in configuration by following the configuration path **Controlling • Profitability Analysis • Structures • Define Operating Concern • Maintain Operating Concern** or by calling Transaction KEA0.

In the **Maintain Operating Concern** window in the **Operating Concern** field, enter a name with four letters or digits, and then press F5 or click on the 🖫 (**Save**) icon. A popup opens asking you to confirm if you want to create the operating concern (see Figure 2.2). Confirm the popup by clicking on **Yes**.

Cross-client organizational object
When you confirm the popup with **Yes**, the popup displayed in Figure 2.3 will open showing **Caution! You are changing/deleting cross-client settings**. Press Enter. As mentioned earlier, the operating concern is a cross-client organizational object, so you'll see this message again when doing some other configuration settings.

2.1 Creating the Operating Concern

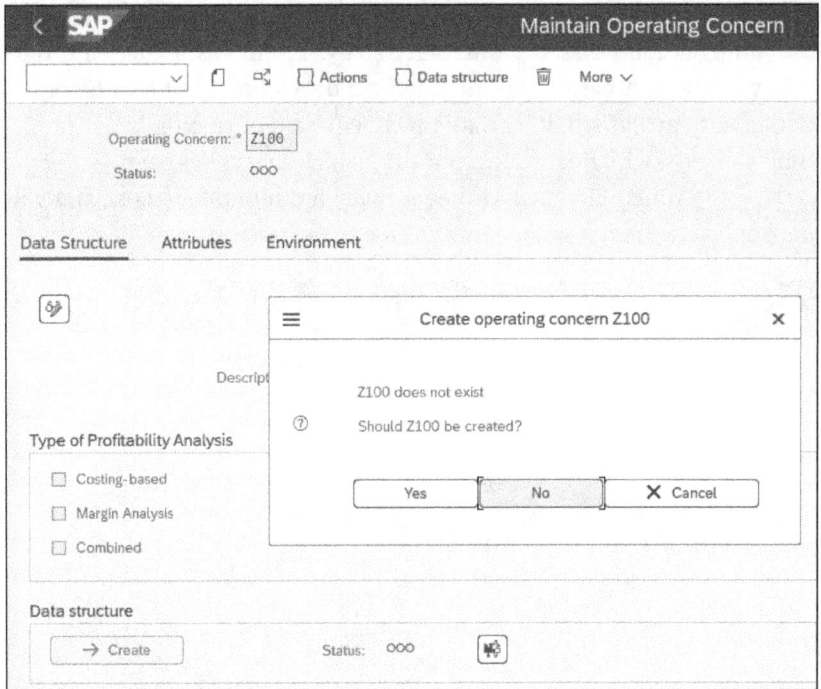

Figure 2.2 Creating an Operating Concern

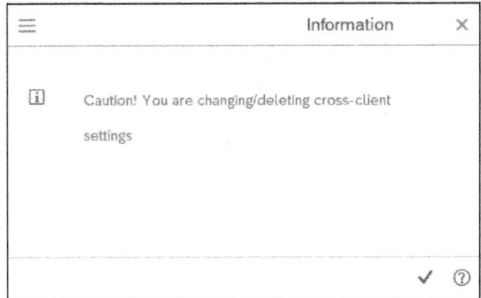

Figure 2.3 Popup with Warning Message before Creation of the Operating Concern

2.1.3 Maintain Further Settings for the Operating Concern

After confirming the warning message, you'll be able to maintain further settings for the operating concern. In Figure 2.4, you start by configuring the **Data Structure** tab. First, give the operating concern a name. In this example, enter "Operating Concern CO-PA" in the **Description** field. In the **Type of Profitability Analysis** section, you define which type of profitability analysis you're going to activate in your operating concern: costing-based, margin analysis, or both. In the example in Figure 2.4 both **Costing-based** profitability analysis and **Margin Analysis** are activated. You can also see

Type of profitability analysis

43

the **Combined** checkbox. Combined profitability analysis was developed by combining account-based profitability analysis with costing-based profitability analysis. It basically is an extended development of costing-based profitability analysis that warrants reconciliation of costing-based profitability analysis with financial accounting, which always has been a struggle. Combined profitability analysis never made it out of pilot status, and you were only allowed to use the same with permission from SAP.

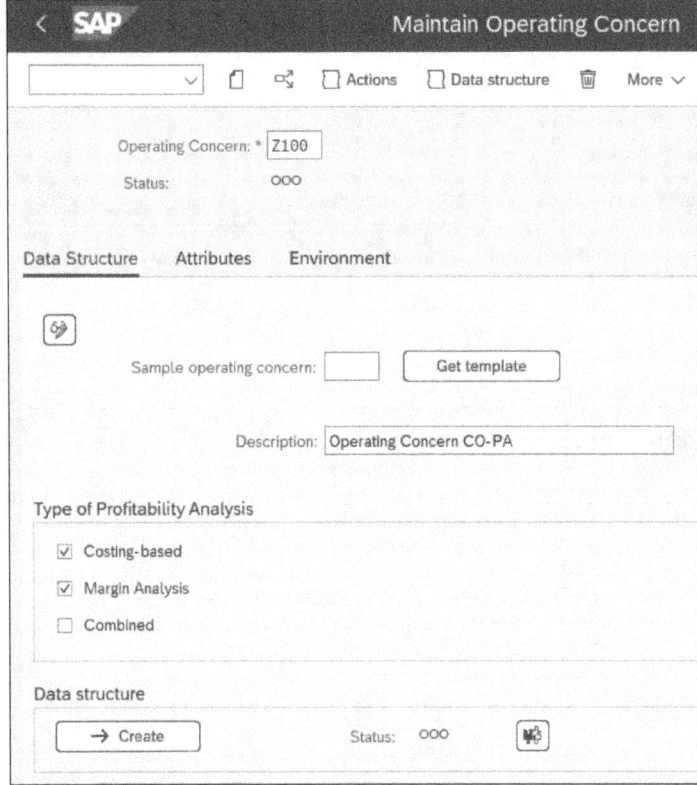

Figure 2.4 Maintaining Data Structure in the Operating Concern

A sample operating concern

In Figure 2.4, note the **Sample operating concern** field. The SAP S/4HANA system includes several sample operating concern templates. Using a sample operating concern has advantages and disadvantages. It can save you a lot of configuration work because the system copies all configuration settings of the sample operating concern. However, making changes to settings of an operating concern is a rather complex task. You'll usually work without a sample operating concern to remain flexible and to avoid the time-consuming task of deleting characteristics or other settings relevant to the operating concern after it's created on the basis of the sample operating concern.

2.1 Creating the Operating Concern

Last but not least, you can see the **Data Structure** section (Figure 2.4). Click on → (Create Data Structure) to go to a screen where you can assign characteristics and value fields to the operating concern. We'll first look at the other tabs of the operating concern before we come back to the **Data Structure** section.

Creating a data structure

> **Recommendation for SAP S/4HANA Finance**
>
> With SAP S/4HANA Finance, SAP explicitly recommends activating margin analysis because the characteristics are also stored in the Universal Journal—the new central table for the actual data of financial accounting and controlling—which facilitates the reconciliation with financial accounting. Note that account-based profitability analysis was renamed to margin analysis. In configuration, you'll still see account-based profitability analysis because not all configuration settings have been renamed yet.

2.1.4 Define Operating Concern Currency

On the **Attributes** tab shown in Figure 2.5, you define the operating concern currency, which is usually maintained in the group currency. In this example, the **USD** currency in selected in the **Operating concern currency** field.

Defining the operating concern currency

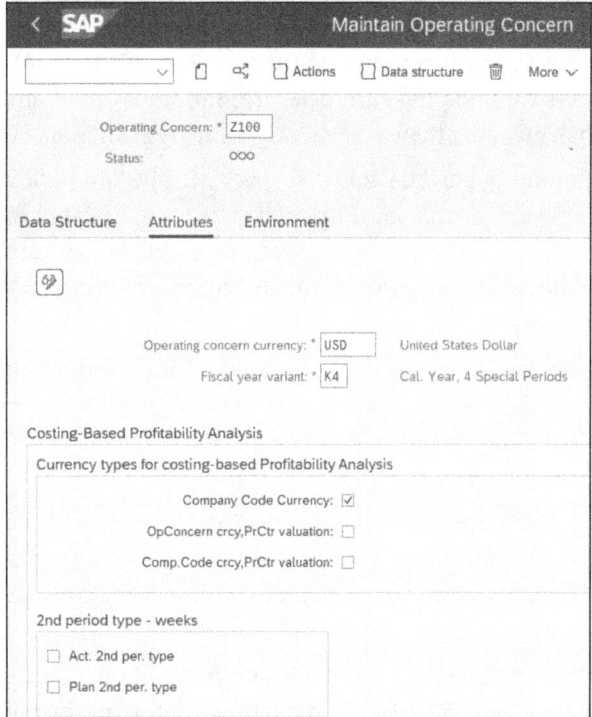

Figure 2.5 Maintaining Attributes in the Operating Concern

45

2   Configuring the Operating Concern and Basic Settings for Profitability Analysis

Assigning a fiscal year variant

Apart from the operating concern currency, you need to assign a fiscal year variant to the operating concern. In Figure 2.5, the fiscal year variant **K4** is assigned, which is the fiscal year variant for the calendar year. Remember, all controlling areas that will be assigned to the operating concern need to have the same fiscal year variant.

Additional currencies for costing-based profitability analysis

In the **Costing-Based Profitability Analysis** section, you can define additional currencies for costing-based profitability analysis. As values of costing-based profitability analysis are stored in different data tables, you need to maintain the currency settings for the operating concern. In costing-based profitability analysis, the actual values are updated in the operating concern currency and additionally in the currency combinations that you can specify on the **Attributes** tab.

In the example in Figure 2.5, the currency is updated in addition to the operating concern currency in the **Company Code Currency**. The costing-based profitability analysis plan values, on the other hand, are updated in the operating concern currency only. If you change the operating concern currency, it will also be changed for historical values in costing-based profitability analysis. However, the values aren't converted. If you change the **Operating concern currency** field from **EUR** to **USD**, for example, a value of 10 EUR is displayed as 10 USD.

Currencies in planning costing-based profitability analysis

In addition to the company code currency, you can update the profit center valuation in costing-based profitability analysis if you use parallel valuations and transfer prices in profit center accounting. Release 1611 of SAP S/4HANA Finance allows for updating currencies from parallel valuations in the Universal Journal, meaning they are available for margin analysis.

Activating profit center valuation

For costing-based profitability analysis, you can specify in the **2nd period type - weeks** section that the actual and/or plan data is to be updated in weeks. Therefore, in addition to the month and year, there will be an extra characteristic that will update the number of the calendar week the profitability segment has been created in.

Status of the operating concern

Save the changes by pressing [Ctrl]+[S] or by following **More • Operating Concern • Save**. Now, let's look at the **Environment** tab in Figure 2.6 where you can check the status of the operating concern or activate the operating concern after you've made changes to its data structure.

The operating concern is activated and ready to use if both the **Cross-client part** and the **Client-specific part** have a green **Status**. In our example, you first need to assign characteristics and value fields to the operating concern before you can activate it, so go back to the **Data Structure** tab.

Maintain the data structure of the operating concern

Click on ➡ (Create Data Structure) in the **Data Structure** tab of the operating concern to open **Edit Data Structure: Characteristics Screen**. In the **Chars** tab, you can assign up to a maximum of 60 characteristics. All characteristics that are available to assign to the operating concern are in the list on the lower-right side of the **Transfer from** area. The characteristics here are either provided by SAP and can be assigned to your operating concern, or they are custom-specific characteristics that are available in the client. There are a lot of characteristics assigned to any operating concern by default, such as company code, cost center, and so on. You'll learn more about those fixed characteristics in Chapter 3.

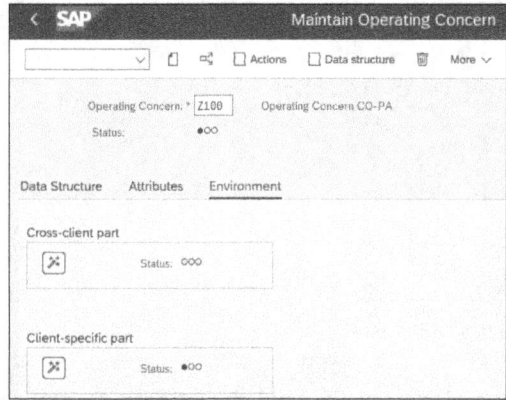

Figure 2.6 Check the Operating Concern Environment

To assign a characteristic to an operating concern, the characteristic has to be created and activated, which you'll learn about in Chapter 3 as well. Now, let's choose a few characteristics from the list in Figure 2.7 and assign them to the data structure. You can scroll in the list and mark the characteristics you want to assign to your data structure.

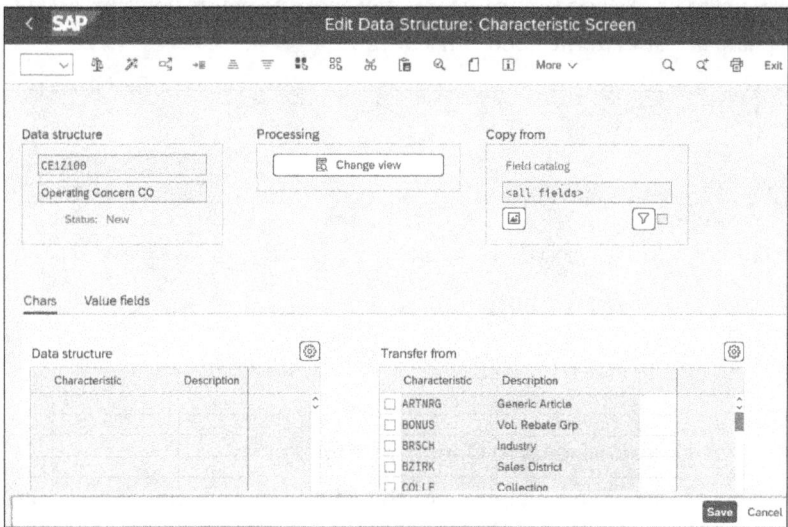

Figure 2.7 Assigning Characteristics to the Operating Concern

2 Configuring the Operating Concern and Basic Settings for Profitability Analysis

Choose characteristics Figure 2.8 shows three characteristics (**KUNWE**, **LAND1**, **MATKL**) selected by clicking on the checkbox in the first column.

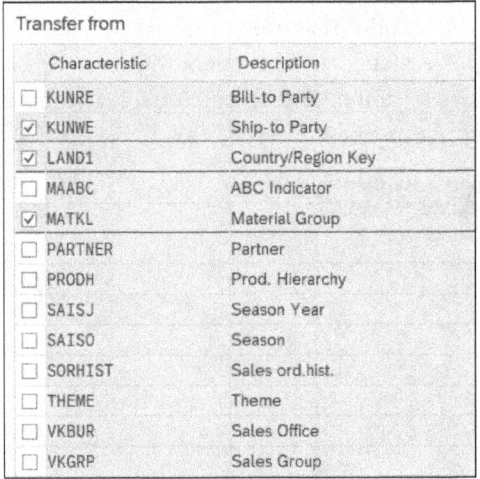

Figure 2.8 Choosing Characteristics to Transfer

Assign characteristics With ◁ (Transfer Fields), you can transfer the fields in the data structure. The three marked characteristics from Figure 2.8 are now assigned to the operating concern in Figure 2.9. When you click on **Edit • Change View**, you get more details of the characteristics such as **Length** and **Origin table**, as displayed in Figure 2.9. With ▷ (Reset Fields), you can remove characteristics from your data structure. It's hard to remove characteristics from the data structure after the operating concern is already activated. You can't remove assigned characteristics for which a data record already exists, so you need to have a good concept for your reporting before you start assigning characteristics to the operating concern.

Chars	Value fields					
Data structure						
Characteristic	Description	Cat.	Length	Check table	Origin table	Domain
☐ KUNWE	Ship-to Party	CHAR	10	KNA1	PAPARTNER	KUNNR
☐ LAND1	Country/Region Key	CHAR	3	T005	KNA1	LAND1
☐ MATKL	Material Group	CHAR	9	T023	MARA	MATKL

Figure 2.9 Characteristics Assigned to the Operating Concern

2.1.5 Assign Value Fields to Operating Concern

Choose value fields You can only save the assignment of characteristics after you have at least one value field assigned to the operating concern. The assignment of value fields to the operating concern is necessary because both costing-based profitability analysis and margin analysis have been activated.

The **Edit Data Structure: Value Field Screen** showing the **Value fields** tab appears in Figure 2.10, which looks very similar to the **Edit Data Structure: Characteristic Screen**.

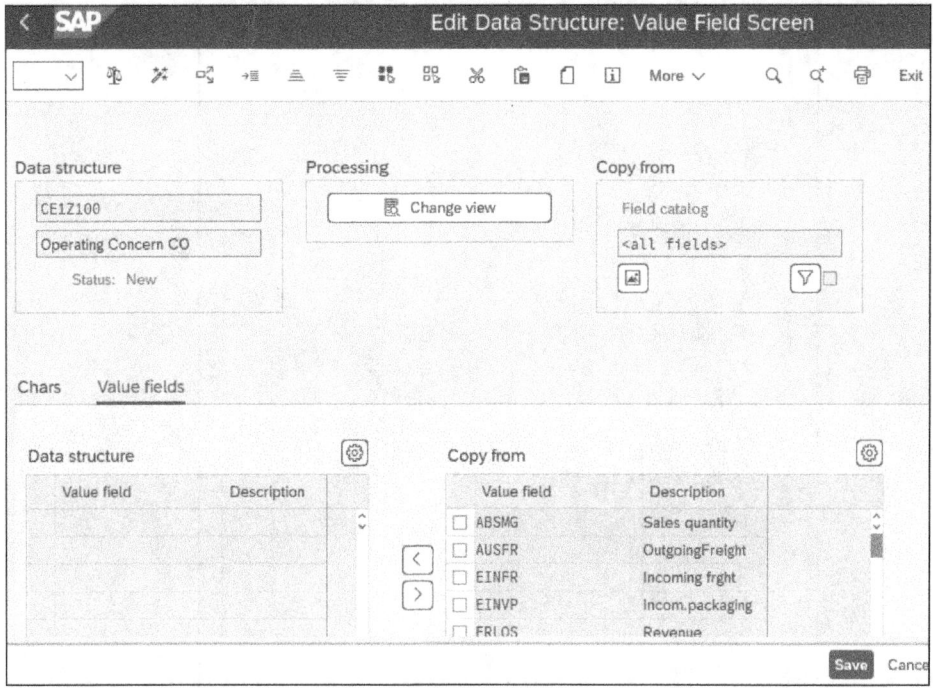

Figure 2.10 Assign Value Fields to the Operating Concern

Clicking on ▽ (**All Fields**) in the **Copy from** section on the upper-right side of the screen opens the popup shown in Figure 2.11 where you can filter for **Quantity** fields in the **Value field** column in the **Copy from** section in Figure 2.10. There are different ways to filter or search for value fields by setting a filter. You can also use the filter conditions for characteristics. You assign value fields to the data structure the same way you assigned characteristics to it: mark the value fields you want to assign, and then copy them to the data structure with < (Transfer Fields).

Value fields assignment to the operating concern

After you select all the characteristics and value fields you want to assign to the operating concern, save by clicking on **Save**. The saving of the data structures can take some time because the data tables will be generated in the background for costing-based profitability analysis. Refer to Figure 2.7 to see the **Data structure** section in the upper-left screen. The table name **CE1Z100** shown here is the database table for the data structure in costing-based profitability analysis.

Save and activate the data structure

2   Configuring the Operating Concern and Basic Settings for Profitability Analysis

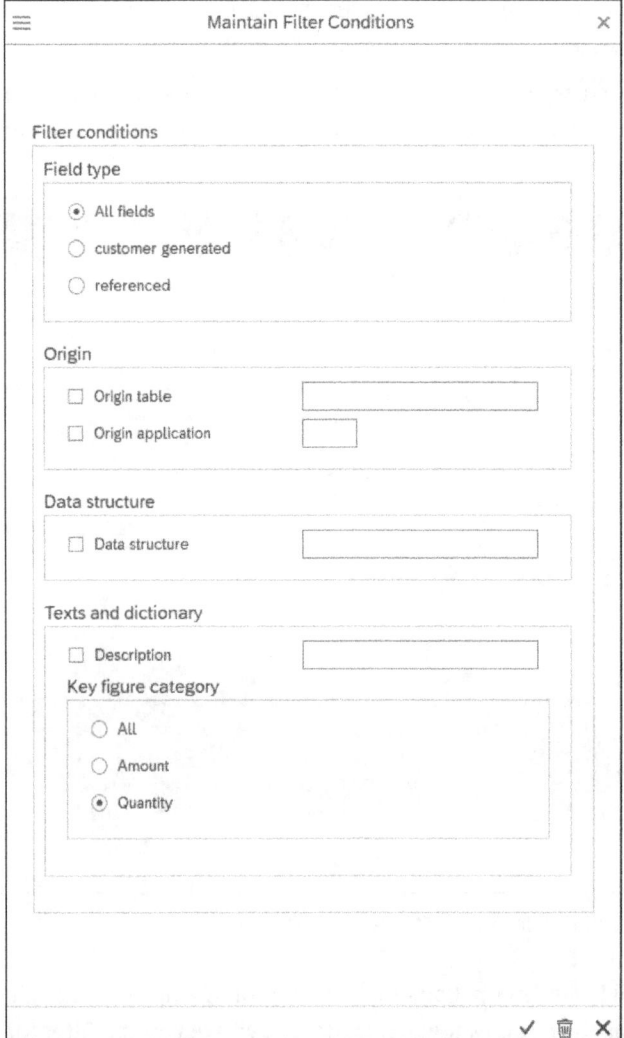

Figure 2.11 Filter Value Fields to Copy to the Operating Concern

Generate the operating concern

After the data structure is saved, you need to activate it with [✱] (Activate). After the activation, go back to the operating concern by pressing [F3]. You'll see the popup in Figure 2.12 asking if you want to generate the operating concern environment. Click **Yes** to activate the operating concern.

Operating concern status

After the operating concern is generated, go back to the **Environment** tab to see that both the **Cross-client part** and the **Client-specific part** are green, as is the overall **Status** of the operating concern right under the **Operating Concern** number in Figure 2.13.

The operating concern is now ready for use, but first let's look at other tasks that need to be completed before we start working with profitability analysis.

2.1 Creating the Operating Concern

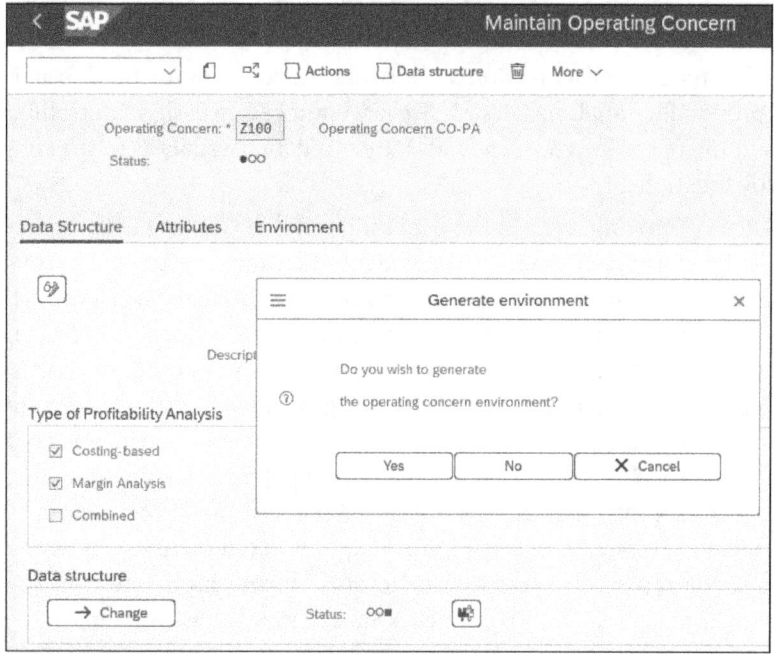

Figure 2.12 Generate the Operating Concern

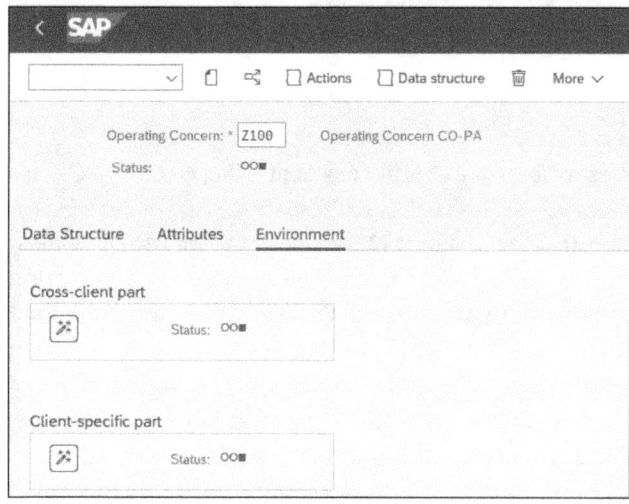

Figure 2.13 Operating Concern Status

> **Number of Characteristics and Value Fields in the Operating Concern**
>
> You can assign a maximum of 60 characteristics and 200 value fields to the operating concern. See SAP Note 160892 to extend the assignment. However, this modification isn't recommended because standard code is changed, and problems may occur regarding release changes.

2.2 Activating Profitability Analysis

After the creation and activation of the operating concern, you need to activate profitability analysis. First, however, you need to assign controlling areas to the operating concern to activate profitability analysis in the controlling functionality.

Assigning controlling areas

To assign controlling areas to the operating concern, follow the configuration path **Enterprise Structure • Assignment • Controlling • Assign Controlling Area to Operating Concern**. The transaction shows an overview of all controlling areas in the system in Figure 2.14 in the **COAr** (controlling area) column. In the **OpCo** (operating concern) column, you can see which controlling areas have an operating concern assigned. In our example in Figure 2.14, controlling area **KW01** has been assigned for operating concern **Z100** created earlier. You can assign several controlling areas to an operating concern if the controlling areas are assigned the same fiscal year variant as the operating concern. After confirming the assignment by pressing [Enter], the system displays the name of the operating concern in the **Name** column. Press [Ctrl]+[S] or click **Save** to save the assignment.

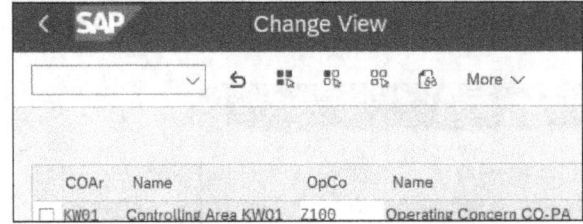

Figure 2.14 Assigning the Operating Concern to a Controlling Area

Activating profitability analysis: Transaction KEKE

After the controlling areas are assigned to the operating concern, you need to activate profitability analysis in the controlling area via Transaction KEKE or by following the configuration path **Controlling • Profitability Analysis • Flows of Actual Values • Activate Profitability Analysis**. As shown in Figure 2.15, the left **COAr** (controlling area) column provides an overview of the controlling areas available in the system. The **From FY** (from fiscal year) column indicates as of which year the controlling area has been valid. The **Op.co** (operating concern) column indicates whether and, if so, which operating concern is assigned to the controlling area. Figure 2.15 controlling area **Z100** is activated for both costing-based and account-based profitability analysis by selecting the corresponding checkboxes (**costing-based** and **Margin Analysis**). With account-based profitability analysis, **Margin Analysis** will be activated. SAP didn't change the naming in all configuration transactions, yet. You can also see a column for **combined**, which can't be checked, as we didn't activate combined profitability analysis in our operating concern. Save the changes with [Ctrl]+[S] or **Save**.

2.2 Activating Profitability Analysis

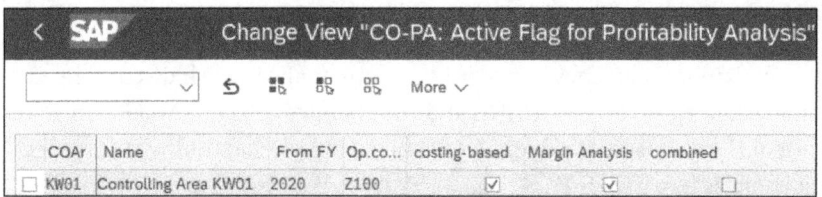

Figure 2.15 Activating Profitability Analysis in the Controlling Area

After you've saved the activation, you can see in Figure 2.16 that the **costing-based** and **Margin Analysis** checkboxes are checked under **Activate Components**.

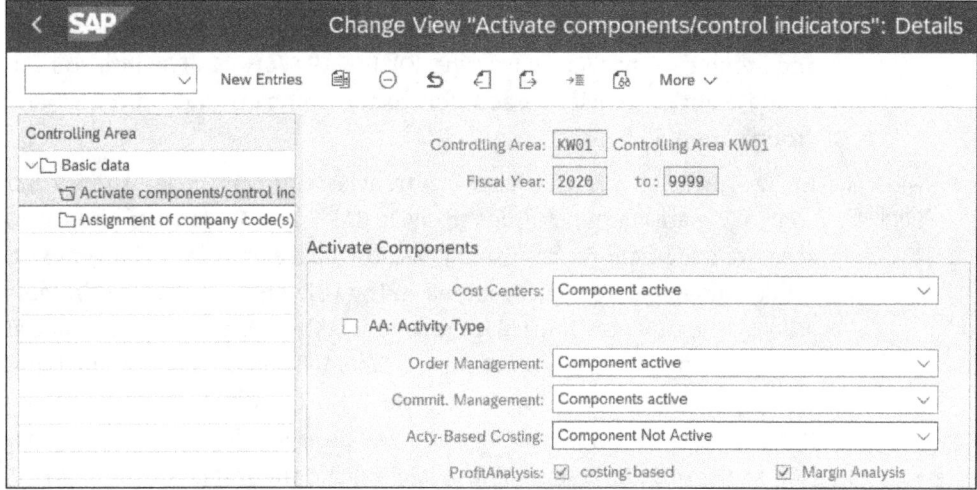

Figure 2.16 Activated Components in the Controlling Area

With its activation completed, profitability analysis is almost ready to use. The following sections describe which steps you have to perform to finally use profitability analysis.

Activated components in the controlling area

> **Operating Concern**
>
> In the operating concern, you define which type of profitability analysis you want to use. You can activate margin analysis, costing-based profitability analysis, or both types of profitability analysis.
>
> If you use SAP S/4HANA Finance, it's recommended to activate margin analysis. If you're working with SAP S/4HANA, you can activate both types of profitability analysis, whereas in SAP S/4HANA Cloud, you can only activate margin analysis.

53

2.3 Currencies in the Operating Concern

Local currencies in financial accounting

To explain the currency concept in margin analysis, we'll discuss the currency concept in SAP S/4HANA and the changes to FI of SAP ERP. In traditional FI (new general ledger [G/L]), you can use a maximum of three local currencies in parallel:

- Company code currency
- Group currency
- Global company currency

If you post a document, the document currency is converted into all three local currencies. In SAP S/4HANA Finance, on the other hand, the system converts all documents into the company code currency (currency type 10) and group currency (currency type 30). In SAP S/4HANA Finance, you can define eight currencies in addition to the company code currency and group currency.

Currencies in controlling

In CO of SAP ERP, you can use two currencies: company code currency and controlling area currency. Conversely, in SAP S/4HANA Finance, there's no separate document for controlling because cost elements no longer exist. All documents are primarily stored in the Universal Journal (i.e., in table ACDOCA). The Universal Journal is the central table for storing actual financial accounting and controlling data in SAP S/4HANA Finance. The original controlling database tables, such as table COEP (CO Object: Line Items [by Period]) still exist in the background and are also filled with data so that SAP standard reports are still available in the traditional CO in SAP ERP. Consequently, all original currency settings from CO are kept.

Controlling area currency

In controlling, you define the controlling area currency in the controlling area. To determine the currency in the controlling area, call Transaction OKKP, or use the following Customizing path: **Controlling • General Controlling • Organization • Maintain Controlling Area • Maintain Controlling Area**.

In the **Change View "Basic data": Details** screen shown in Figure 2.17, you can see that **Currency Type 30** is assigned for the **Group currency** in **Controlling Area KW01** in the **Currency Setting** section.

Group currency

SAP recommends assigning the group currency to the controlling area, as the group currency is usually used in all company codes, minimizing rounding differences. In addition, select the **Diff. CCode Currency** (different company code currency) checkbox to make the controlling documents available in the group currency and company code currency.

2.3 Currencies in the Operating Concern

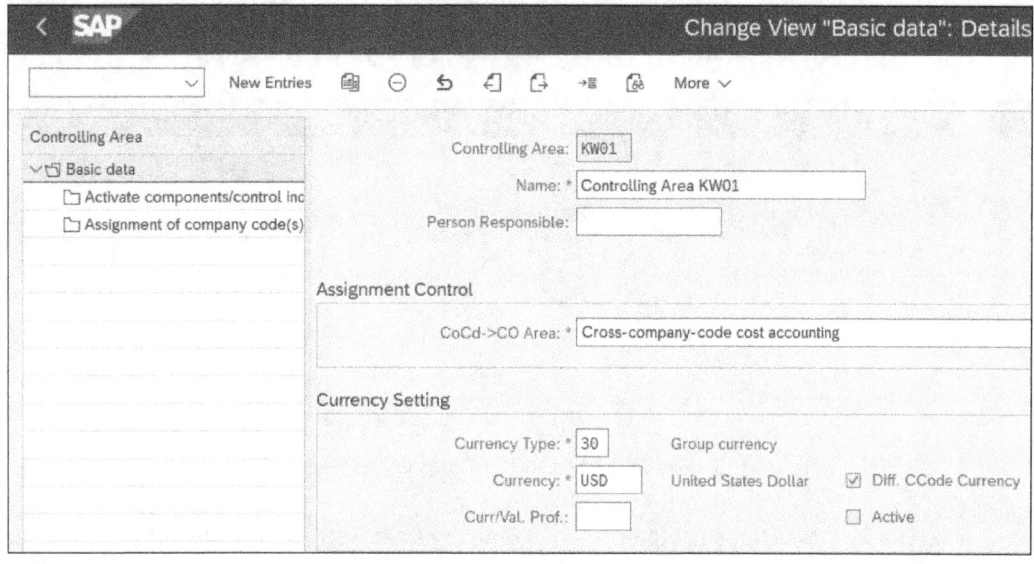

Figure 2.17 Maintaining Currencies in the Controlling Area

In SAP S/4HANA Finance, you implement the settings for the currency types in financial accounting for each ledger separately. To do so, follow the configuration path **Financial Accounting • Financial Accounting Global Settings • Ledgers • Ledger • Define Settings for Ledgers and Currency Types**. In Figure 2.18, you can see an overview of all ledgers in the system. You can check whether the leading ledger **0L** has the group currency 30 assigned, which is the controlling area currency for this example.

Overview of ledgers

Figure 2.18 Overview of Ledgers in the System

Navigate to the **Currency Types** folder in the **Dialog Structure** section. You see in Figure 2.19 an overview of all currency types in the system. **Currency Type 30** for **Group Currency** is highly recommended to use as the controlling area currency when using margin analysis.

Overview of currency types

Currency Type 20 is a currency type for the **Controlling Area**. If you're working with a group currency that is merely used for consolidation, and you

Currency type 20

55

2 Configuring the Operating Concern and Basic Settings for Profitability Analysis

want to display your controlling area data in a different currency, you can choose **Currency Type 20** and assign it to your controlling area.

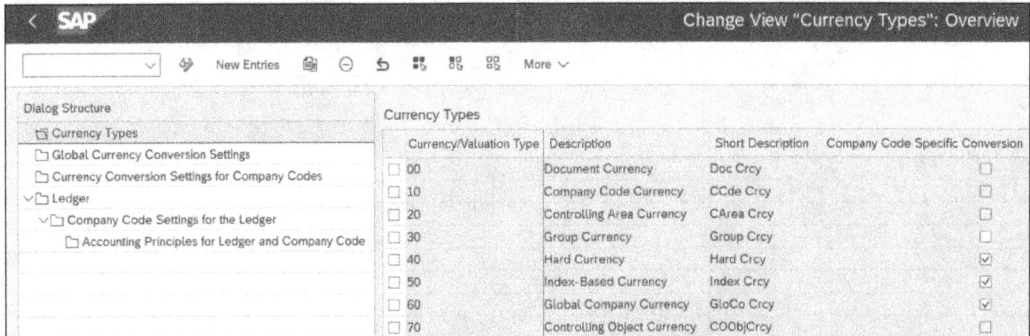

Figure 2.19 Overview of Currency Types in the System

Check currency settings for company codes

Now, let's navigate to the **Company Code Settings for the Ledger** folder in the **Dialog Structure**. In Figure 2.20, you can see that company code **1000** has both the local currency type **10** and the global currency type **30** assigned. All settings are as expected.

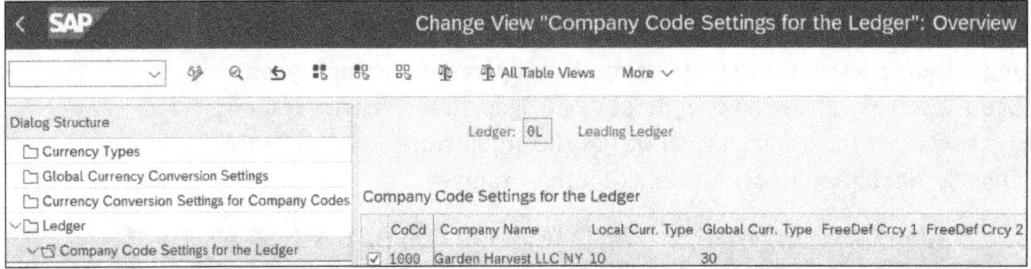

Figure 2.20 Display Company Code Settings for the Ledger

> **Currencies**
>
> Two currencies are available in the controlling area: company code currency and controlling area currency. In addition, you can define an operating concern currency in the operating concern.

2.4 Creating an Extension Ledger for Profitability Analysis

Definition: extension ledger

For some functionalities in margin analysis, you'll need an extension ledger to store predictive data. An extension ledger sits on top of a ledger, as shown in Figure 2.21. Let's consider an example: When you post a document in leading ledger 0L, it won't be posted in extension ledger EL. Likewise, when you post a document in extension ledger EL, it won't be posted in leading ledger 0L. When you run a P&L and balance sheet report for leading ledger 0L, you'll only see documents that are posted in leading ledger 0L; however, when you run a P&L and balance sheet report for extension ledger EL, you'll see both postings from leading ledger 0L and extension ledger 0L. All non-generally accepted accounting principles (GAAP)–relevant postings occur in extension ledger 0L. Anything GAAP-relevant will be posted in leading ledger 0L. There's a strict separation of these two ledgers.

An extension ledger can't live on its own; it always references a standard ledger. The extension ledger inherits the basic settings of the underlying ledger, such as assigned currency types. Let's see how extension ledgers are created.

Reference ledger

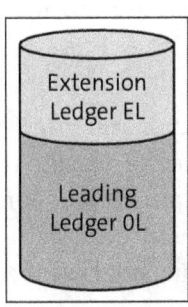

Figure 2.21 Relationship between Extension Ledger and Leading Ledger

Extension ledgers are created in financial accounting via the configuration path **Financial Accounting • Financial Accounting Global Settings • Ledgers • Ledger • Define Settings for Ledgers and Currency Types**. By pressing F5 or clicking **New Entries**, you can create an extension ledger, as shown in Figure 2.22. For this example, we created extension ledger **EL** by maintaining the following criteria:

Create an extension ledger

- **Ledger Name**
  Enter the name of the ledger. The name has two characteristics (for this example, extension ledger **EL**).

- **Leading**
  If you're creating a leading ledger, you need to check the checkbox in this column.

- **Ldgr Type**
  Choose from two ledger types:
  - **Standard ledger**: A standalone ledger, for example, 0L or LC for local accounting.
  - **Extension ledger**: A ledger that references a standard ledger (chosen for this example).

2 Configuring the Operating Concern and Basic Settings for Profitability Analysis

- **ExtLdgrTyp**
  Defines how the extension ledger can be used. To use the extension ledger for predictive accounting in margin analysis, choose **Line items with technical numbers/no deletion possible**.

- **Underlying Ledger**
  Defines which standard ledger the extension ledger is referencing (in this example, ledger **OL** is the reference ledger).

- **Man.Pstgs Not Allwd**
  Select this checkbox to prevent the extension ledger from accepting manual postings. This setting isn't mandatory for using predictive accounting in margin analysis. It's selected in this example to make sure that all data in the extension ledger is derived from transactions in SAP S/4HANA and that there are no accruals or predictions entered manually.

- **AcctgPrinc of Ledger**
  The accounting principle of the ledger specifies which accounting principle the ledger is assigned to. All company codes assigned to the extension ledger EL will be assigned to the same accounting principle. In our example, we assign the **Extension Ledger** to the accounting principle **USAP** (Local GAAP for USA), which is the same accounting principle the ledger OL is assigned to.

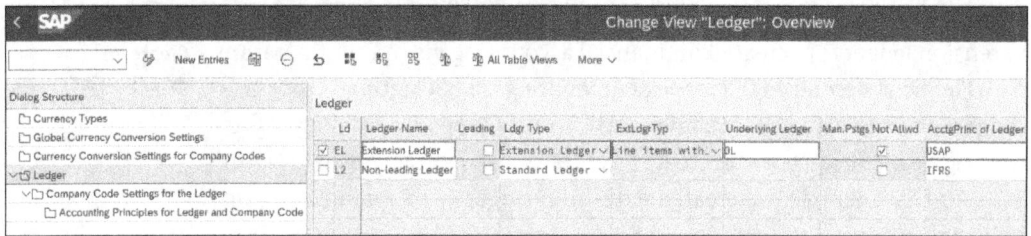

Figure 2.22 Creating the Extension Ledger

Company code settings in an extension ledger

After creating the extension ledger, you need to check the company code settings. In the **Dialog Structure** on the left side of the screen shown in Figure 2.23, open the **Company Code Settings for the Ledger** folder. Because the reference ledger for the extension ledger is ledger OL, the extension ledger is automatically assigned to all company codes in the system that have ledger OL. Therefore, you don't have to do anything except check in Figure 2.23 that company code **2000** has an entry for **Ledger EL**.

Check the prediction ledger

Before you can use the extension ledger in margin analysis, you need to define it as a prediction ledger by following the configuration path **Financial Accounting • Predictive Accounting • Check Prediction Ledger**. Press F5 or click **New Entries** to assign the extension ledger **EL** in Figure 2.24 as the prediction ledger. Only if the extension ledger is created correctly will you be able to assign it as a prediction ledger for margin analysis. You can also define the extension ledger as **Relevant for Commitment Management**, but this can only be activated for one extension ledger, not multiple.

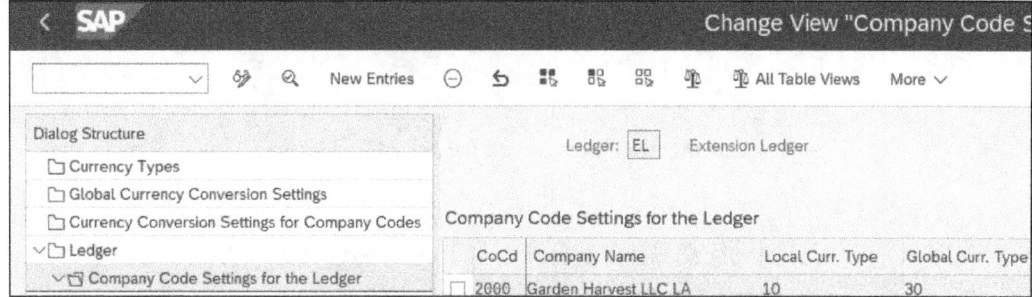

Figure 2.23 Check Company Code Settings for the Extension Ledger

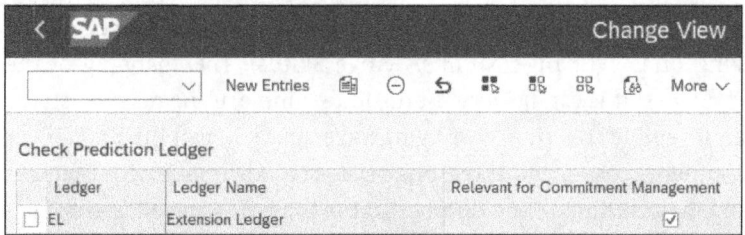

Figure 2.24 Define the Prediction Ledger

With the extension ledger, you can use predictive accounting in margin analysis. You'll learn how to set up predictive accounting in Chapter 4.

2.5 Creating Number Ranges for Profitability Analysis

In this section, you'll learn how to create number ranges for costing-based profitability analysis. For margin analysis, you don't have to create number ranges.

To maintain number ranges, call Transaction KEN1 or follow the configuration path **Controlling • Profitability Analysis • Flows of Actual Values • Initial Steps • Define Number Ranges for Actual Postings**.

Maintaining number ranges

In the screen shown in Figure 2.25, you enter the **Operating concern** for which you want to maintain the number range. In our example, number ranges are maintained for operating concern **Z100**. To maintain intervals, click on 🖉 **Intervals** (Change Intervals).

2   Configuring the Operating Concern and Basic Settings for Profitability Analysis

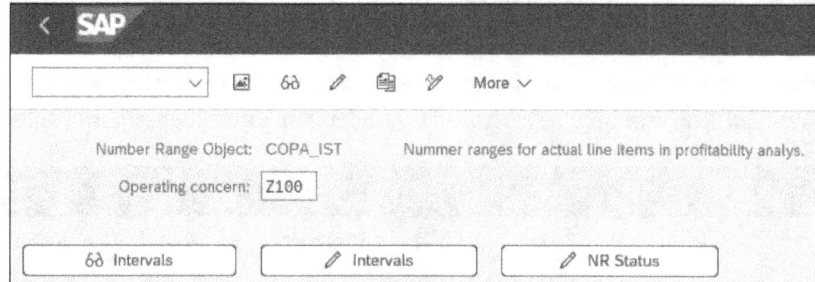

Figure 2.25 Maintaining Intervals for Actual Line Items

Maintain intervals   In Figure 2.26, you can see that interval **01** has been created. You can adjust the number range, but for this example, leave everything as it is.

The **NR Status** column displays the number status that indicates how many documents have already been posted. In Figure 2.26, the number status is **0**, which means that there are no documents posted in costing-based profitability analysis yet. You can change the number status in Transaction KEN1 by clicking on 🖉 (Change Number Range Status). The changing of the number range status can be very useful, for example, if the number status has been reset due to a transport by mistake, and you can thus no longer save documents for technical reasons because a document with the next free number according to the number status is already stored in the database. If numbers are supposed to be assigned externally, the checkbox in the **External** column will be selected. This checkbox is obsolete in SAP S/4HANA as it's no longer possible to assign an external number range to costing-based profitability analysis in SAP S/4HANA.

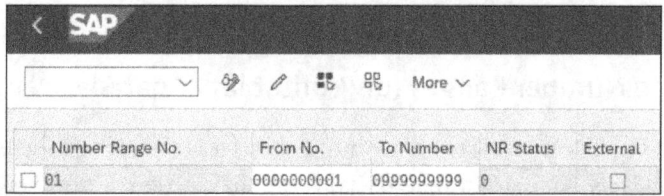

Figure 2.26 Maintain the Number Interval for Costing-Based Profitability Analysis

Record types   The transaction to maintain number intervals for costing-based CO-PA looked different in SAP ERP. You could maintain a number range by record type. Record types are only available in costing-based profitability analysis and describe the business transaction from which the transferred data is obtained. Record types can be compared to document types in financial accounting. You can use transaction types in analyses.

2.5 Creating Number Ranges for Profitability Analysis

The SAP system provides the record types listed in Table 2.1 in costing-based profitability analysis.

Overview of record types

Element	Element Text
A	Incoming Sales Orders
B	Direct Account Assignment
C	Order/Project Settlement
D	Overhead Costs
E	Single Transaction Costing
F	Billing Data
G	Customer Arrangements
H	Statistical Key Figures
I	Customer Order Project
Y	External Data

Table 2.1 Overview of Transaction Types

Save your settings. The maintenance of number range intervals isn't linked to automatic transport recording. You have to add changes to number range intervals to the transport request manually.

To do so, select **More • Interval • Transport** in the menu bar. The system displays a warning message that indicates the number range intervals in the target system are deleted during the transport. The general recommendation is to maintain number ranges directly in the target system. You can access the transactions for the maintenance of number ranges in the production system, irrespective of the corresponding configuration. If you still want to transport number range intervals, confirm the warning message by choosing **Yes**, and select a transport request or create a new transport request via which you want to transport the settings.

Transporting number ranges

If you use the planning function in costing-based profitability analysis, you also have to create number range intervals to store planning data. To maintain number ranges for planning, call Transaction KEN2, or follow the configuration path **Controlling • Profitability Analysis • Planning • Initial Steps • Define Number Ranges for Planning Data**.

Number ranges for planning data

You maintain number range intervals for planning data in the same way as number range intervals for actual data. In Figure 2.27, for example, enter the **Operating concern** "Z100" to maintain number range intervals for.

Number range intervals

2   Configuring the Operating Concern and Basic Settings for Profitability Analysis

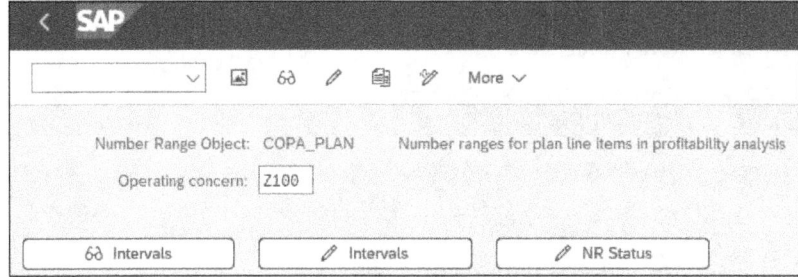

Figure 2.27 Maintaining Intervals for Plan Line Items

Maintaining number ranges

To maintain number ranges, click on ✎ **Intervals** (Change Intervals), and you'll see the number range interval that has already been created by the system in Figure 2.28.

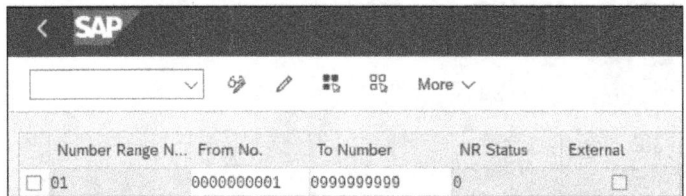

Figure 2.28 Maintaining Number Intervals for Plan Line Items in Costing-Based Profitability Analysis

Transporting number range intervals

Like the number range intervals for actual line items, the number range intervals for plan line items aren't linked to automatic transport recording. You should maintain number range intervals directly in the target system to avoid overwriting number ranges due to transports.

> **Number Ranges**
>
> You have to create number ranges only in costing-based profitability analysis for both actuals and plan.

2.6   Assigning Versions to the Operating Concern

In controlling, you use versions that enable you to store actual data and different variants of planning data. When controlling areas are created, the system automatically creates version 0 to store planning and actual data.

Version for actual data

Version 0 is the only version that is able to store actual data. Every version has operating concern–specific settings, which you maintain via configuration path **Controlling** • **Profitability Analysis** • **Planning** • **Initial Steps** • **Maintain Versions**. In Figure 2.29, you see an overview of all versions that

2.6 Assigning Versions to the Operating Concern

are available in the system. Select **Version 0** by clicking on the beginning of the row and navigating to the **Dialog Structure** on the left side of the screen to the **Settings in Operating Concern** folder.

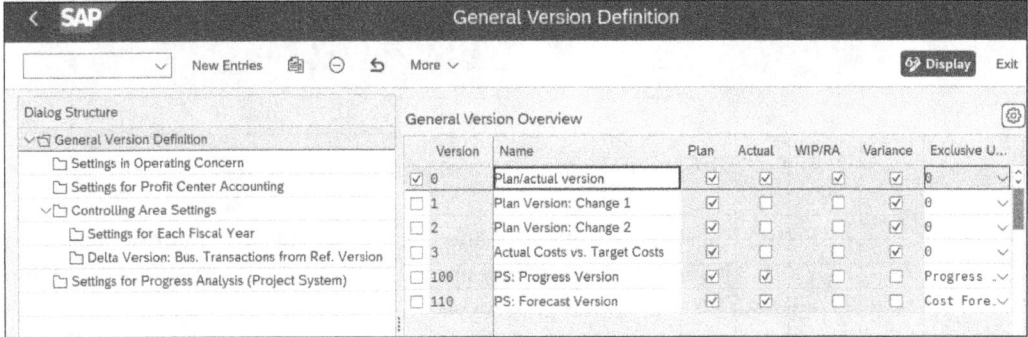

Figure 2.29 Maintaining Versions

After you've clicked on the **Settings in Operating Concern** folder, the system displays the popup shown in Figure 2.30 asking you to enter the operating concern for which you want to maintain the version. You also need to choose the type of profitability analysis. In the example in Figure 2.30, **costing-based** has been chosen, but your choice doesn't matter because the version will be maintained for both types of profitability analysis, so there's no need to go back and make the same entries for margin analysis. Confirm your entries with Enter.

Set the operating concern

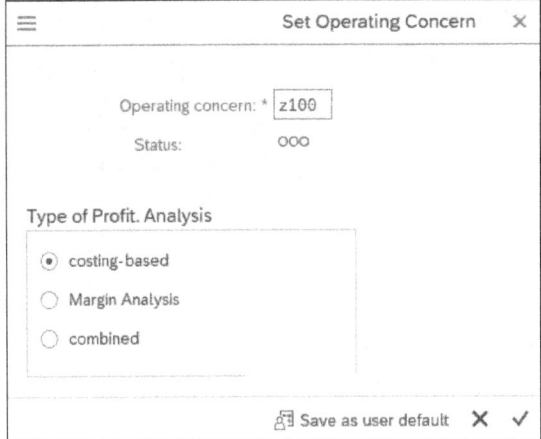

Figure 2.30 Set Operating Concern for Version Maintenance

After confirming your operating concern, you'll see another popup, as shown in Figure 2.31, asking if you want to create version 0 for the operating concern. Confirm the popup by clicking on **Yes**.

63

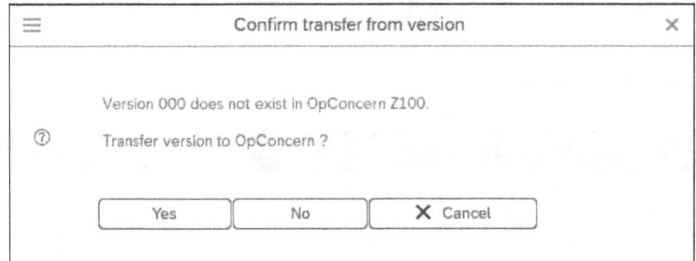

Figure 2.31 Confirm Transfer from Version 0

Figure 2.32 shows the maintenance view of **Version 0** in **Operating concern Z100**. All fields are already prepopulated. You can make the following entries in the version maintenance.

Locking versions
In the **Attributes** section, you can lock the version by selecting the **Version Locked** checkbox. Locking the version means that neither actual data nor planning data can be stored in version 0 for the operating concern, thus no postings are possible in profitability analysis.

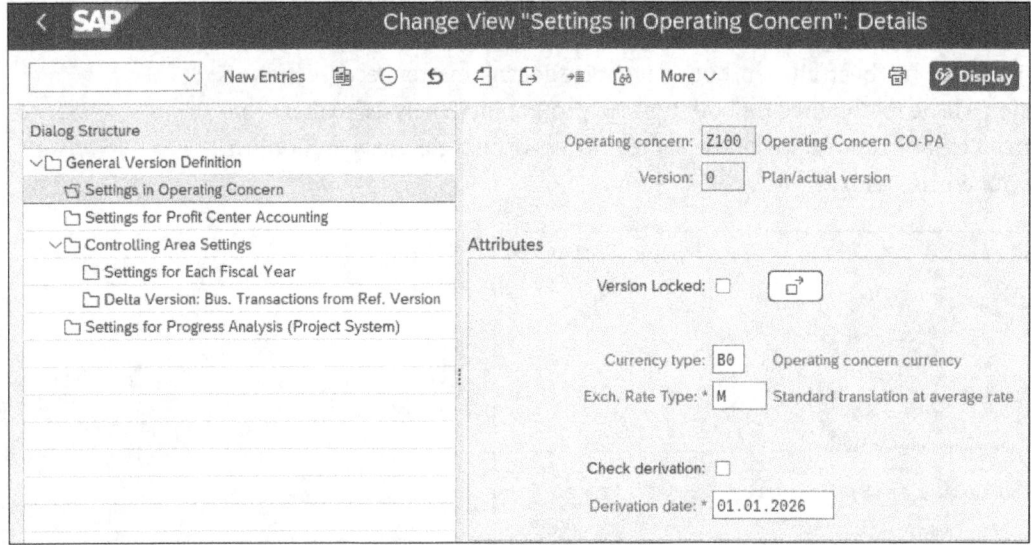

Figure 2.32 Maintain Version 0 for the Operating Concern

Maintaining currency types
You enter the currency in which values in the operating concern can be displayed and planned in the **Currency type** field ("B0"). Usually, currency type B0 (**Operating concern currency**) is maintained here.

In the **Exch. Rate Type** field ("M"), you define the exchange rate type with which foreign currencies are converted in profitability analysis planning.

Checking the derivation
If you select the **Check derivation** checkbox, you determine that characteristic derivations that are defined in the system are checked when planning

data is entered. For example, if a derivation for the **Domestic/International** characteristic exists, and the **Domestic** characteristic value is maintained for the country **Germany**, you can't maintain planning data for the **Germany/International** characteristic combination. The **Derivation date** field is mandatory but only relevant if the **Check derivation** checkbox is selected. In this case, the system checks the derivation of the characteristic values according to the derivation rules that are valid on this date.

Save your changes after your version maintenance is complete.

Saving the version

> **Versions**
>
> Version 0 for storing actual data and planning data always needs to be activated for the operating concern, irrespective of whether you use costing-based profitability analysis or margin analysis.
>
> In addition to version 0, you have to activate all other plan versions in the operating concern that you're planning to use.

2.7 Setting the Operating Concern

When working with profitability analysis, you'll be asked every time you call a transaction to enter the operating concern. To avoid this, you can set the operating concern in your user parameters by calling Transaction KEBC or following the menu path **Accounting** • **Controlling** • **Profitability Analysis** • **Environment** • **Set Operating Concern**.

Set the operating concern

In Figure 2.33, enter your operating concern. In this example, **Operating concern Z100** has been set. You also have to choose the type of profitability analysis.

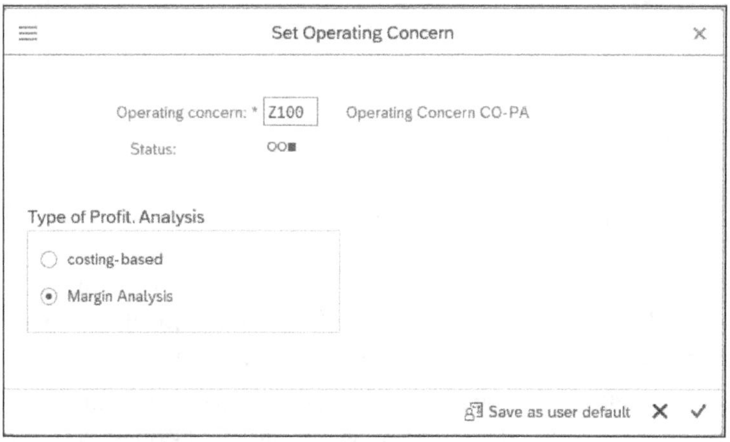

Figure 2.33 Set the Operating Concern

65

We chose **Margin Analysis** for this example, although both types of profitability analysis are activated in the operating concern. We just assume that we'll use margin analysis more often than costing-based profitability analysis. Click on 📋 **Save as user default** to save your changes in your user parameter.

You can go back to Transaction KEBC and change your settings anytime.

2.8 Transporting the Operating Concern

After you've finished the configuration for the operating concern, you can transport it, including all related settings.

Transporting the operating concern

Call Transaction KE3I or follow the configuration path **Controlling** · **Profitability Analysis** · **Tools** · **Production Startup** · **Transport**. In the screen shown in Figure 2.34, you can choose the objects you want to transport (**Operating concern** in this example).

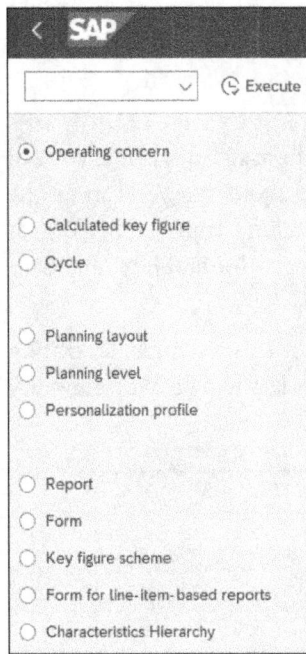

Figure 2.34 Transporting the Operating Concern

Click on 🕒 **Execute**, or press [F8]. The system then displays the dialog box in Figure 2.35 that prompts you to choose or create a **Customizing Request** or **Workbench Request**. Press [Enter] to confirm your entries.

The system provides an overview of all operating concerns that are available for transport. In this example, as shown in Figure 2.36, operating concern **Z100** has been selected by clicking on the beginning of the row.

2.8 Transporting the Operating Concern

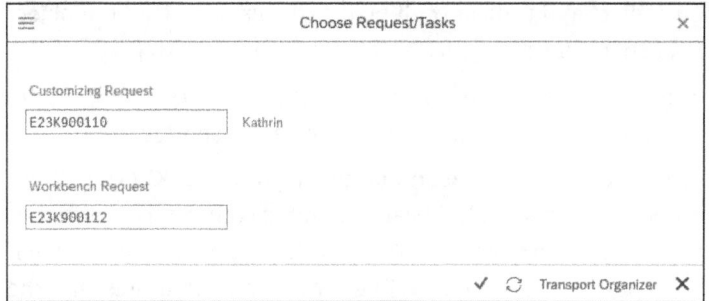

Figure 2.35 Selecting Transport Requests

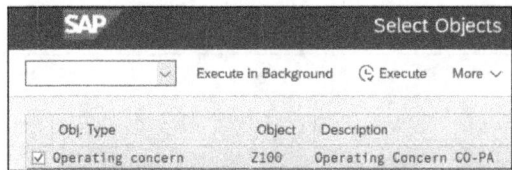

Figure 2.36 Selecting the Operating Concern for Transport

After choosing your operating concern for transport, click on ⊕ **Execute** to choose the objects you want to include in your transport.

In Figure 2.37, you see an overview of all dependent object classes that are available for transport.

Transporting object classes

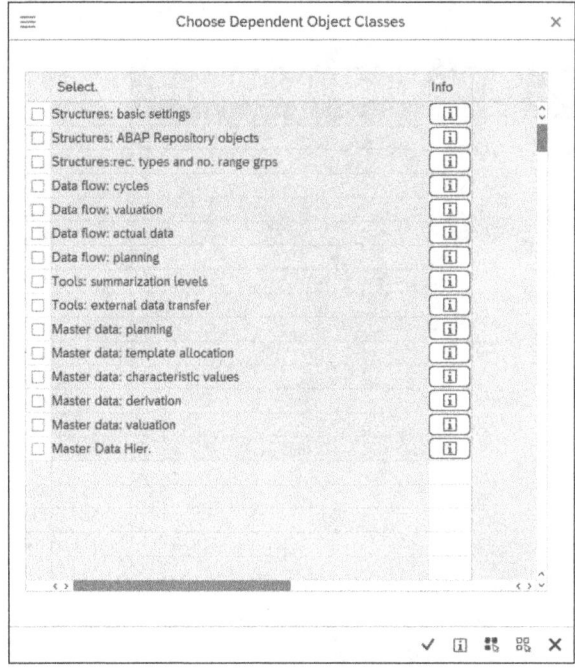

Figure 2.37 Choosing Object Classes for Transport

You can either select and transport all object classes or restrict the object classes to those that have changed. Confirm your entries with [Enter].

You get a confirmation as shown in Figure 2.38 that the selected objects are now included in the transport requests and can be transported.

After importing the requests, you should use Transaction KEAO to check whether the operating concern has been generated completely or whether you have to generate it retroactively. You can check the status in the **Environment** tab, as shown in Figure 2.39. The green status indicates that the operating concern is generated completely. You can activate the **Cross-client part** and **Client-specific part** independently via their [✗] (Activate) buttons.

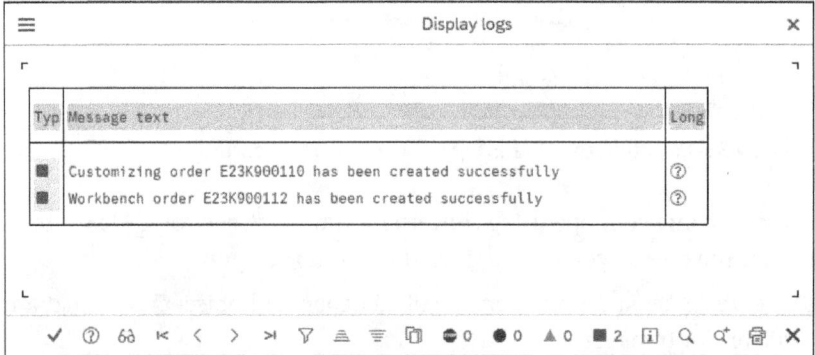

Figure 2.38 Created Transport Requests for Operating Concern

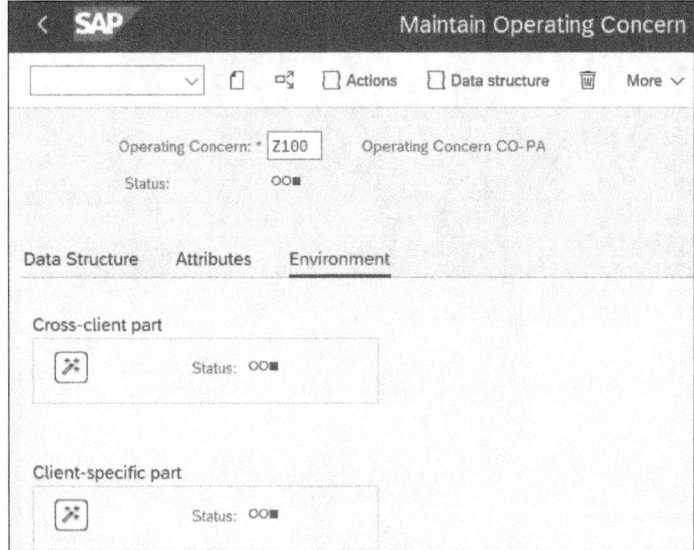

Figure 2.39 Check the Operating Concern Status

> **Transporting the Operating Concern**
> You can transport the operating concern for margin analysis and costing-based profitability analysis to the target system using the corresponding Customizing transaction.

2.9 Summary

This chapter explained how to create the organizational structure of profitability analysis and all initial settings to activate it. Most settings are valid for costing-based profitability analysis and margin analysis, but when this wasn't the case, we pointed that out in the text. After the settings in this chapter are completed, you can start posting to the operating concern, at least in margin analysis. For costing-based profitability analysis, you first have to configure some of the interfaces described in Chapter 5.

The next chapter provides detailed information on how to create additional characteristics and characteristic derivations.

Chapter 3
Configuring Characteristics

Characteristics define on which level you can analyze your data. Both costing-based profitability analysis and margin analysis use characteristics to structure reporting.

Characteristics are required to define the basic structure of costing-based profitability analysis and margin analysis. You should put some time into defining characteristics and thinking about strategic goals when you set up your characteristics. It isn't easy to change characteristics after you're productive with costing-based profitability analysis or margin analysis. You can't delete characteristics that have postings at all. In addition, you're limited to 60 characteristics that you can assign to your operating concern. There are some fixed characteristics that are assigned to every operating concern by default. The 60 characteristics can be assigned in addition to the fixed characteristics.

Characteristics: basic structure of profitability analysis

All characteristics in margin analysis are available in the Universal Journal. The characteristics in costing-based analysis are stored in the separate characteristics table CE4xxxx. The xxxx can be replaced with the technical name of your operating concern.

Storing characteristics

The fixed characteristics in SAP standard can be accessed from the data structure that you can display via the operating concern. To see the data structure, use Transaction KEA0 to display the operating concern Z100, and then click on →| **Display** (Display Data Structure) in the **Data structure** section (see Figure 3.1). By choosing **Extras** • **Display fixed fields**, you can display the fixed fields that are assigned to every operating concern by default.

Fixed characteristics

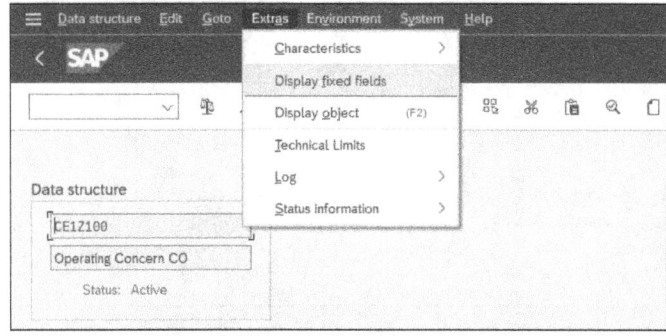

Figure 3.1 Display Fixed Fields

3   Configuring Characteristics

Types of fixed characteristics

The fixed fields are divided into three categories:

- **Charact.**
  In this tab, you'll find characteristics for the organizational structure, such as company code, profit center, and so on, as well as characteristics for the main master data, such as customer, vendor, article, and so on.
- **Techn. fields**
  In this tab, you'll find the document numbers and all characteristics related to the period, such as fiscal year, posting period, and so on.
- **Acctg amounts**
  In this tab, you'll find all the quantity fields and amount fields.

The fixed fields have more relevancy for costing-based profitability analysis as all those fields are available in the Universal Journal by default and will be updated in margin analysis after the operating concern is activated. Before you assign characteristics to your operating concern, you should check the fixed fields to see which ones are assigned to your operating concern by default. Figure 3.2 shows an extract of the characteristics that are fixed fields. We won't list all the fixed fields here, but you now know how to display them.

Field name	Description	Use	DTyp	Length	Orig.Table	Data element
ARTNR	Product	F	CHAR	40	MARA	ARTNR
BUKRS	Company Code	F	CHAR	4		BUKRS
COPA_KOSTL	Cost center	F	CHAR	10		COPA_KOSTL
FKART	Billing Type	F	CHAR	4		FKART
FKBER	Functional Area	F	CHAR	16		FKBER
GSBER	Business Area	F	CHAR	4		GSBER
KAUFN	Sales Order	F	CHAR	10		KDAUF
KDPOS	Sales Ord. Item	F	NUMC	6		KDPOS
KNDNR	Customer	F	CHAR	10	KNA1	KUNDE_PA
KOKRS	CO Area	F	CHAR	4		KOKRS
KSTRG	Cost Object	F	CHAR	12		KSTRG
LIEFNR	Delivery	F	CHAR	10		VBELN_VL
LIEFPOS	Item	F	NUMC	6		POSNR_VL
PPRCTR	Partner PC	F	CHAR	10		PPRCTR
PRCTR	Profit Center	F	CHAR	10	MARC	PRCTR
PSPNR	WBS Element	F	NUMC	8		PS_PSP_PNR
RKAUFNR	Order	F	CHAR	12		AUFNR
SEGMENT	Segment	F	CHAR	10		FB_SEGMENT
SERVICE_DOC_ID	Service Document	F	CHAR	10		FCO_SRVDOC_ID
SERVICE_DOC_ITEM_ID	Service Doc. Item	F	NUMC	6		FCO_SRVDOC_ITEM
SERVICE_DOC_TYPE	Service Doc. Type	F	CHAR	4		FCO_SRVDOC_TYPE
SOLUTION_ORDER_ID	Solution Order	F	CHAR	10		FCO_SOLUTION_OF
SOLUTION_ORDER_ITE..	Solution Order Item	F	NUMC	6		FCO_SOLUTION_OF

Figure 3.2 Fixed Fields of an Operating Concern

In addition to the fixed fields, you can assign characteristics to the operating concern from a list of already available characteristics in the client. You also can create customer-specific characteristics and assign them to the operating concern. The following types of customer-specifics characteristics are available in profitability analysis:

- **Creating characteristics from SAP tables**
  You can copy characteristics from existing SAP standard tables.
- **Creating characteristics with your own value maintenance**
  You can create your own values for your user-defined characteristic.
- **Creating characteristics without value maintenance**
  You can assign values to characteristics that don't have specific values. These characteristics are often created for technical reasons.
- **Creating characteristics with reference to existing values**
  You can copy an existing characteristic, including its values.

Types of customer-specific characteristics

In this chapter, we'll give you step-by-step instructions on how to create the different types of characteristics and give meaningful examples for when to create each type of characteristic. Then, you'll learn how to create characteristic derivations and how to analyze them. Finally, we'll show you how to realign characteristics when you, for example, decide to add a new characteristic to your operating concern and want to derive that characteristic for all existing documents.

3.1 Characteristics

In this section, we'll introduce the various forms of user-defined characteristics. We'll show how you can create user-specific characteristics and add them to the operating concern.

3.1.1 Creating, Changing, and Updating Characteristics

Transaction KEA5 can be used to create, change, and display characteristics. You can access Transaction KEA5 by following the configuration path: **Controlling • Profitability Analysis • Structures • Define Operating Concern • Maintain Characteristics**.

Using Transaction KEA5

Figure 3.3 shows the selection screen to create/change or display characteristics transactions. The screen is divided into two sections. In the **Choose Characteristics** area, you have the following different functions for changing/displaying a characteristic:

- **All Characteristics**
  All characteristics that have been created in the system are displayed/available in change mode.

- **Chars from operating concern**
  Only the characteristics of a predefined operating concern are displayed/available in change mode.

- **Characteristics that are not used in operating concerns**
  Characteristics that aren't used in any operating concern existing in the system are displayed/available in change mode.

Create new characteristics

In the **Create Characteristic** area of the screen, you can create a new characteristic. We explain how to create the different types of customer-specific characteristics in Section 3.1.2 and the following.

Change characteristics

In this example, we'll execute Transaction KEA5 to view all characteristics that are assigned to operating concern Z100 in change mode. Therefore, select **Chars from operating concern**, enter "Z100" in the text box, and click on [pencil icon] **Change**.

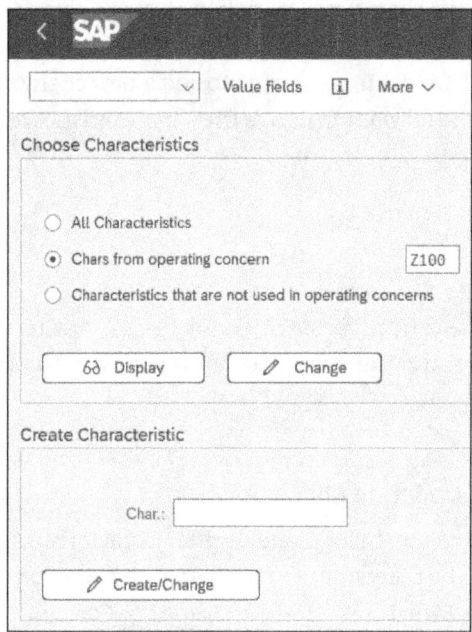

Figure 3.3 Selection Screen to Create/Change and Display Characteristics

Characteristics of the operating concern

In Figure 3.4, you see an overview of all characteristics assigned to operating concern Z100. You can change the name of the customer-specific characteristics in change mode. Otherwise, no changes to characteristics are possible. You recognize customer-specific characteristics by their technical name because they must start with "WW". In Figure 3.4, no customer-specific characteristics have been assigned to the operating concern yet, but the following information about the characteristics is available:

- **Char.**
  The technical name of the characteristic. Customer-specific characteristics start with "WW".
- **Description**
  The description of the characteristic.
- **Short text**
  The short text of the characteristic.
- **DTyp**
  Whether the characteristic is a numeric characteristic (**NUMC**) or whether other characters are permitted as values (**CHAR**).
- **Lgth.**
  The lengths of the characteristic values.
- **Origin Table**
  For all noncustomer-specific characteristics, the origin table columns from which table they are originating.
- **Origin field d**
  The origin field if the characteristic is created with reference to an existing characteristic.

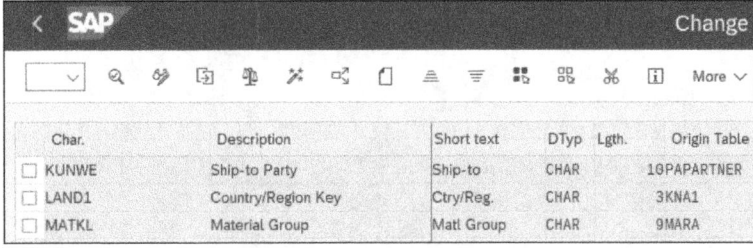

Figure 3.4 List of Characteristics Assigned to the Operating Concern

Let's look at a characteristic's details. Double-click on one of the characteristics, and you'll be forwarded to the detailed view. For example, Figure 3.5 shows the details of **Characteristic LAND1** (Land).

Details of characteristics

The detail screen is divided into several sections:

Characteristic detail screen

- **Texts**
  Defines the description/name of the characteristic.
- **ABAP Dictionary**
  Displays the technical name of the characteristic.
- **Origin**
  Contains the origin table and origin field if the characteristic has been created with reference to an existing SAP table.

3 Configuring Characteristics

- **Further Properties**
  Provides information on whether the characteristic is already active.

- **Validation**
  Shows which characteristic type the characteristic has and if the characteristic has a check table that is storing its values. You'll learn more about the different characteristic types in Section 3.1.2 and following.

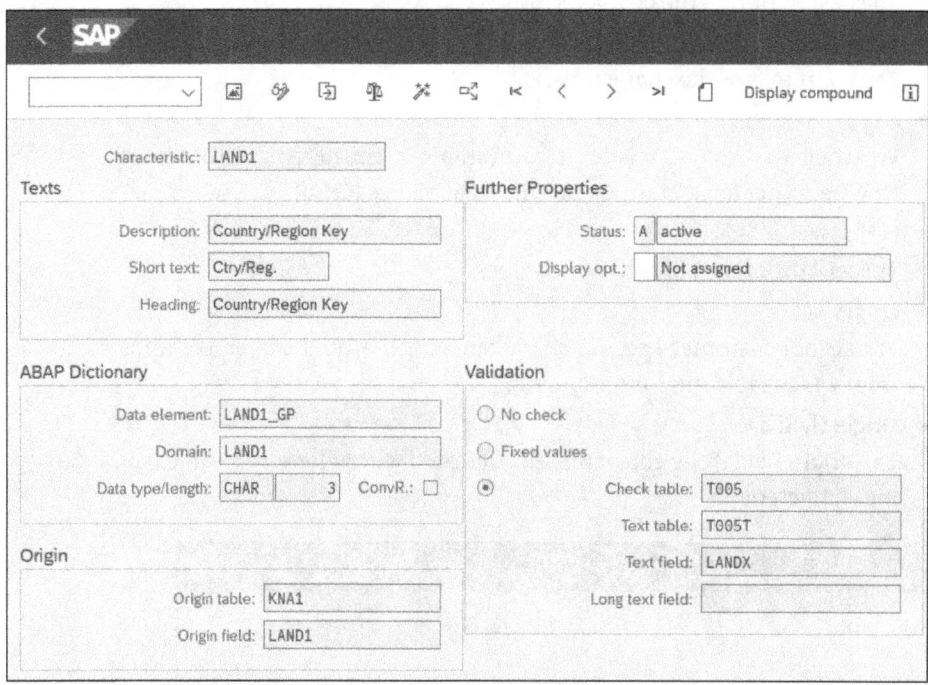

Figure 3.5 Details of a Characteristic

3.1.2 Creating a Characteristic from an SAP Table

Using Transaction KEA5

To create a characteristic with reference to an SAP table, go to Transaction KEA5, or follow the configuration path **Controlling • Profitability Analysis • Structures • Define Operating Concern • Maintain Characteristics**.

You create a new characteristic by clicking on 🖉 **Create/ Change** in the **Create Characteristic** section of the screen shown in Figure 3.6.

Choose a characteristic

In the popup shown in Figure 3.7, choose the type of characteristic to create (e.g., **Transfer from SAP table**). With the F4 Help, you can display all the reference tables that are available for transferring a characteristic, as shown in Table 3.1. You need to know the technical name and table from which you want to transfer the characteristic.

3.1 Characteristics

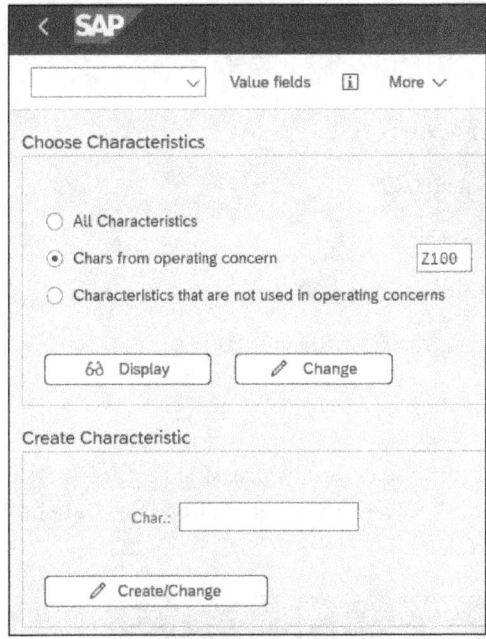

Figure 3.6 Create a Characteristic

Table	Description
KNA1	General Data in Customer Master
KNB1	Customer Master (Company Code)
KNVV	Customer Master Sales Data
MARA	General Material Data
MARC	Plant Data for Material
MVKE	Sales Data for Material
PACPROJECT	Characteristics from cProjects That Can Be Used in CO-PA
PACRMIPM	Characteristics from CRM IPM That Can Be Used in CO-PA
PACRMSLS	Characteristics from CRM Sales That Can Be Used in CO-PA
PACRMSRV	Characteristics from CRM Services That Can Be Used in CO-PA
PAPARTNER	SD Partners That Can Be Used in CO-PA
PARETAIL	Retail Fields That Can Be Used in CO-PA
PRPS	WBS (Work Breakdown Structure) Element Master Data

Overview of reference tables

Table 3.1 Overview of SAP Tables

3   Configuring Characteristics

Table	Description
T001W	Plants/Branches
VBAK	Sales Document: Header Data
VBAP	Sales Document: Item Data
VBKD	Sales Document: Business Data
VIAUFKST	Generated Table for View VIAUFKST
WTY_COPA	Warranty Processing: Characteristics for CO-PA

Table 3.1 Overview of SAP Tables (Cont.)

Choosing a reference table

After you select the table, confirm the selection by pressing `Enter`. (In the example shown in Figure 3.7, table **KNVV – Customer Master Sales Data** has been selected as the reference table.)

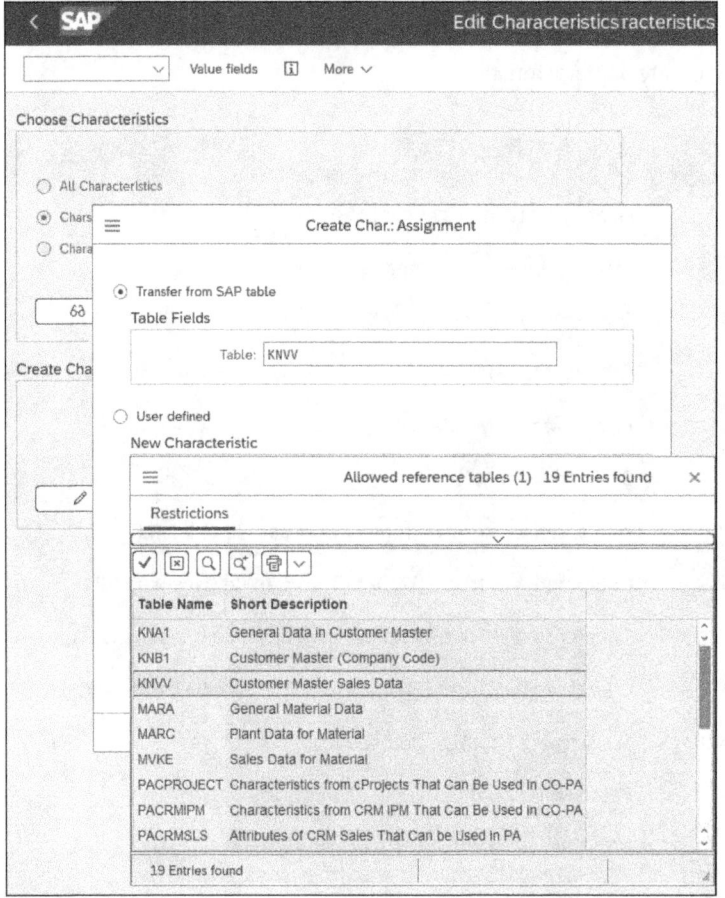

Figure 3.7 Choose the Reference Table for the Characteristic

In Figure 3.8, you see all fields that are available in table KNVV (Customer Master Sales Data). The characteristics that are grayed out can't be transferred to the operating concern. They may already be transferred, they may be part of the fixed fields, or a transfer is technically not possible. All characteristics with white background can be transferred. Select the checkbox next to **INCO1 – Incoterms**, and then press `Enter`.

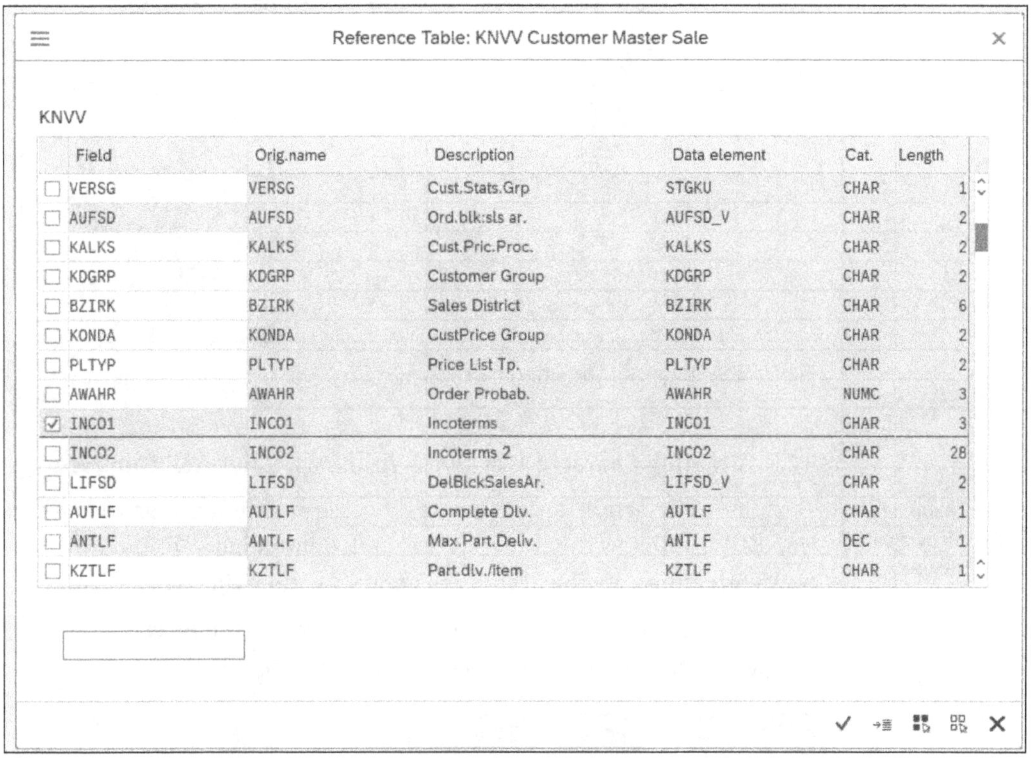

Choosing a characteristic

Field	Orig.name	Description	Data element	Cat.	Length
☐ VERSG	VERSG	Cust.Stats.Grp	STGKU	CHAR	1
☐ AUFSD	AUFSD	Ord.blk:sls ar.	AUFSD_V	CHAR	2
☐ KALKS	KALKS	Cust.Pric.Proc.	KALKS	CHAR	2
☐ KDGRP	KDGRP	Customer Group	KDGRP	CHAR	2
☐ BZIRK	BZIRK	Sales District	BZIRK	CHAR	6
☐ KONDA	KONDA	CustPrice Group	KONDA	CHAR	2
☐ PLTYP	PLTYP	Price List Tp.	PLTYP	CHAR	2
☐ AWAHR	AWAHR	Order Probab.	AWAHR	NUMC	3
☑ INCO1	INCO1	Incoterms	INCO1	CHAR	3
☐ INCO2	INCO2	Incoterms 2	INCO2	CHAR	28
☐ LIFSD	LIFSD	DelBlckSalesAr.	LIFSD_V	CHAR	2
☐ AUTLF	AUTLF	Complete Dlv.	AUTLF	CHAR	1
☐ ANTLF	ANTLF	Max.Part.Deliv.	ANTLF	DEC	1
☐ KZTLF	KZTLF	Part.dlv./item	KZTLF	CHAR	1

Figure 3.8 Choose the Characteristic for Transfer

In Figure 3.9, you see the details of **Characteristic INCO1 – Incoterms**. Because it was created from a reference field, you can't make any changes to the details. Activate the characteristic by clicking (Activate), and save the characteristic with **Save** or `Ctrl`+`S`.

Details of the characteristic

To make the characteristic available in costing-based profitability analysis and margin analysis, you have to assign it to the operating concern. We explain how to assign characteristics to the operating concern in Section 3.1.6.

Assignment to an operating concern

3 Configuring Characteristics

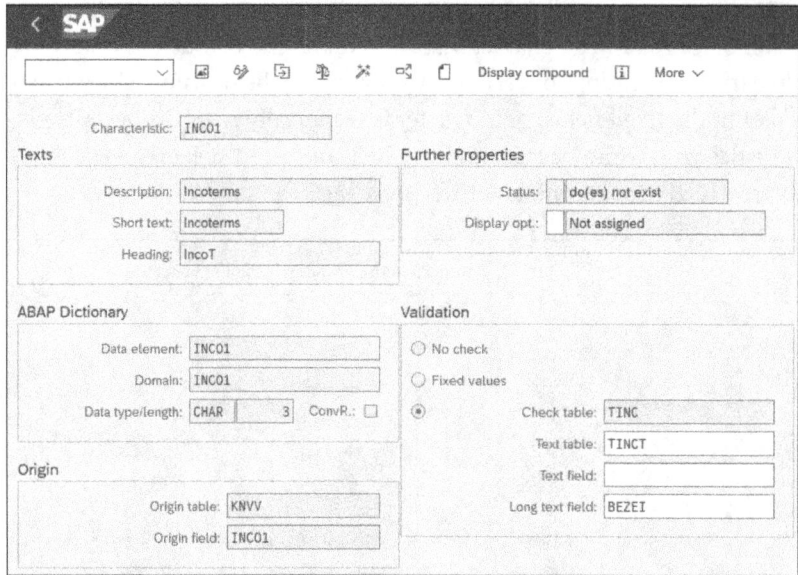

Figure 3.9 Details of the Characteristic

3.1.3 Creating Characteristics with Your Own Value Maintenance

Characteristic with your own value maintenance

To create a characteristic with your own value maintenance, go to Transaction KEA5, or follow the configuration path **Controlling • Profitability Analysis • Structures • Define Operating Concern • Maintain Characteristics**.

In Figure 3.10, for example, create a new characteristic by clicking on 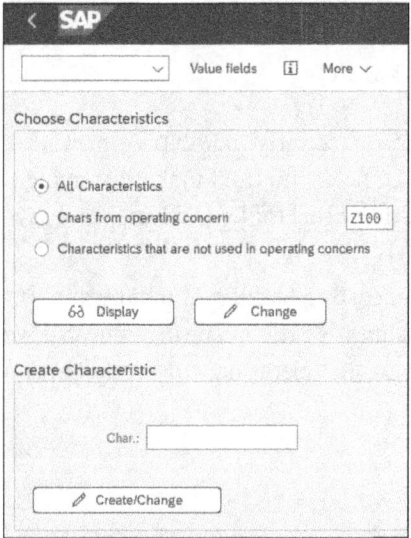 **Create/Change** in the **Create Characteristic** section.

Figure 3.10 Create Characteristics

80

3.1 Characteristics

In the screen that appears, choose to create a **User defined** characteristic (see Figure 3.11) because you don't want to reference any existing SAP table with the characteristic. Next, enter the technical name "WWCON" (remember, user-defined characteristics must start with "WW") and the description "Continent" for this example. The objective is to create a characteristic with your own value maintenance that will assign every country to a continent using a derivation you'll create in Section 3.2.1.

Choose the characteristic type

We're assuming that the sales districts are divided by continents, and to create a margin contribution report by continent, you can assign every country to a continent to facilitate reporting. Choose **With own value maintenance**, and press Enter.

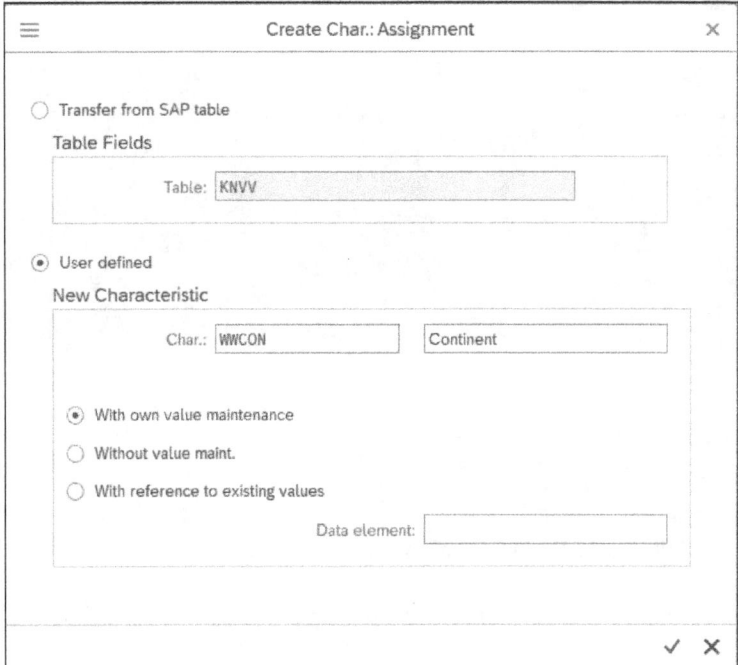

Figure 3.11 Creating Characteristics with Your Own Value Maintenance

In Figure 3.12, you maintain the details of the characteristic WWCON – Continent in the following fields:

Maintaining details of the characteristic

- **Short text**
  Enter a short text for characteristic WWCON ("Continent" for this example).

- **Heading**
  Enter a heading for characteristic WWCON ("Continent" for this example). The **Heading** is displayed in the column name when you maintain values for characteristic WWCON.

81

3 Configuring Characteristics

- **Data type/length**
  Define if the values for characteristic WWCON are numeric or a mix of letters and numbers (in this example, **NUMC** for numeric data type and a field length of **4**).

Generating check tables
Activate your characteristic with (Activate), and you'll see the popup shown in Figure 3.13. The system asks if you want to generate the check tables automatically or manually. Choose **Automatic**.

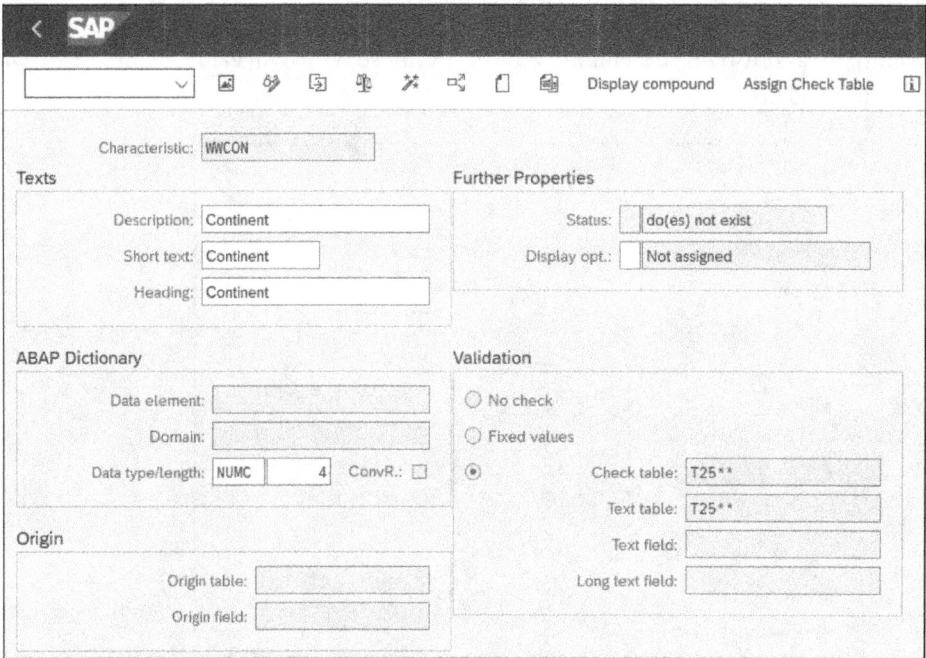

Figure 3.12 Maintaining the Details of the Characteristic

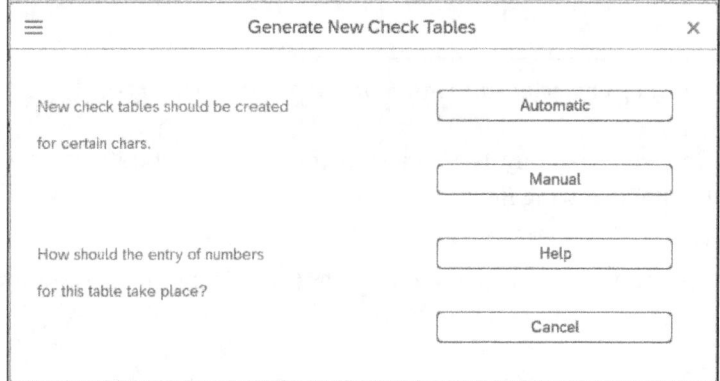

Figure 3.13 Generating New Check Tables for Characteristics with Value Maintenance

3.1 Characteristics

After confirming, the system will generate the check tables in the background and enter their technical names in the **Validation** section in the characteristic details of **WWCON** (see Figure 3.14).

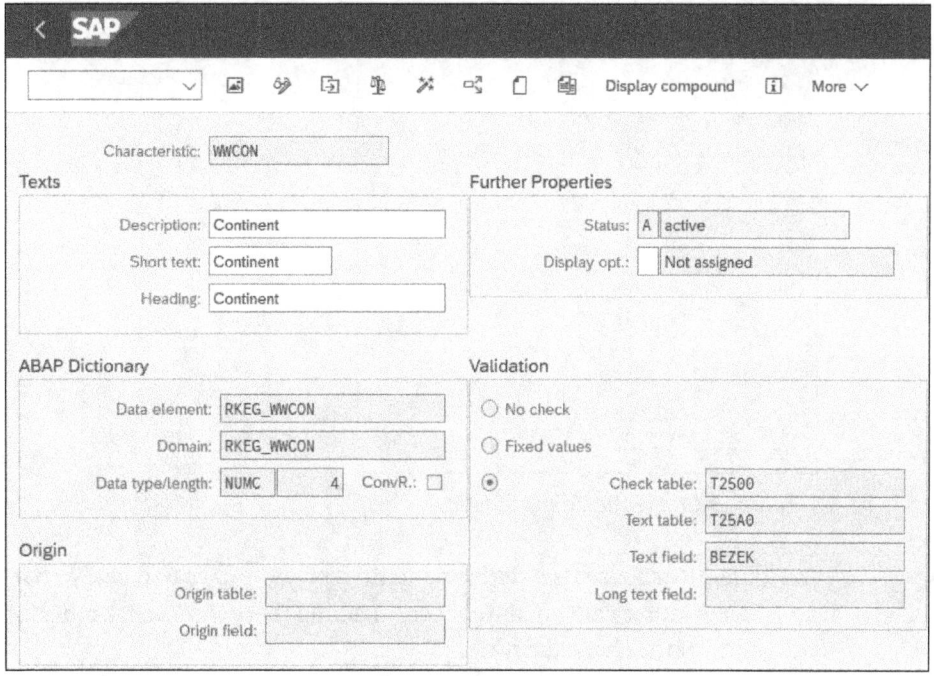

Figure 3.14 Details of the Characteristic with Generated Check Tables

Save the characteristic with **Save** or `Ctrl`+`S`. The next step is to assign the characteristic to the operating concern, which you'll do in Section 3.1.6. After the characteristic is assigned to the operating concern, you can maintain values for characteristic WWCON. You learn how to maintain characteristic values in Section 3.1.7.

Save characteristic

3.1.4 Creating Characteristics Without Value Maintenance

To create a characteristic without value maintenance, go to Transaction KEA5, or follow the configuration path **Controlling • Profitability Analysis • Structures • Define Operating Concern • Maintain Characteristics**.

Create characteristics without value maintenance

You create a new characteristic by clicking on 🖉 **Create/Change** in the **Create Characteristic** section shown in Figure 3.15.

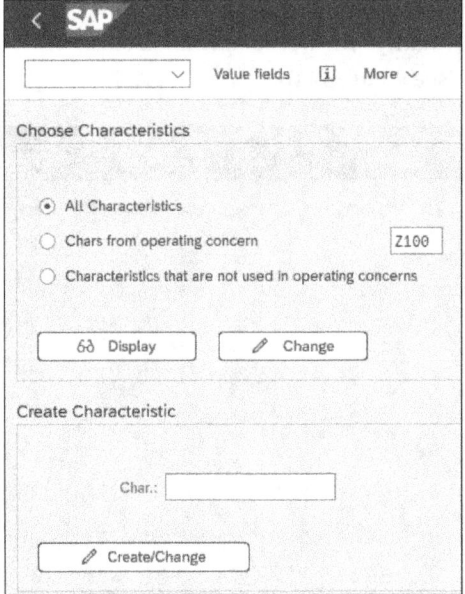

Figure 3.15 Create the Characteristic

In Figure 3.16, choose **User defined** and then press [Enter] to create a characteristic without value maintenance, meaning there is no check table for the values of the characteristic.

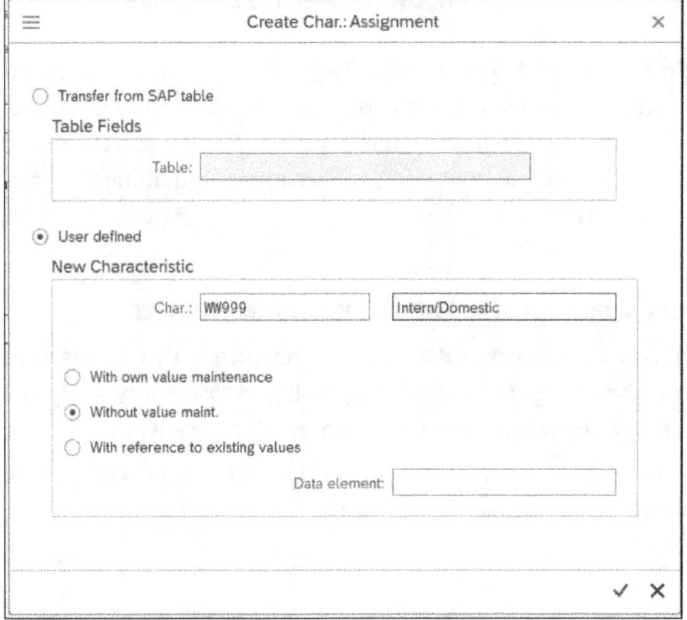

Figure 3.16 Creating Characteristics without Value Maintenance

There will be no set of allowed values or texts assigned to the characteristic. Consequently, there will be no validity check taking place when postings occur on that characteristic. Enter the technical name "WW999" (must start with "WW") and the description "Intern/Domestic". The objective is to create a characteristic without your own value maintenance. The characteristic for this example will define whether a sale is international or domestic. You'll assign that characteristic to a characteristic derivation that you'll create in Section 3.2.3 to all sales transactions. In Figure 3.16, in the **New Characteristic** section, choose **Without value maint.**, and press Enter.

In Figure 3.17, enter the details of characteristic **WW999** as follows: **Short text** ("Intern/Dom") and **Heading** ("Intern/Domestic"). In the **ABAP Dictionary** section, enter **Data type/length** as "CHAR" (alphanumeric) and "2", which means values or text in characteristic **WW999** can be moved with a length of 2.

Maintain details of the characteristic

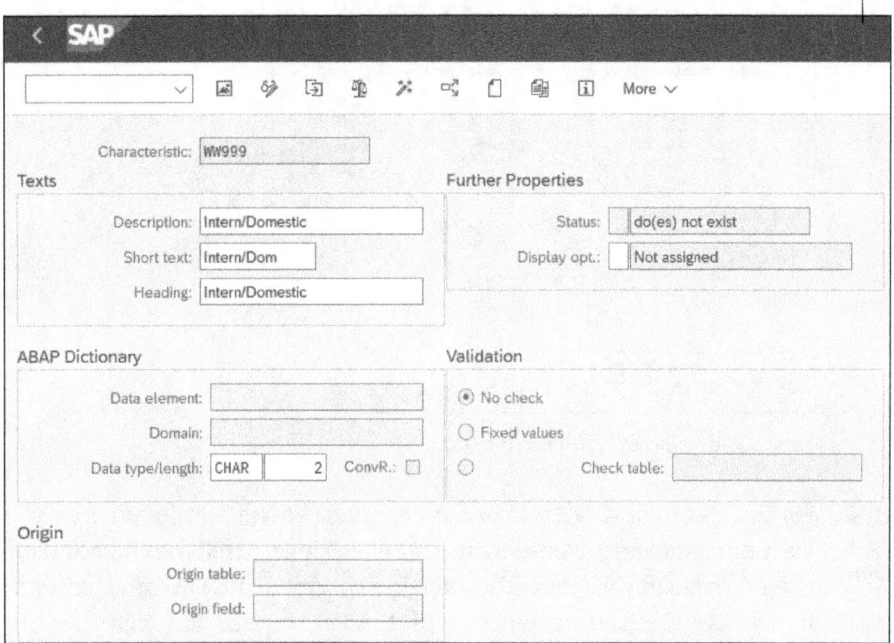

Figure 3.17 Maintain Details of the Characteristic without Value Maintenance

Activate the characteristic with (Activate), and save your settings with **Save** or Ctrl+S. For the characteristic to be available in costing-based profitability analysis and margin analysis, you have to assign it to the operating concern, which you'll learn about in Section 3.1.6.

Activate and save the characteristic

To enable the derivation of the characteristic just created, you have to create a derivation rule to let the system know which documents have to be assigned to this characteristic and which criteria have to be met beforehand. You'll learn how to derive characteristic WW999 to documents in Section 3.2.3.

3 Configuring Characteristics

3.1.5 Creating Characteristics with Reference to Existing Values

Characteristic with reference to existing values

To create a characteristic with reference to existing values, go to Transaction KEA5 or follow the configuration path **Controlling • Profitability Analysis • Structures • Define Operating Concern • Maintain Characteristics**.

You create a new characteristic by clicking on ✏ **Create/Change** in the **Create Characteristic** section, as shown in Figure 3.18.

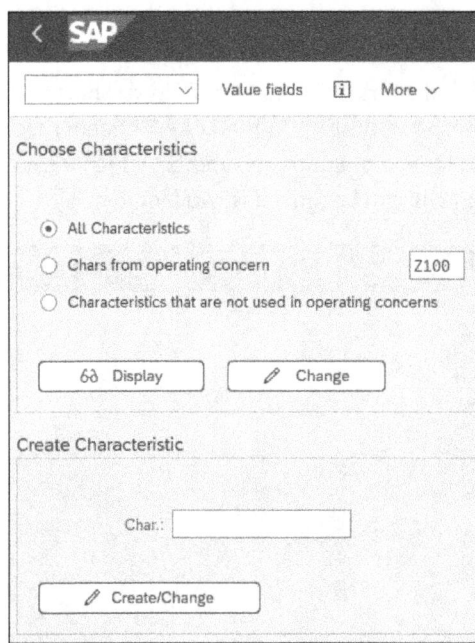

Figure 3.18 Create the Characteristic

Create a characteristic for ship-to country

For this example, create a **User defined** characteristic with reference to an existing characteristic (see Figure 3.19), which means that the characteristic takes on the same values as the reference characteristic. Enter the technical name "WWLAN" and the description "Land of Ship-to". The objective is to create a characteristic with reference to the country key. In addition to the country of the sold-to customer, you can derive the country of the ship-to customer. It doesn't make sense to create a characteristic and maintain all the values manually, so you create a characteristic with reference to existing values. You'll assign that characteristic with a characteristic derivation that you'll create later in Section 3.2.1 to all sales transactions.

Next, choose **With reference to existing values** in the **New Characteristic** section, and enter "BELAND" as the **Data element** to reference. To find the correct data element to reference, you can use F4 help in the **Data element** field. Confirm the selection by pressing Enter.

3.1 Characteristics

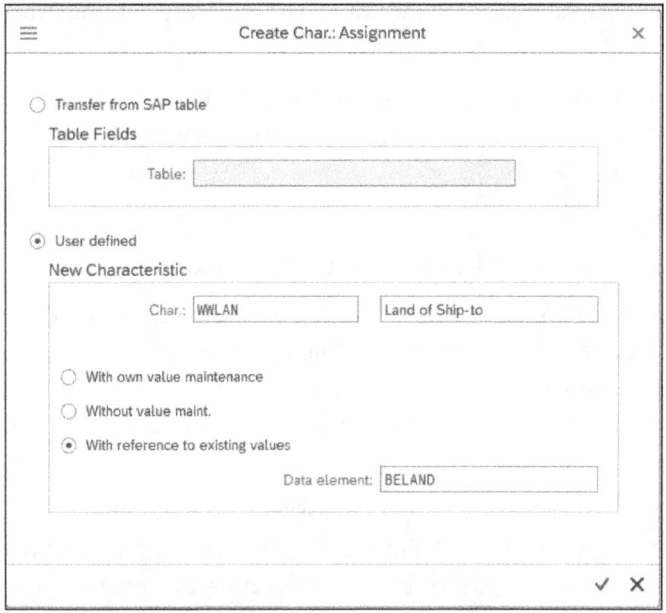

Figure 3.19 Create the Characteristic with Reference to Existing Values

Figure 3.20 shows the details of characteristic **WWLAN**. There are no fields for you to maintain because all attributes are copied from the reference characteristic **BELAND**. Even the **Description** is copied from the reference field and changed to **Dest. C/R.** (Destination Country).

Maintain details of the characteristic

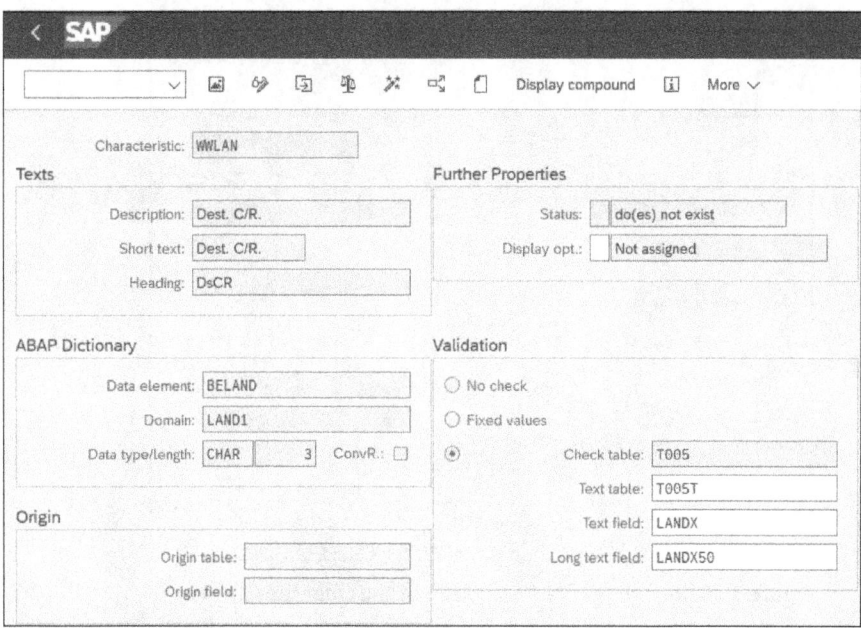

Figure 3.20 Maintain Details of the User-Defined Characteristic

3 Configuring Characteristics

Activate and save the characteristic

Activate the characteristic with [Activate icon] (Activate), and save your entries with **Save** or `Ctrl`+`S`.

In the next section, you'll learn how to assign the characteristic to the operating concern to make it available for use in costing-based profitability analysis and margin analysis.

3.1.6 Assigning Characteristics to the Operating Concern

Assign characteristics to the operating concern

By using Transaction KEAO or by following the configuration path **Controlling • Profitability Analysis • Structures • Define Operating Concern • Maintain Operating Concern**, you can assign the characteristics to the operating concern and make them available for use in costing-based profitability analysis and margin analysis.

In the screen shown in Figure 3.21, you can assign the characteristics just created to operating concern Z100. Enter the operating concern to which you want to assign your characteristics in the **Operating Concern** field, and click on [Display/Change icon] (Display/Change) to change the data structure. After you do this, you can assign new characteristics to the data structure.

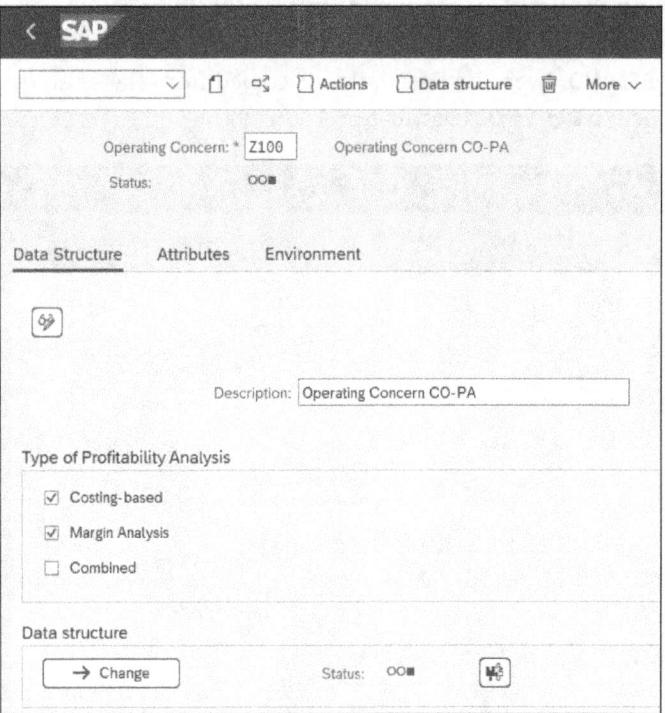

Figure 3.21 Change the Data Structure of the Operating Concern

In the **Transfer from** section, select all the characteristics created in this chapter, that is, **WW999 – Intern/Domestic, WWCON – Continent**, and **WWLAN – Dest. Country** (see Figure 3.22). With [<] (Transfer Fields), you can move those characteristics to the **Data structure** section to assign them to the operating concern.

Changing the data structure

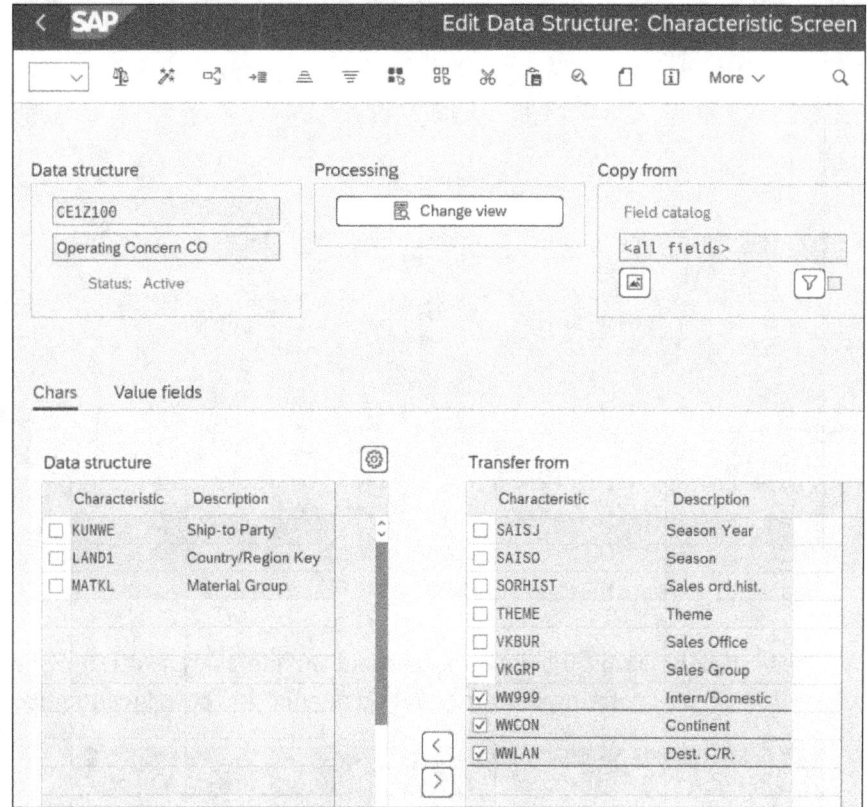

Figure 3.22 Transfer Characteristics to the Data Structure

Next, activate the data structure with [✱] (Activate). At the bottom of the screen shown in Figure 3.23, you can see that the data structure was successfully activated, and the characteristics in the **Data structure** section changed their color to blue. Now, save your changes with **Save** or [Ctrl]+[S].

Activate the data structure

Now, go back to the operating concern by clicking [<] (Back) or pressing [F3]. You might get a popup that asks if you want to generate the environment of the operating concern. Confirm the popup by pressing [Enter]. If not, then follow Figure 3.24 and go to the **Environment** tab of the operating concern. Activate the **Cross-client part** and **Client-specific part** manually with [✱] (Activate).

3 Configuring Characteristics

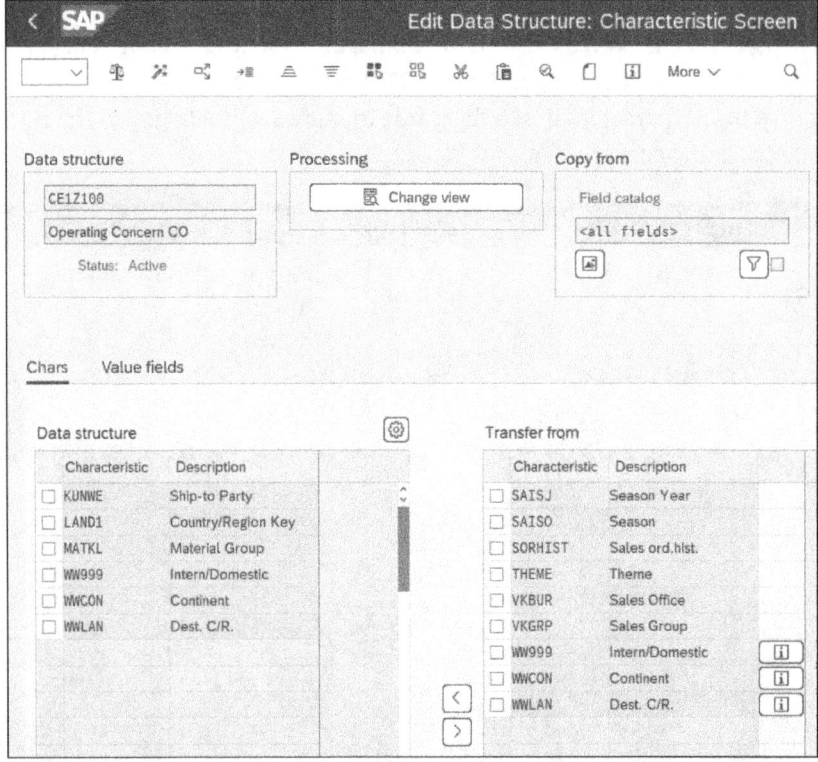

Figure 3.23 Activate the Data Structure

The costing-based profitability analysis and margin analysis can only be used when the operating concern is fully generated and has a green status.

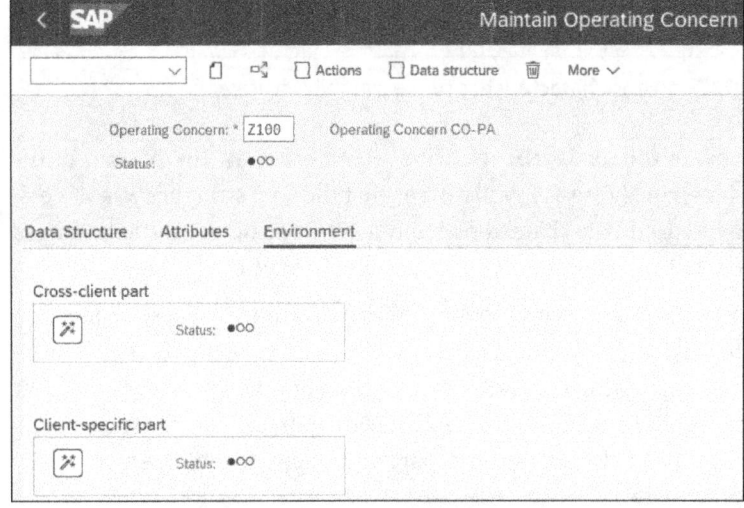

Figure 3.24 Generating the Environment of the Operating Concern

3.1.7 Maintaining Characteristic Values

For the WWCON – Continent characteristic with own value maintenance created in Section 3.1.3, you now need to assign values to it via Transaction KES1 or the configuration path **Controlling • Profitability Analysis • Master Data • Characteristic Values • Maintain Characteristic Values**.

Maintain characteristic values

Figure 3.25 provides an overview of all characteristics with your own value maintenance that are assigned to the operating concern. In this example, only the WWCON – Continent characteristic with own value maintenance is shown. Double-click on **Continent** to maintain values for characteristic WWCON.

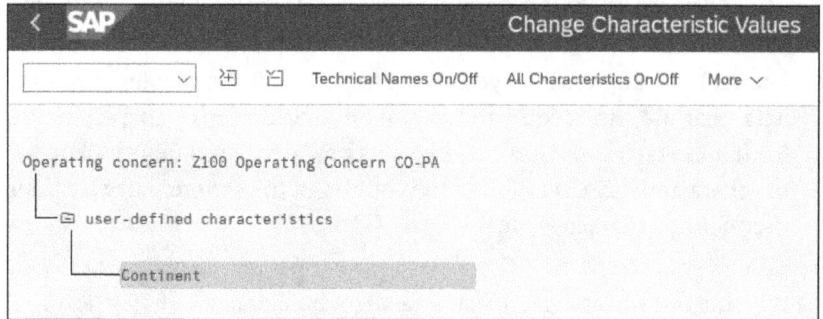

Figure 3.25 Maintain User-Defined Characteristics

In Figure 3.26, you can maintain values with **New Entries** or [F5]. In the **Continent** column, you specify the technical name of the characteristic, and in the **Descriptn** (description) column, you maintain the description for the created characteristics. In this example, create the characteristic value **0001 – North America**. Save the changes with **Save** or [Ctrl]+[S].

Creating user-defined characteristic values

Now, you need to create a characteristic derivation that the characteristic values get derived from in the document. You'll learn how to create a characteristic derivation for characteristic WWCON – Continent in Section 3.2.1.

Deriving characteristics

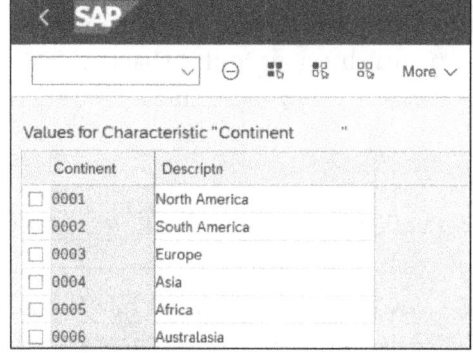

Figure 3.26 Maintain Values for the Characteristic

3.2 Characteristics Derivations

Types of characteristic derivation

You've already learned about the different characteristic types, but for the system to be able to derive the characteristics when the profitability segment is created, you need to create a characteristic derivation. The SAP system provides the following types of characteristic derivation, which are explained further in this section:

- **Derivation rule**
  The derivation rule is used most in profitability analysis because it enables you to derive characteristic values on the basis of various combinations of characteristics. Section 3.2.1 explains how to create characteristic derivations for your user-defined characteristic WWCON - Continent.

- **Table lookup**
  The table lookup enables you to access tables, for example, of master data, that are included in the document and to derive characteristics from these tables. Section 3.2.2 explains how to create a table lookup for the characteristic derivation of the country of the ship-to party for your user-defined characteristic WWLAN.

- **Move**
  With moves, you assign a specific value to the characteristic. Section 3.2.3 explains how to create a characteristic derivation for your user-defined characteristic WW999.

- **Clear**
  With clears, you can reset characteristics. Characteristic values might have to be reset for technical reasons before another value is assigned to the characteristic. Section 3.2.4 covers creating derivations to clear fields.

- **Enhancement**
  Enhancements enable you to create characteristic derivations via user exits. This is useful for the creation of complex characteristic derivations when the SAP standard reaches its limits. Section 3.2.5 covers derivations with enhancements.

In the next subsections, you'll learn how to create characteristic derivations for the user-defined characteristics created earlier in this chapter. You'll also learn how to test your derivations and how to execute a derivation analysis.

3.2.1 Creating Characteristic Derivations with Derivation Rules

Characteristic derivation with derivation rules

The characteristic derivation with derivation rules is the first and most used derivation. For this example, you'll create a derivation rule for the user-defined characteristic WWCON – Continent. To create a characteristic derivation with derivation rule, go to Transaction KEDR, or follow the con-

3.2 Characteristics Derivations

figuration path **KEDR Controlling • Profitability Analysis • Master Data • Define Characteristic Derivation**.

To create a characteristic derivation, click on [icon] (Display <-> Change) to switch to the change mode of the characteristic derivation. You'll then see the possibility to create a new characteristic derivation with [icon] (Create Step) or `F5`. In Figure 3.27, choose **Derivation rule**, and press `Enter`.

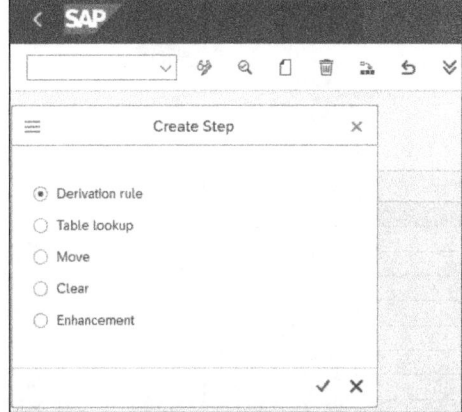

Figure 3.27 Create the Derivation Rule

In Figure 3.28, define **Step Description** as "Derivation WWCON – Continent" before defining the rules for the derivation. The screen is divided into two sections:

- **Source Fields**
  Define the fields that are required to derive the user-defined characteristic. You can enter multiple fields. In this example, where characteristic **Continent** is supposed to be derived, maintain characteristic **WWLAN (Country/Region of Destination of Goods)** as the source field. When the source field is filled, the target field will be derived.

- **Target Fields**
  Define the fields that will be filled with characteristic values. In the example in Figure 3.28, maintain the user-defined characteristic with your own value maintenance **WWCON (Continent)**.

Defining rule definitions

Navigate to the **Condition** tab in Figure 3.29. Here, you can define additional conditions that have to be met to trigger the derivation. You can define the billing type and a specific value here, for example. In this case, characteristic WWCON – Continent would only be derived for this specific billing type. In the example in Figure 3.29, you don't need to add any further conditions for a successful characteristic derivation for characteristic WWCON – Continent for every document that has the characteristic WWLAN – Country of Destination of Goods filled.

Maintain condition in the derivation rule

3　Configuring Characteristics

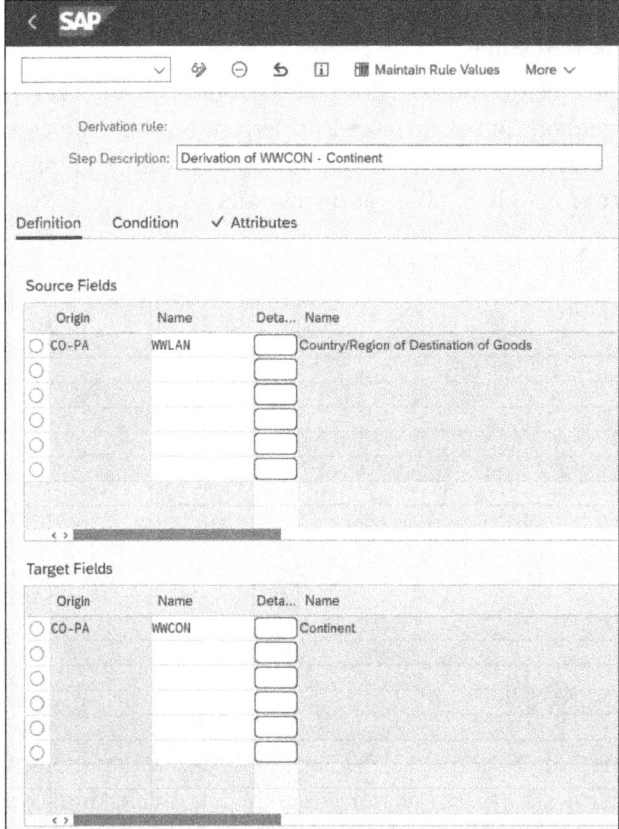

Figure 3.28　Maintain the Definition in the Derivation Rule

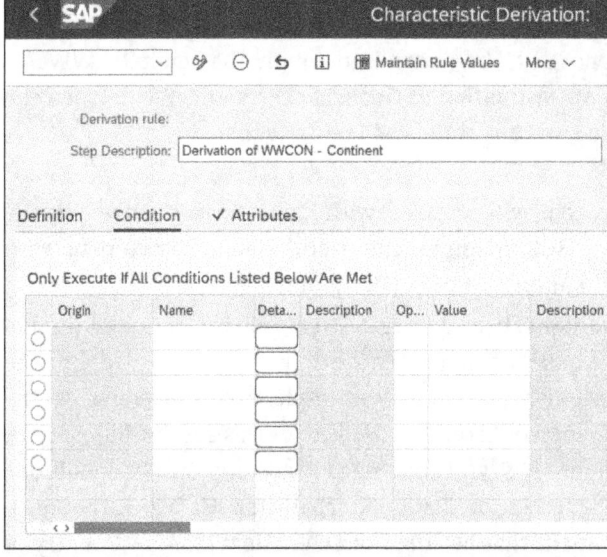

Figure 3.29　Maintain the Condition in the Derivation Rule

Navigate to the **Attributes** tab in Figure 3.30. Here, you can maintain the following settings:

- **Issue error message if no value found**
  If this setting is enabled, the system issues an error message when a document is created that contains the source field but for which characteristic WWCON – Continent can't be derived. You should consider carefully whether to select this checkbox or not because its selection disrupts the process. You have the possibility to re-derive characteristics or correct falsely derived characteristics with the realignment of characteristics. You learn how to use the realignment in Section 3.2.8.

- **Maintain entries using validity date**
  If this setting is enabled, you can maintain the derivation of the characteristic values using a validity date. This is useful to obtain some kind of history if characteristic values change frequently.

- **Optimize access, no from-to values can then be maintained**
  To improve the performance, you can optimize accesses to the characteristic derivation by not maintaining from-to values. For example, if you maintain a value from customer 1 to customer 2000, the system includes all 2,000 customers in the characteristic derivation check. This might take some time; consequently, you can improve the performance by preventing interval maintenance. Set this attribute to optimize the access in Figure 3.30.

- **User-defined step ID (optional, see F1 help)**
  In this field, you can enter a step ID, which makes sense if you use user exits. In addition, the step ID enables you to directly navigate to the rule maintenance in the initial screen.

Maintain attributes in the derivation rule

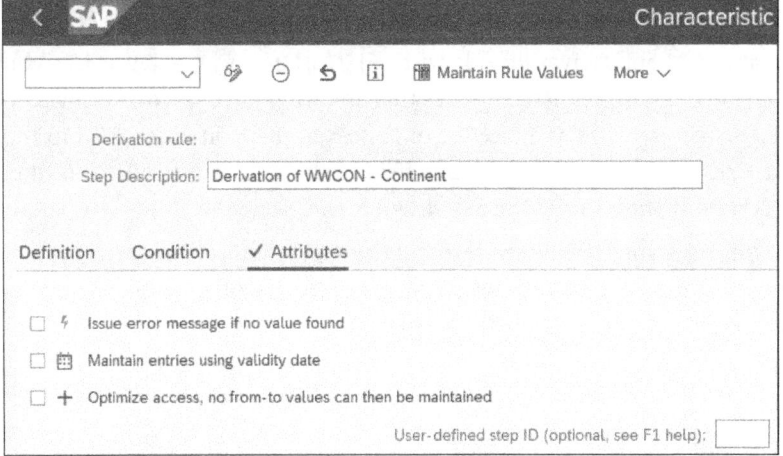

Figure 3.30 Maintain Attributes in Derivation rule

3   Configuring Characteristics

Save your settings with **Save** or [Ctrl]+[S]. To maintain rule values, click on [☰] **Maintain Rule Values**.

Transporting rule entries

In Figure 3.31, in the **Country/Region of Destination of Goods** column, select the source values for the source characteristic WWLAN, and select the target values from the **Continent** column for the target characteristic WWCON. The example shows the rule that when the country of destination is **US** (**USA**), the continent **1** (**North America**) will be derived. Save your rules with **Save** or [Ctrl]+[S].

Go back to the Transaction KEDR screen and add rule entries to a transport request with [🚚] (**Transport**) and transport them to the target system. However, if the target system is a production system, you can maintain rule entries directly in the target system via Transaction KEDR. This makes sense because derivation rules often use master data that isn't available in the source system.

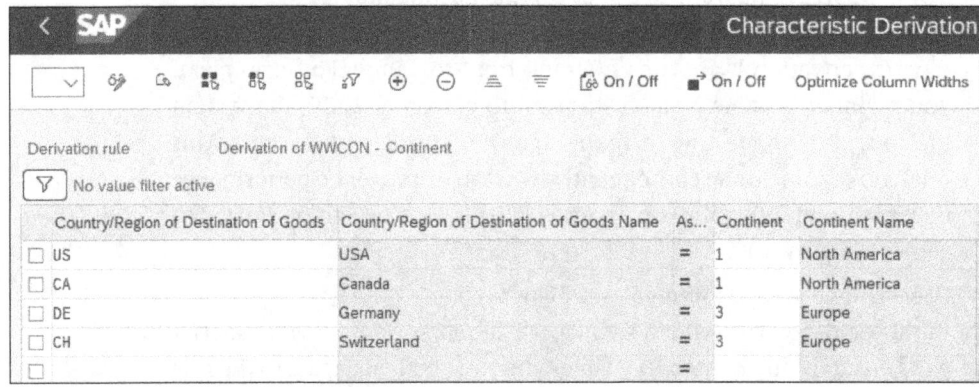

Figure 3.31 Maintain the Derivation Rule

Test derivation rule

Going back to the overview of the derivation rules with [F3] or [◀] (Back), you can test your derivation rule with [▼] (Test) or [F8]. In Figure 3.32, you see an overview of all characteristics that are assigned to the operating concern (custom-specific and fixed). For this example, you enter "US" in the **Dest. C/R.** field, as this is the source field for the derivation rule, and then click on **Derivation** to test the derivation rule.

Display result of the derivation rule test

In Figure 3.33, you can see the result of the derivation rule test. The system derived **Continent 1 – North America** for the **Dest. C/R. US**, which matches with the rule maintained earlier in Figure 3.31.

3.2 Characteristics Derivations

Figure 3.32 Test the Derivation Rule

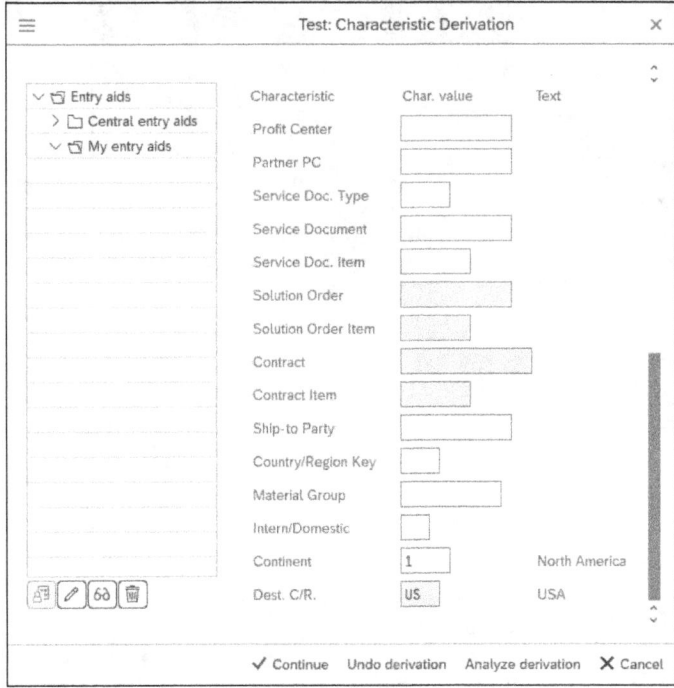

Figure 3.33 Display the Result Test Derivation Rule

3 Configuring Characteristics

3.2.2 Creating Characteristic Derivations with Table Lookup

Characteristic derivation with table lookup

The characteristic derivation with table lookup allows you to access characteristics that are available in standard SAP tables but can't get created as a user-defined characteristic from an SAP table, such as the one created in Section 3.1.2. The characteristic to derive with the derivation with table lookup should be part of the upstream process; it isn't possible to derive a characteristic that is totally unrelated to the business process.

In this example, you'll create a characteristic derivation with table lookup for the country of the ship-to customer. In practice, the sold-to customer is often in a different country than the ship-to customer, which means you might be interested to know where the goods are shipped to, as this is most likely the market in which they will be sold. A characteristic for the country of destination for goods was created in Section 3.1.5, which will be used in the characteristic derivation with table lookup. To create a characteristic derivation with table lookup, go to Transaction KEDR, or follow the configuration path **Controlling • Profitability Analysis • Master Data • Define Characteristic Derivation**.

Selecting the type of characteristic derivation

To create a characteristic derivation, click on 🖉 (Display <-> Change) to switch to the change mode of the characteristic derivation. You'll then see the possibility to create a new characteristic derivation with 🗇 (Create Step) or F5. In Figure 3.34, choose **Table lookup**, and then press Enter.

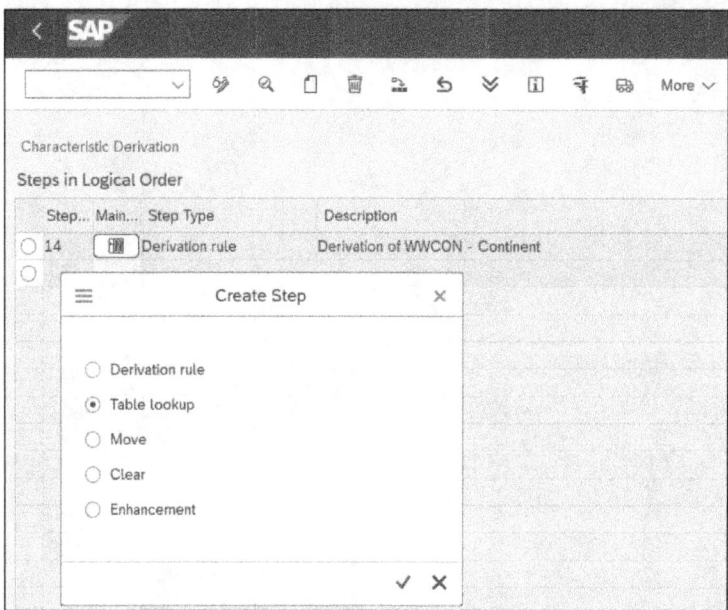

Figure 3.34 Create the Table Lookup

Choosing the table name for lookup

Figure 3.35 shows a popup that asks you to enter a table name. This is the table in which the characteristic you want to derive is stored. In this case, this is the ship-to country.

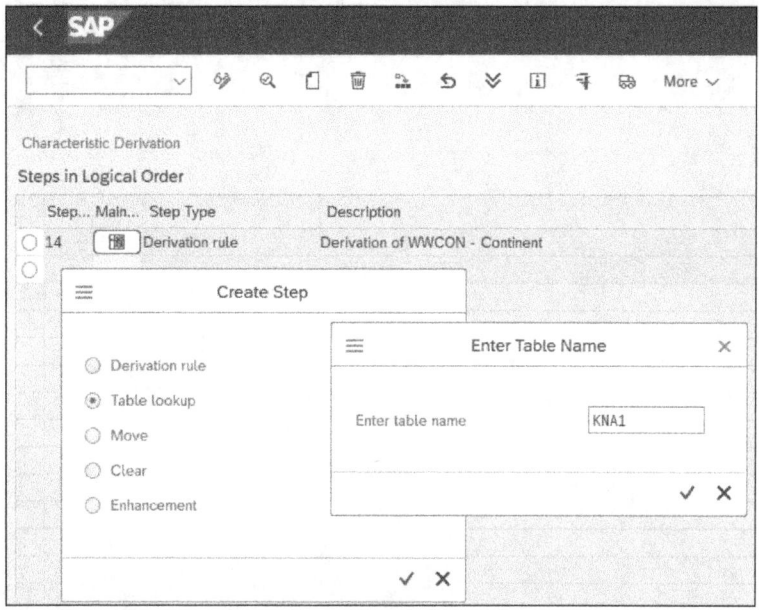

Figure 3.35 Entering the Table for Table Lookup

The country of a customer is stored in table KNA1 (General Data in Customer Master). That is why you select **KNA1** as the table for the creation of the characteristic derivation with table lookup earlier in Figure 3.7 Confirm your selection with [Enter].

Next, you maintain a name for the characteristic derivation. In Figure 3.36, choose **Derivation of WWLAN – Country of Dest..** The definition of the characteristic derivation with table lookup is divided into two sections:

Defining source and target fields for the table lookup

- **Source Fields for Table Lookup**
  The source fields for the table lookup are the fields required to derive our country of destination. In Figure 3.36, the **KNA1 – KUNNR – Customer Number** is entered by default. This is the customer number the system will read from table KNA1. The system is instructed to read KNA1 – KUNNR with the entry of the KUNWE field, which is the ship-to party characteristic you assigned to the operating concern.

- **Assignment of Table Fields to Target Fields**
  In the target fields, you tell the system which field to derive from the customer number. In Figure 3.36, the country of the customer needs to be derived. In this example, the ship-to country is entered in the **WWLAN – Country/Region of Destination of Goods** user-defined characteristic from Section 3.1.5.

3   Configuring Characteristics

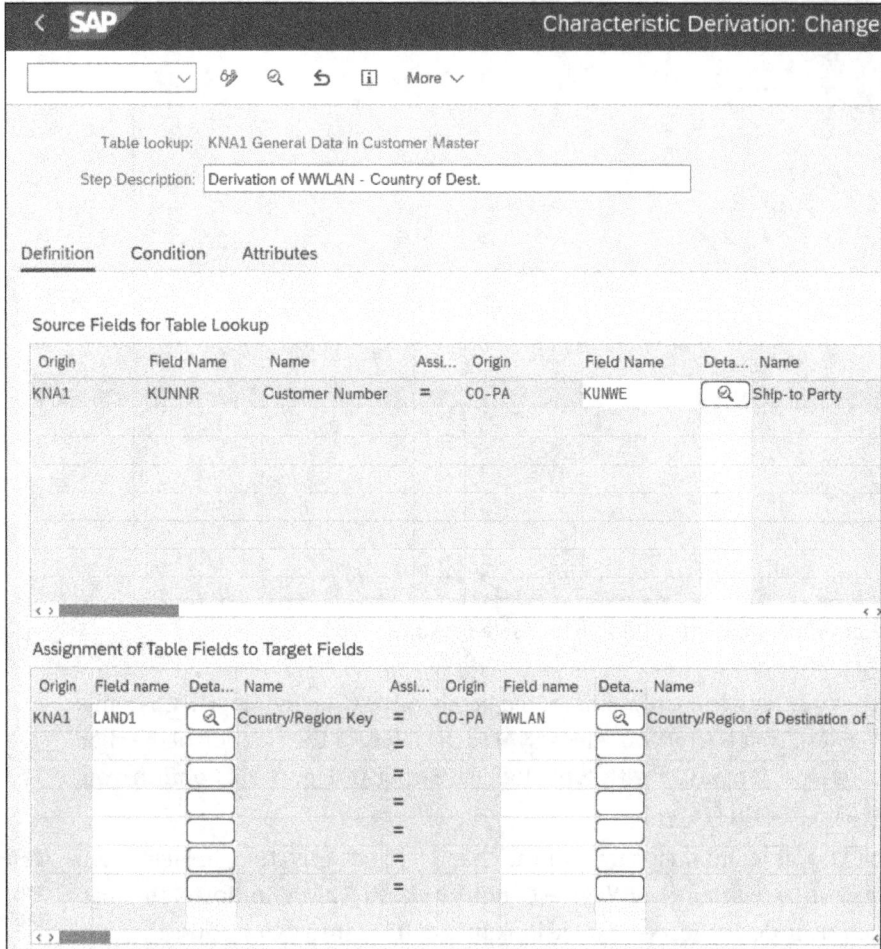

Figure 3.36 Define Source and Target Fields for the Table Lookup

Basically, in one sentence, the table lookup is saying: take the ship-to customer and check in table KNA1 for the country of the ship-to customer, and transfer the entry to WWLAN – Country/Region of Destination of Goods.

Maintain conditions for the table lookup

Figure 3.37 shows the next tab, **Condition**, in which you can maintain conditions/rules for which the characteristic derivation will be executed. For this example, no conditions need to be maintained because the characteristic derivation needs to be executed for any document that is being transferred to costing-based profitability analysis and margin analysis.

Maintain attributes for the table lookup

Figure 3.38 shows the **Attributes** tab with the **Issue error message if no value found** checkbox. If you select this checkbox, all documents that are transferred to costing-based profitability analysis/margin analysis where no characteristic value can be derived for WWLAN won't be transferred, and you'll receive an error message.

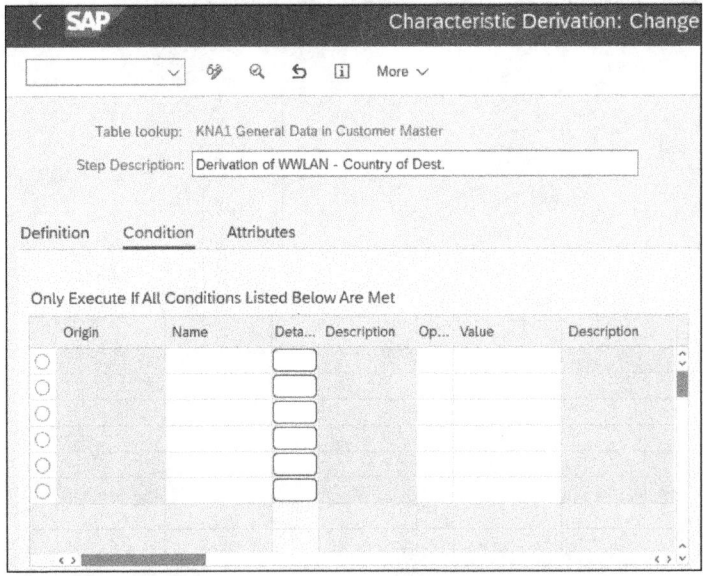

Figure 3.37 Maintain Conditions for the Characteristic Derivation

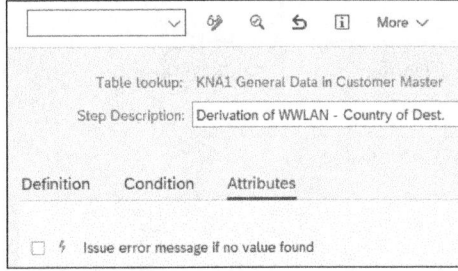

Figure 3.38 Maintain Attributes for the Characteristic Derivation

For this example, leave the checkbox empty so you can execute a realignment if some characteristics are missing or not derived correctly. In Section 3.2.8, you'll see how to create and execute the realignment for characteristics.

Now, let's test the characteristic derivation in Transaction KEDR with 🔲 (Test) or F8. In Figure 3.39, enter the **Customer** number "190" and the sales area data (i.e., enter "0001" for **Sales Org.** and "01" for both **Distr. Channel** and **Division**) as the partner roles for the customer are maintained depending on the sales area. The system needs to know in which sales area it has to check for the ship-to information.

After you enter all the source data, execute the test of the derivation with **Derivation**. In Figure 3.40, scroll down to characteristic **Dest. C/R.** to see that **US** for USA has been derived. This is also the country of the ship-to customer that is assigned to customer 190. This is evidence that the characteristic derivation with table lookup is working!

3   Configuring Characteristics

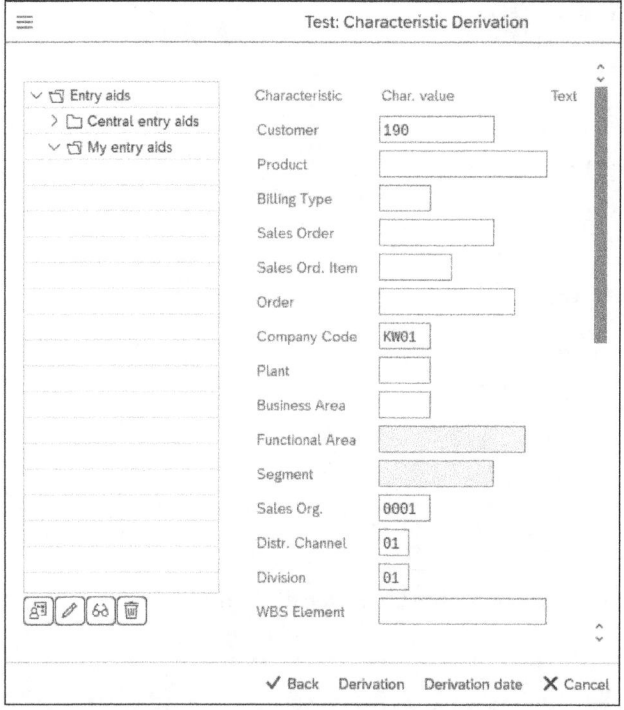

Figure 3.39 Testing the Characteristic Derivation with Table Lookup

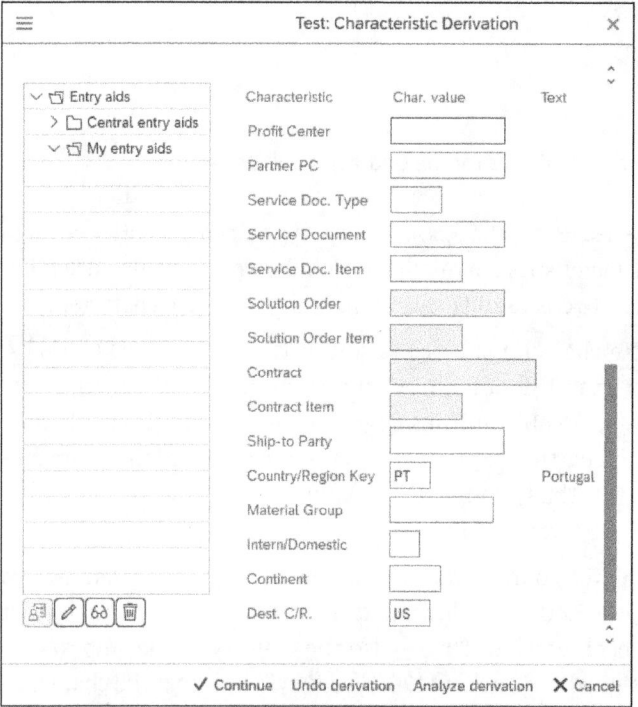

Figure 3.40 Checking the Characteristic Derivation with Table Lookup

3.2.3 Creating Characteristic Derivations with Moves

Now, let's look at how you can create a characteristic derivation with moves. With the move, you can assign a value to a characteristic, for example, the characteristic without value maintenance created in Section 3.1.4. You can either copy a value in that characteristic with the move or copy part of a characteristic value in the new characteristic. This example shows how you assign the **Constant** I for international or **D** for domestic to the characteristic WWIAD – International or Domestic. With this characteristic, the user can easily see if a sale has occurred domestically or internationally.

Characteristic derivation with moves

To create a characteristic derivation with move, click on 🗒 (Display <-> Change) to switch to the change mode of the characteristic derivation. You'll then be able to create a new characteristic derivation with 📄 (Create Step) or [F5]. In Figure 3.41, choose **Move**, and then press [Enter].

Selecting the type of characteristic derivation

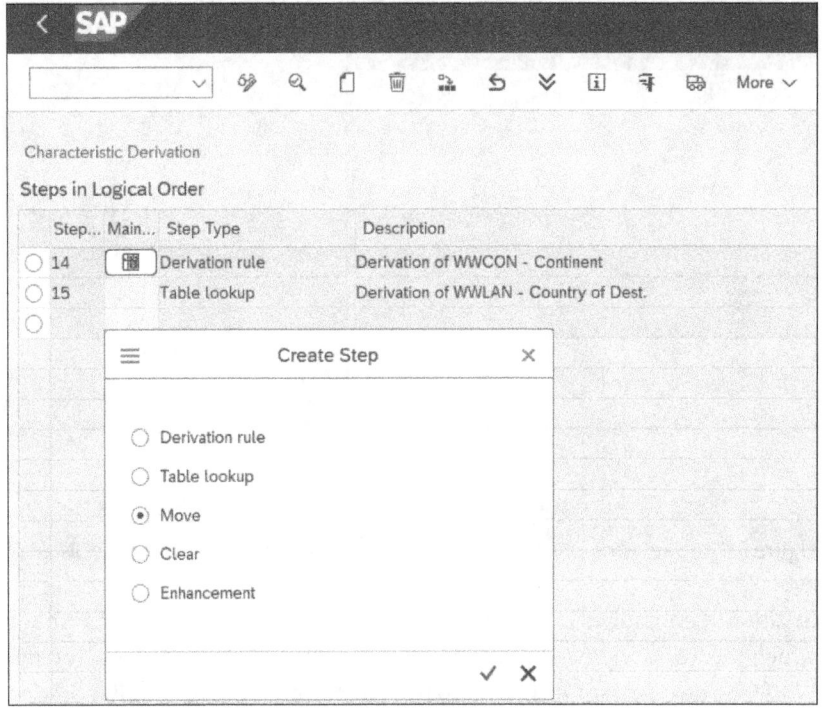

Figure 3.41 Create Characteristic Derivation with Move

In the screen shown in Figure 3.42, enter the characteristic derivation **Step Description** as "Derivation of WW999 - International". You have to actually create two characteristic derivations to (1) fill the characteristic with a **Constant IN** for international and (2) create a characteristic derivation to fill characteristic **WW999** with a **Constant CO** for domestic.

Maintain characteristic derivation with the Move option

3 Configuring Characteristics

On the **Definition** tab, you define with which value you want to supply the characteristic. The following two options are available:

- **Source field**
  If you enable the **Source field** option, you can select a characteristic that is assigned to the operating concern. Some global characteristics are also available, such as PLIKZ (Planned/Actual Indicator).

- **Constant**
  If the **Constant** option is enabled, you can maintain a characteristic value with up to 10 digits as a constant.

Defining target fields

In the **Target field** line, you specify the characteristic to which the value of the source field or constant will be transferred. In the example in Figure 3.42, select **Constant IN** for international to be transferred to characteristic **WW999** (Intern/Domestic).

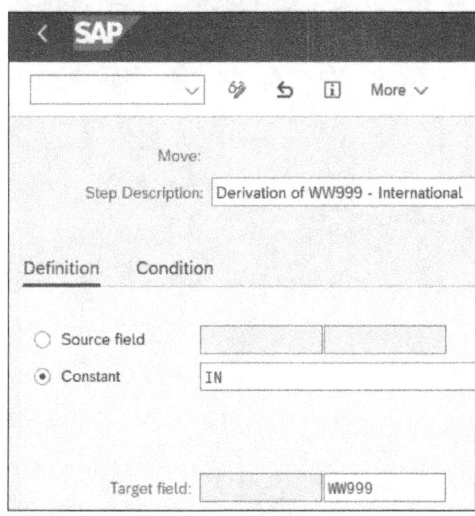

Figure 3.42 Create the Characteristic Derivation with Move for Constant IN

Define conditions

In the screen shown in Figure 3.43, you maintain the conditions under which the Constant IN – International is to be transferred to characteristic WW999. The Constant IN is only to be transferred if the field **BZIRK – Sales District** isn't filled in the document. In this example, the **Sales District** is only filled for customers in the United States, which is why you can use this criterion to distinguish between domestic and international sales.

Save the characteristic derivation

Save your characteristics derivation with **Save** or [Ctrl]+[S]. Now, you have to repeat the steps from Figure 3.44 to create another characteristics derivation with move for characteristic WW999 – Intern/Domestic to assign **Constant DO** for domestic to WW999 for domestic sales. Figure 3.44 shows the characteristic derivation description **Derivation of WW999 -**

Domestic and **Constant DO** for domestic selected to be transferred to characteristic **WW999** (Intern/Domestic).

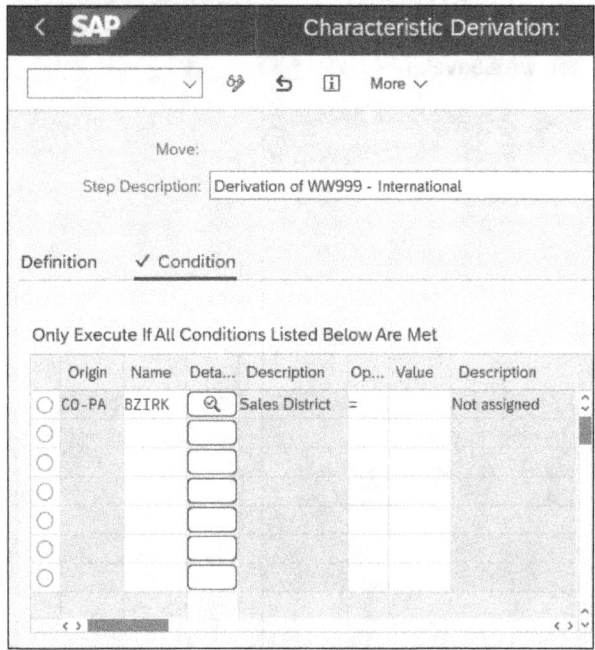

Figure 3.43 Maintain Conditions for the Characteristic Derivation with Move for Constant IN

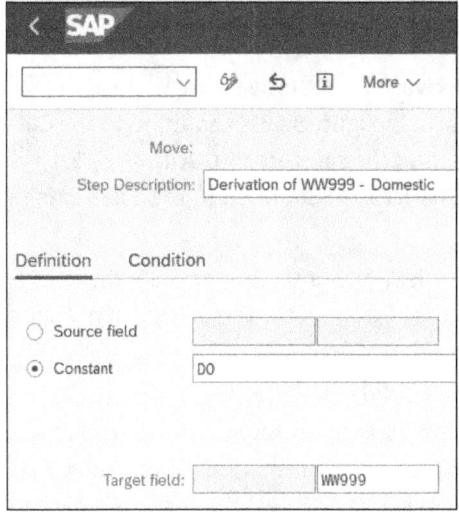

Figure 3.44 Create the Characteristic Derivation with Move for Constant DO

In the screen shown in Figure 3.45, you maintain the conditions under which the constant DO for domestic is to be transferred to characteristic

Maintain conditions

WW999. The constant DO is only to be transferred if the **field BZIRK – Sales District** is filled in the document. In this example, the **Sales District** is only filled for customers in the United States, which is why you can use this criterion to distinguish between domestic and international sales. Save your characteristics derivation with **Save** or Ctrl+S.

Figure 3.45 Maintain Conditions for the Characteristic Derivation with Move for Constant DO

Test characteristic derivation for domestic customer

Now, let's test our characteristic derivation with move with (Test) or F8 for domestic **Customer 190**, as shown in Figure 3.46. In addition to the **Customer** (not visible on the screen), enter the sales area ("0001", "01", "01") as the characteristic **BZIRK – Sales District** is maintained in the sales data in the customer master.

Check characteristic derivation for domestic customer

After you enter the customer and sales area for testing, execute the derivation with **Derivation**, and scroll to the bottom of the screen shown in Figure 3.47. There, you see that the system derived the **Sales District** as well as the value **DO** for domestic in characteristic **Intern/Domestic**.

Test characteristic derivation for international customer

The characteristic derivation for domestic customers seems to work. Now, let's look at the characteristic derivation for international customers. Execute the test with (Test) or F8. In Figure 3.48, test the characteristic derivation with move for international **Customer 163**. In addition to the customer, enter the sales area ("0001", "01", and "0491") as the characteristic **BZIRK – Sales District** is maintained in the sales data in the customer master.

3.2 Characteristics Derivations

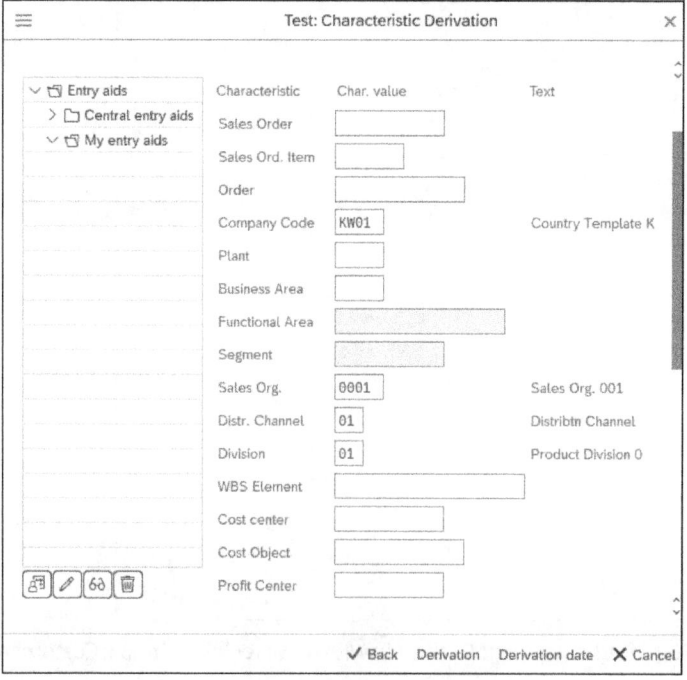

Figure 3.46 Test Characteristic Derivation with Move for Domestic Customer

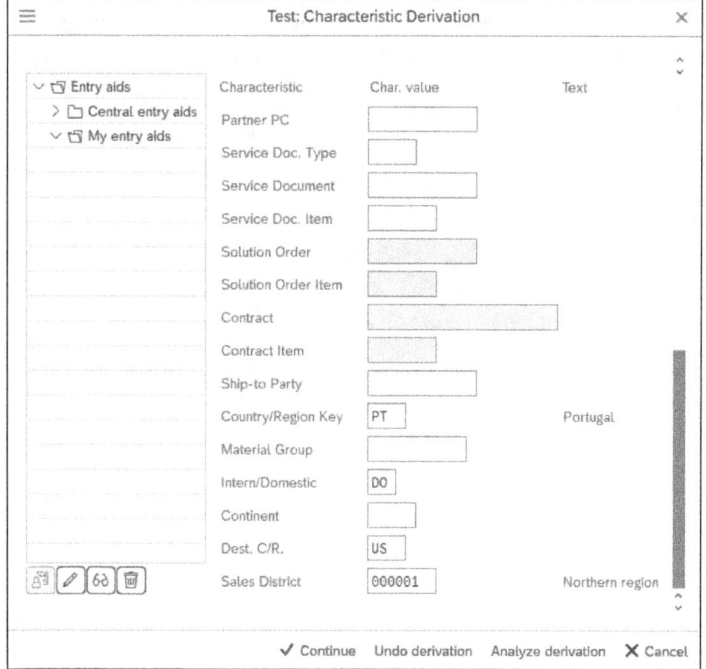

Figure 3.47 Check Result of the Characteristic Derivation with Move for Domestic Customer

3 Configuring Characteristics

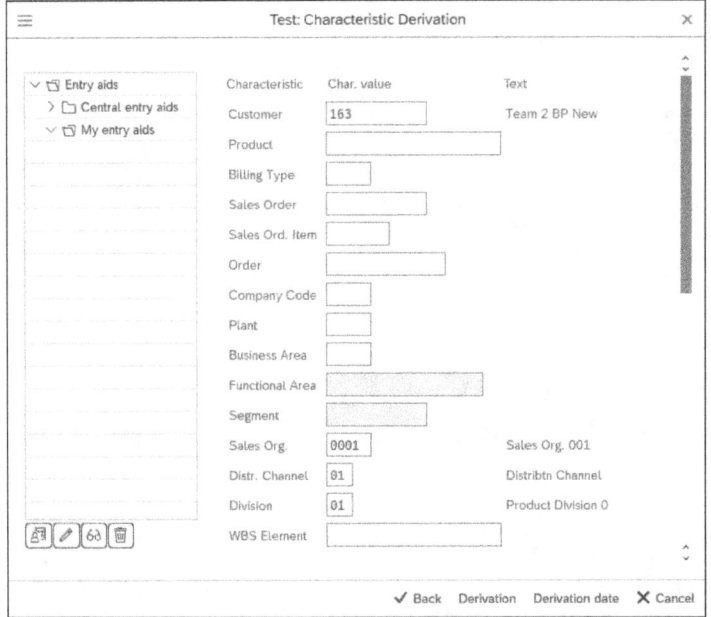

Figure 3.48 Test the Characteristic Derivation with Move for International Customer

Check the derivation for international customer

Let's execute the derivation with **Derivation**, and then scroll to the bottom of the screen in Figure 3.49. There, you see that the system didn't derive a **Sales District** and that the value **IN** for international is filled in characteristic **Intern/Domestic**. Both characteristic derivations with move seem to work.

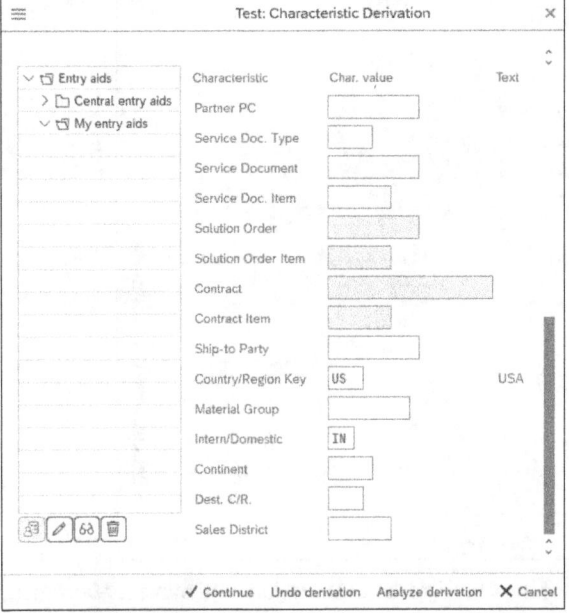

Figure 3.49 Check the Result of Characteristic Derivation with Move for International Customer

3.2 Characteristics Derivations

3.2.4 Creating Characteristic Derivations to Clear Fields

Let's look at how to create a characteristic derivation to clear fields. With this type of characteristic derivation, you can initialize characteristics and clear any type of characteristic. In this example, you'll clear the sales district out of every profitability segment with customer 190 and then later assign a new sales district with the realignment.

Characteristic derivation to clear fields

To create a characteristic derivation to clear fields, click on 🔄 (Display <-> Change) to switch to the change mode of the characteristic derivation. You'll then see the option to create a new characteristic derivation with 📄 (Create Step) or [F5]. In the screen shown in Figure 3.50, choose **Clear**, and then press [Enter].

Selecting the characteristic derivation type

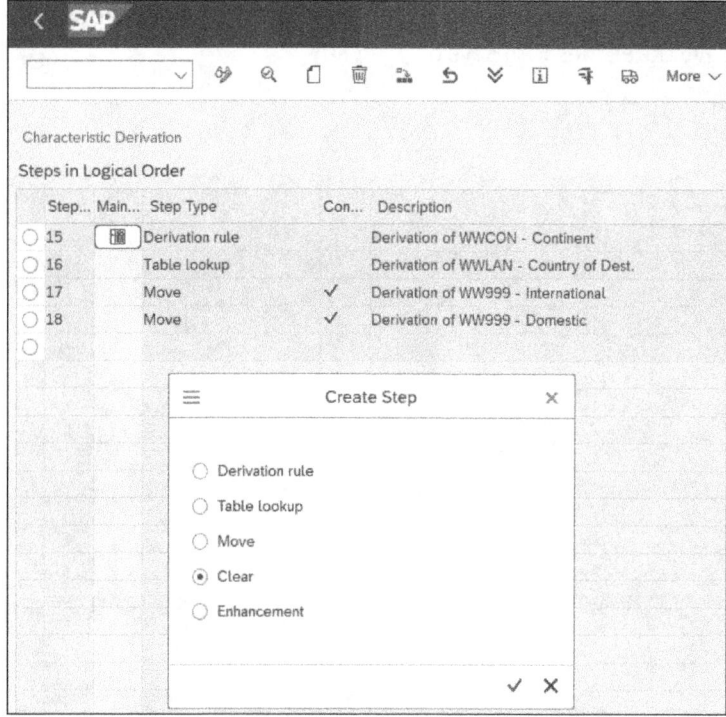

Figure 3.50 Create the Characteristic Derivation to Clear Fields

In Figure 3.51, you can enter a name for the derivation as "Clear BZIRK" and enter "BZIRK" (for sales district) in **Field** as the characteristic to clear with this derivation.

Enter the characteristic derivation name

3   Configuring Characteristics

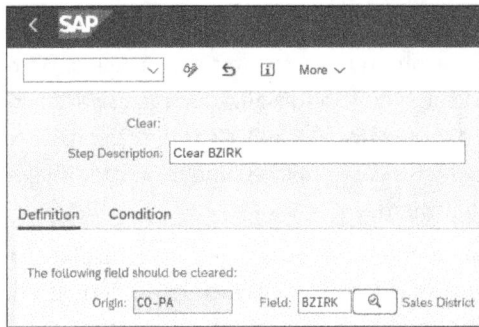

Figure 3.51 Define the Characteristic Derivation to Clear Fields

Maintain the characteristic derivation condition

In Figure 3.52, you maintain the condition/rule when the **BZIRK** field will be cleared. Maintain the rule that all documents with **Customer 190** will be cleared. Save the entries with **Save** or [Ctrl]+[S].

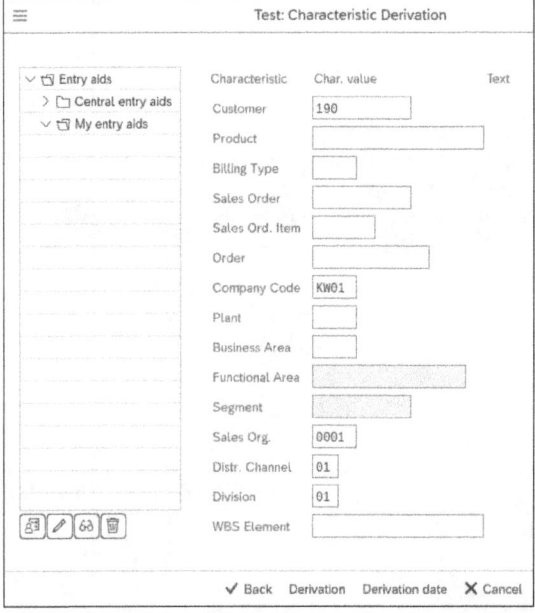

Figure 3.52 Define the Condition for the Characteristic Derivation to Clear Fields

Test the characteristic derivation

Let's now use ▣ (Test) or [F8] to test whether the characteristic derivation to clear fields is working as intended. In the screen shown in Figure 3.53, enter the **Customer** "190", the **Company Code** "KW01", and sales area data (i.e., enter "0001" for **Sales Org.** and "01" for both **Distr. Channel** and **Division**) in the **Test: Characteristic Derivation** screen.

Execute characteristic derivation

Execute the derivation with **Derivation**, and you'll see in Figure 3.54 that the **Sales District** field got cleared. The characteristic derivation to clear the sales district for customer 190 is working as intended.

110

Figure 3.53 Test the Characteristic Derivation to Clear Fields

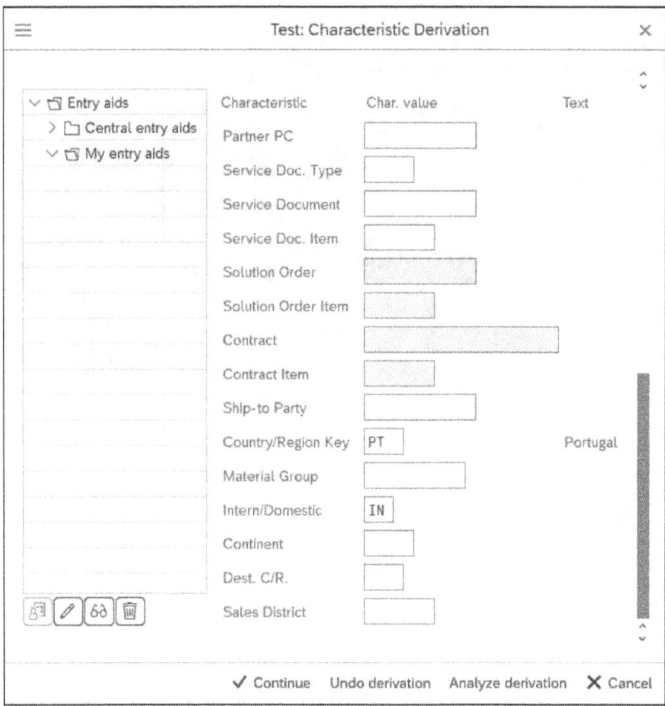

Figure 3.54 Check the Characteristic Derivation to Clear Fields

3 Configuring Characteristics

3.2.5 Creating Characteristic Derivations with Enhancements

Characteristic derivation with enhancement

In costing-based profitability analysis and margin analysis, you can set up an enhancement to derive characteristics; however, you should avoid creating enhancements if possible, so make sure there's no other possibility to derive your characteristic. Usually, with the adjustment of the sequence or the use of technical characteristics, you can derive the characteristic within SAP standard without enhancement.

User exit COPA0001

If you need to create an enhancement, user exit COPA0001 can be implemented to derive characteristics according to your own rules. Note that most fixed characteristic values, such as company code, controlling area, and so on, can't be adjusted with the enhancement, as they are part of the central organizational structure within SAP.

The characteristic derivation is called in many critical places online and in the background, for example, when a financial document is created. That's why there are many restrictions in the enhancement of the characteristic derivation. Before starting to implement the user exit, you need to understand its restrictions.

Create a characteristic derivation with enhancement

To create a characteristic derivation with enhancement, click on [icon] (Display <-> Change) to switch to the change mode of the characteristic derivation. You'll then see the option to create a new characteristic derivation with [icon] (Create Step) or [F5]. In Figure 3.55, choose **Enhancement**, and then press [Enter].

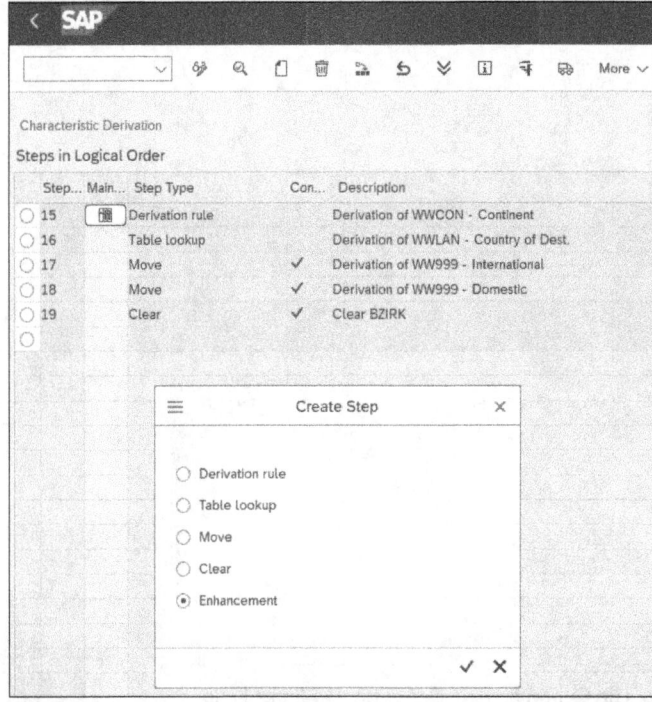

Figure 3.55 Create the Characteristic Derivation for Enhancement

3.2 Characteristics Derivations

In Figure 3.56, you can see that the structure of the characteristic derivation with enhancement is similar to any other characteristic derivation. There is a **Definition** tab in which you define the source and target fields, a **Condition** tab to define rules under which the characteristic derivation is carried out, and an **Attributes** tab to define if an error message should be issued if the derivation wasn't successful. You don't have to make any entries in any of those tabs. The user exit is activated when you create this characteristic derivation as you can see above **Step Description**. By clicking **Source text**, you can directly view the coding of the user exit. Don't forget to save your characteristic derivation with **Save** or Ctrl+S.

Characteristic derivation with enhancement: structure

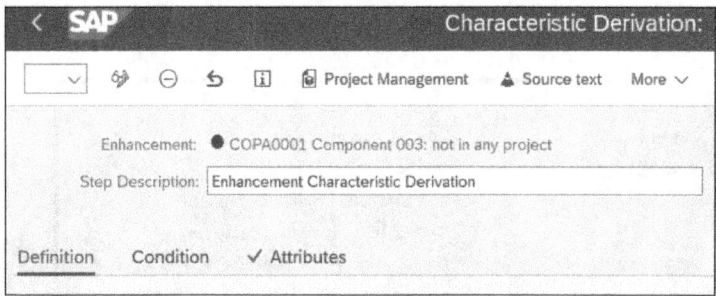

Figure 3.56 Create the Characteristic Derivation with Enhancement

> **Characteristics**
>
> You create characteristics and assign them to the operating concern in costing-based profitability analysis and margin analysis. You can define nearly any type of data field as a characteristic. You can also define characteristic derivations to derive characteristics that can't be assigned to the operating concern directly. The rules for the assignment and derivation of characteristics don't differentiate in the two types of profitability analysis.
>
> SAP S/4HANA allows you to change characteristics after documents have been posted or enrich posted documents with characteristics. It isn't possible to delete characteristics after postings have been made with that characteristic, however.

3.2.6 Sequence of Characteristic Derivation Rules

There is no limitation in the system on how many characteristic derivation rules you can create. Some characteristic derivation rules are created by SAP. When you're in Transaction KEDR, you can display those derivation rules with (Display All Steps).

Sequence of characteristic derivations

113

3 Configuring Characteristics

In Figure 3.57, you can see a number of characteristic derivations that have been created by default. The number of characteristic derivations created by default can change based on the characteristics you assign to your operating concern and the business functionalities you activate/configure. It's recommended that you don't change any of the default characteristic derivations.

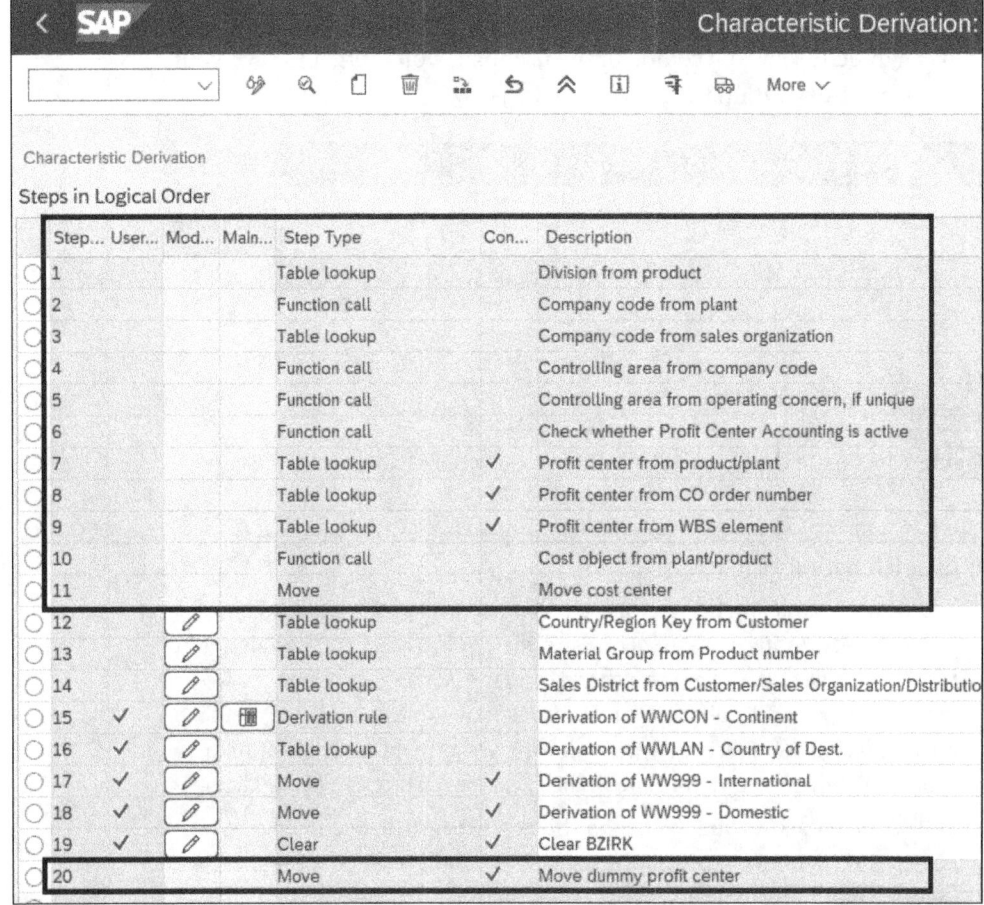

Figure 3.57 Characteristic Derivations Created by SAP

Importance of characteristic derivation sequence

The sequence of characteristic derivations plays a very important role. The system executes the characteristic derivations from top to bottom. You need to make sure that your characteristic derivations are in the correct sequence. Let's look at characteristic derivation **15** and **16** in Figure 3.57:

- Derivation 15 derives WWCON – Continent from WWLAN – Country of Destination of Goods.
- Derivation 16 derives WWLAN – Country/Region of Destination of Goods via table lookup from the ship-to customer.

The system will never be able to derive a continent if the country/region of destination of goods is derived after the system executed the derivation rule of the continent. Therefore, you need to change the sequence here to make sure the characteristic WWCON – Continent can be derived. First, go to change mode of the characteristic derivation with 🖉 (Display <-> Change). After that, mark the characteristic derivation that you want to move. In Figure 3.58, for example, mark characteristic derivation **Step 16 - Table lookup**. Next, you set the cursor in the line you want to move the characteristic derivation, so you set the cursor on **Step 15**. Finally, move the characteristic derivation with 🗒 (Move) or F4.

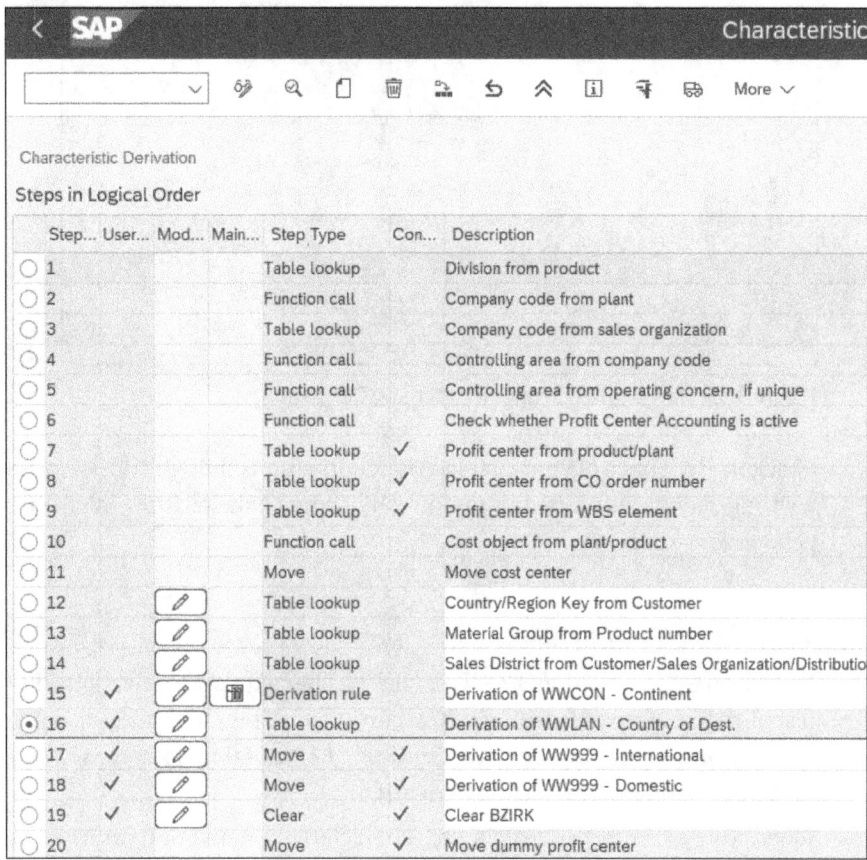

Figure 3.58 Marking the Characteristic Derivation to Move

In Figure 3.59, you can see that the two characteristic derivations **Step 15** and **Step 16** are now in a different order. Save your changes with **Save** or Ctrl+S. After saving your changes, the numbers in Figure 3.59 will be chronological again.

Change the order of characteristic derivation

3   Configuring Characteristics

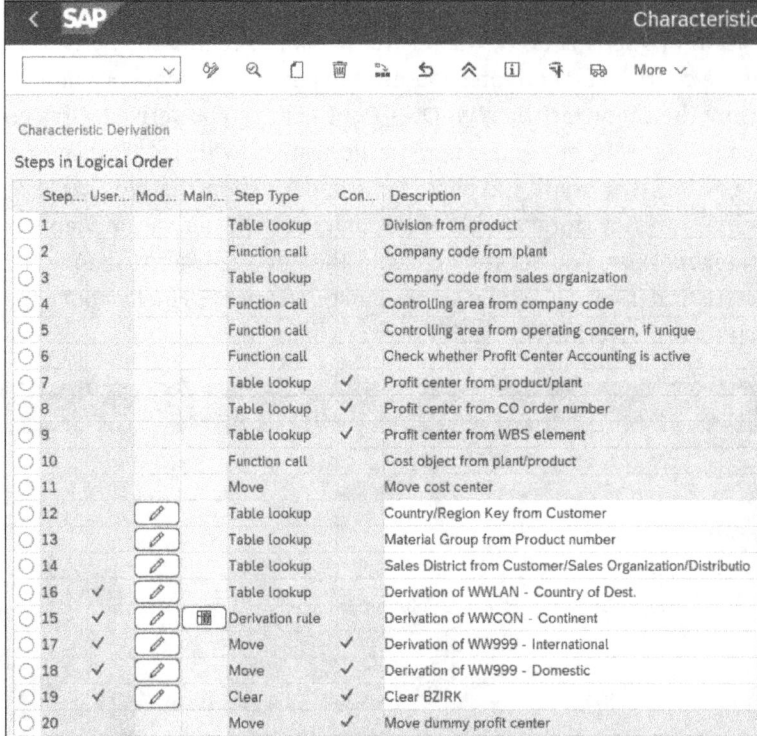

Figure 3.59 New Sequence of Characteristic Derivations

Check the sequence of your characteristic derivations thoughtfully because a wrong sequence can be the reason why some characteristics might not get derived accurately.

3.2.7 Derivation Analysis

In Section 3.2.1 to Section 3.2.5, you created characteristic derivations and tested them afterwards with 🔽 (Test) or [F8] in Transaction KEDR or via the following configuration path: **Controlling** • **Profitability Analysis** • **Master Data** • **Define Characteristic Derivation**.

Testing characteristic derivations

When you execute the test of the characteristic derivation with 🔽 as described in Section 3.2, you see the **Analyze derivation** icon at the bottom of the **Test: Characteristic Derivation** screen in Figure 3.60. This functionality allows you to further analyze the derivation results.

Performing derivation analyses

With **Analyze derivation**, you can see step-by-step why characteristic derivations did or didn't derive a characteristic. In Figure 3.61, you see an overview of all the derivations that exist in your operating concern, including the standard derivations created by SAP and the derivations you created. All highlighted derivations have been executed by the system.

3.2 Characteristics Derivations

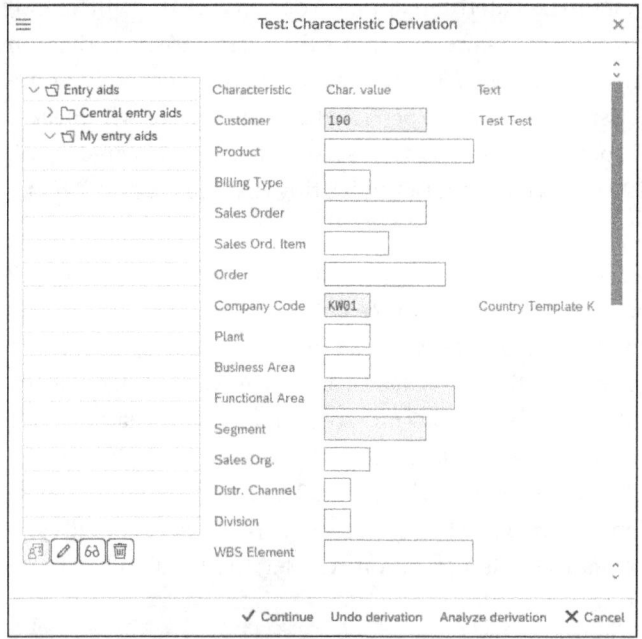

Figure 3.60 Test Characteristic Derivation

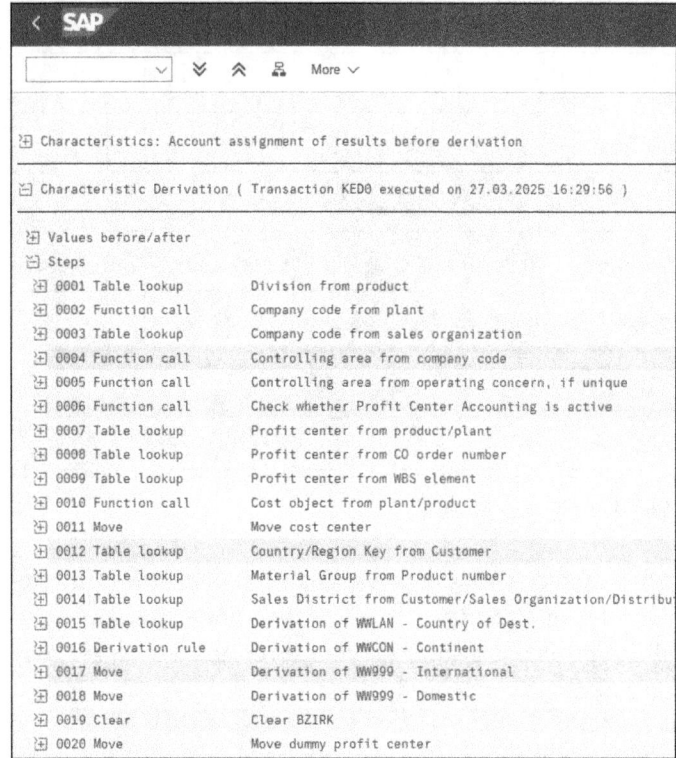

Figure 3.61 Overview of the Analyze Derivation Steps

3 Configuring Characteristics

Analyze characteristic derivation

Let's look at the system-executed derivations. You can click on ⊞ (Expand) at the beginning of the line. Figure 3.62 shows the characteristic derivation **0018 – Move – Derivation of WW999 – Domestic**. If you remember, you created this characteristic derivation in Section 3.2.2. The system will check if the sales district is populated or not. If the sales district is populated, it will fill the field WW999 with a constant DO, indicating that the product is being sold domestically.

Let's take a closer look at what the characteristic analysis in Figure 3.62 reveals. We defined the fields in the **Source fields** and **Target fields** sections when creating the characteristic derivation (refer to Figure 3.36). The source field is the characteristic **BZIRK – Sales District**. The system derived the **Sales District 000001** in this field. The target field is the characteristic **WW999 – International/Domestic** created in Section 3.1.5. The system moved the value DO in the Characteristic WW999, as defined in the characteristics derivation. Figure 3.62 shows that the **Content before** field isn't empty, which means that the derivation will set the constant of DO = Domestic in this field.

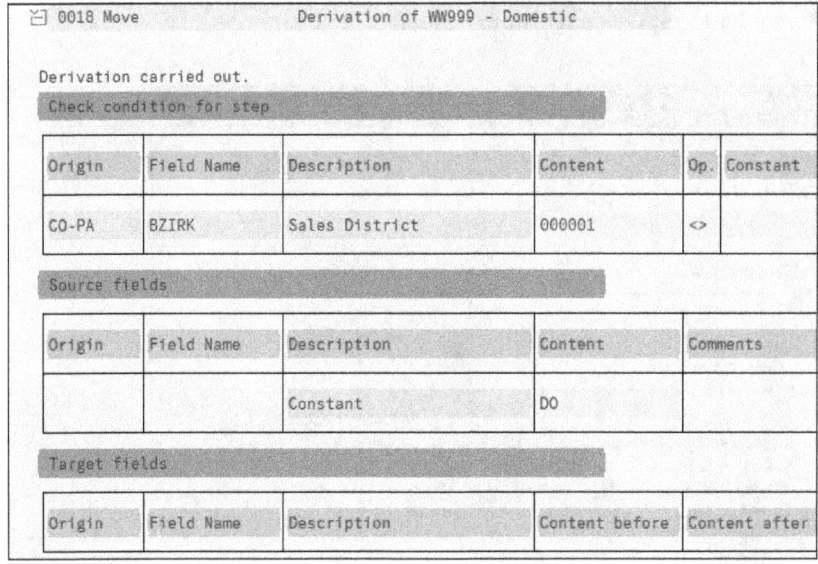

Figure 3.62 Detailed Analysis of Move

The analysis of the characteristic derivation is ideal to test previously created characteristic derivations or analyze characteristic derivations to better understand the results. It also helps your error analysis to see why characteristics did or didn't get derived.

3.2 Characteristics Derivations

> **Derivation Analysis**
> You can use derivation analyses to test derivation rules before they are used in production in both costing-based profitability analysis and margin analysis to help you understand and analyze derivations.

3.2.8 Realignment of Characteristics

What if you realize that your characteristic derivation was set up incorrectly, and all of your profitability segments have the wrong characteristic derived? Or, you're introducing a new characteristic that is missing in all documents that have been posted up to date? There are several cases in which you wish you could change or re-derive characteristics on profitability segments. The realignment of documents allows you to do so. In this section, we'll show you how to change characteristics in existing documents by re-deriving them based on the characteristics derivations maintained in the system.

Realignment of characteristics

Run Realignment

You can change the derivation of existing characteristics in a profitability segment or derive additional characteristics. To do so, go to the Run Realignment (Probability Analysis) app tile, as displayed in Figure 3.63.

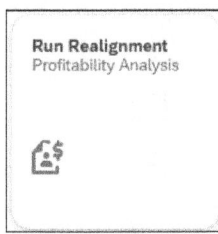

Figure 3.63 Run Realignment App Tile

By clicking on the tile to open the app, you'll see a popup asking you to enter the operating concern and the type of profitability analysis. As margin analysis is saving its data in table ACDOCA and costing-based profitability analysis is saving its data in operating concern–specific tables, we need to specify in which tables we want to update/realign characteristics. If you're using both margin analysis and costing-based profitability analysis, you need to create and execute the realignment run twice.

Set the operating concern for realignment run

By clicking on Realignment Run, you can create a realignment run (Create Realignment Run). After you enter a **Realignment run short text**, as shown in Figure 3.64, confirm your entries with Accept or Enter.

119

3   Configuring Characteristics

Figure 3.64 Create the Realignment Run

Create Request

Create realignment request

Next, you'll create a request with Request (Create Request for Realignment Run, Figure 3.65). In the request, you determine which characteristics are selected for realignment. You can assign multiple requests to one realignment run.

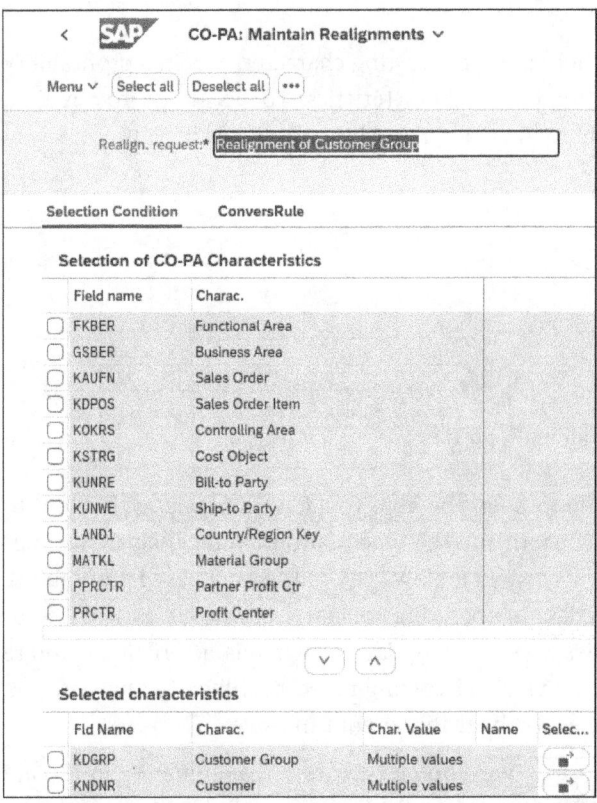

Figure 3.65 Create a Request for Realignment of Customer Groups

3.2 Characteristics Derivations

In the **Selection of CO-PA Characteristics** section of the screen, you can see all the characteristics assigned to the operating concern. You can choose for which characteristics the selection for the documents to be realigned will take place. For this example, choose characteristic **KDGRP – Customer Group** and **KNDNR – Customer** for realignment. You'll likely choose several characteristics because you'll probably have a lot of documents in your productive system. Because this is a test system with a small number of documents, only one characteristic is required here. With ⌵ (Move Field), assign the characteristic KDGRP and KNDNR to the **Selected characteristics** section. You can maintain a value or value range so that only specific material groups will be selected.

Maintain Conversion Rule

After the characteristics are selected, go to the **ConversRule** (conversion rule) tab, as shown in Figure 3.66. Select the characteristic to be derived again from the **Characteristics NOT to be changed** area on the right side of the screen. Here, you see all the characteristics that are assigned to the operating concern.

Conversion rule

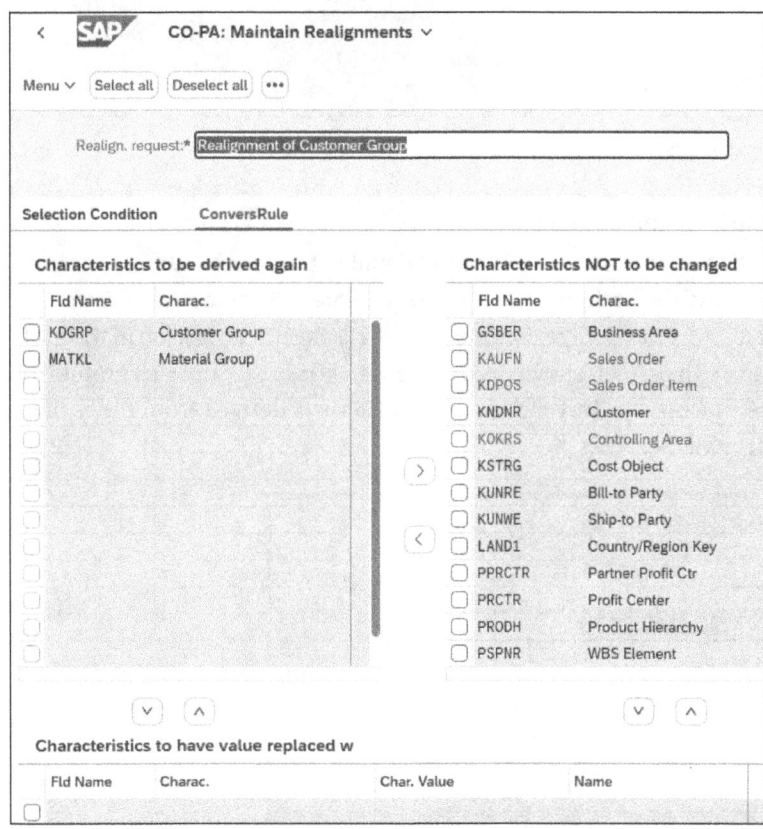

Figure 3.66 Maintain a Conversion Rule for the Characteristic Material Group

3 Configuring Characteristics

A few characteristics appear in light blue here signifying that you can't change them, such as **Controlling Area**, as they are essential characteristics that SAP restricts due to technical reasons.

Mark the characteristics **MATKL** and **KDGRP**, and move them with ⟨ (Move Field) in the **Characteristics to be derived again** section. The **Realignment of Customer Group** request will now derive all material groups and customer groups again when you execute the realignment run.

Test monitor for the realignment run
Before you run the realignment, you need to test whether the realignment will work as expected. To test the realignment, go to ... • **Test monitor** (see Figure 3.67) or press [F8]. This functionality allows you to test the maintained realignment rules for specific documents.

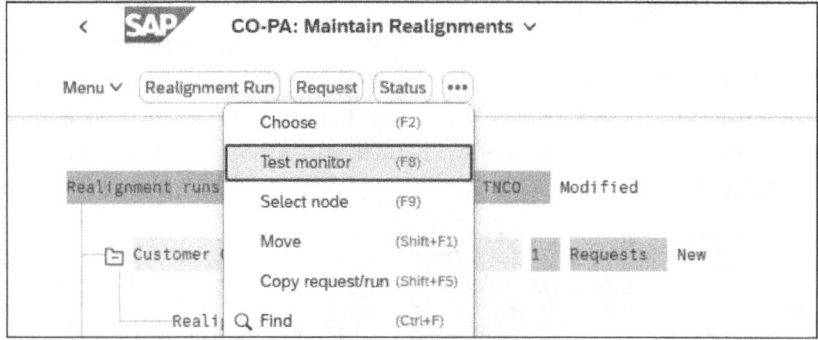

Figure 3.67 Test Realignment Run

Test the Realignment Run

Execute the realignment run in test mode
Now let's test the realignment run in detail before we execute it in the real run. To test the realignment run, go to **More** • **Run/Request** • **Execute** • **Without Start Date**. All realignment runs are executed via a job in the background. In the screen shown in Figure 3.68, choose that the background job will be executed in **Test mode**. The **Job Name** is derived from the realignment run name. Press [Enter].

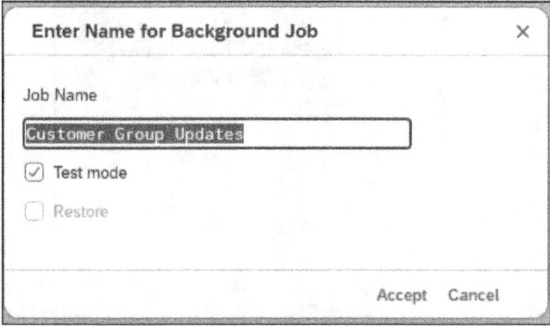

Figure 3.68 Execute the Realignment Run in Test Mode

In Figure 3.69, you can schedule the job to execute the realignment in test mode by clicking on **Immediate**. Save your choices with **Save** or Ctrl+S.

Schedule a job for the realignment run

Figure 3.69 Schedule a Job for the Realignment Run in Test Mode

To review the job status, go to Transaction SM37 (Job Overview). Usually, only IT has access to this transaction. In Figure 3.70, you can see an overview of all the jobs; in this example, there's only one job for the date of **03/28/2025**, and it's finished. In the **Spool** column, you can display the spool list by double-clicking on (Spool List).

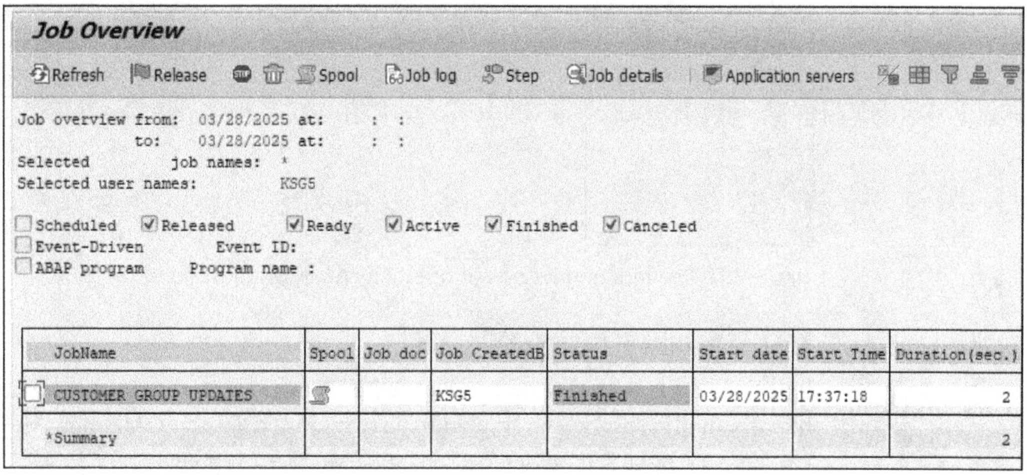

Review the job status

Figure 3.70 Review Job Overview

3   Configuring Characteristics

Spool list   The spool list will show you the result of the job, including whether it ran successfully and the results of the test run.

Execute the Realignment Run

Execute the realignment run in the real run   Now, let's execute the realignment run in the real run. To do so, follow **More • Run/Request • Execute • With Start Date**.

In Figure 3.71, you have to deselect the **Test mode** checkbox and press Enter. Then, the system will display the same screen as shown earlier in Figure 3.69 to schedule the job. Follow the same procedure as for the test run to review the protocol of the job.

Figure 3.71   Execute the Realignment Run in the Real Run

Review profitability segments after realignment run   After the realignment run job finishes, go to the Realignments Results (Profitability Analysis) app tile shown in Figure 3.72 to review the results of your realignment run.

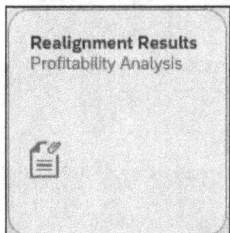

Figure 3.72   Realignment Results (Profitability Analysis) App Tile

Display profitability segments for costing-based profitability analysis

Analyze Results of Realignment Run

In the selection screen shown in Figure 3.73, you can enter various criteria, but as a mandatory selection, you have to choose the ledger. Maintain the selection criteria for **Ledger 0L**, and execute the report with Go. With **Adapt Filters**, you can change the selection criteria and add any characteristic of the operating concern.

3.2 Characteristics Derivations

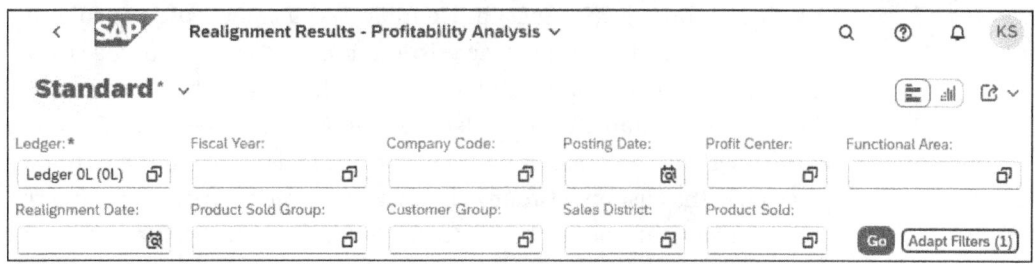

Figure 3.73 Line-Item Browser Selection Screen for Costing-Based Profitability Analysis

In Figure 3.74, you can see an overview of documents that has been updated with the realignment run that we executed.

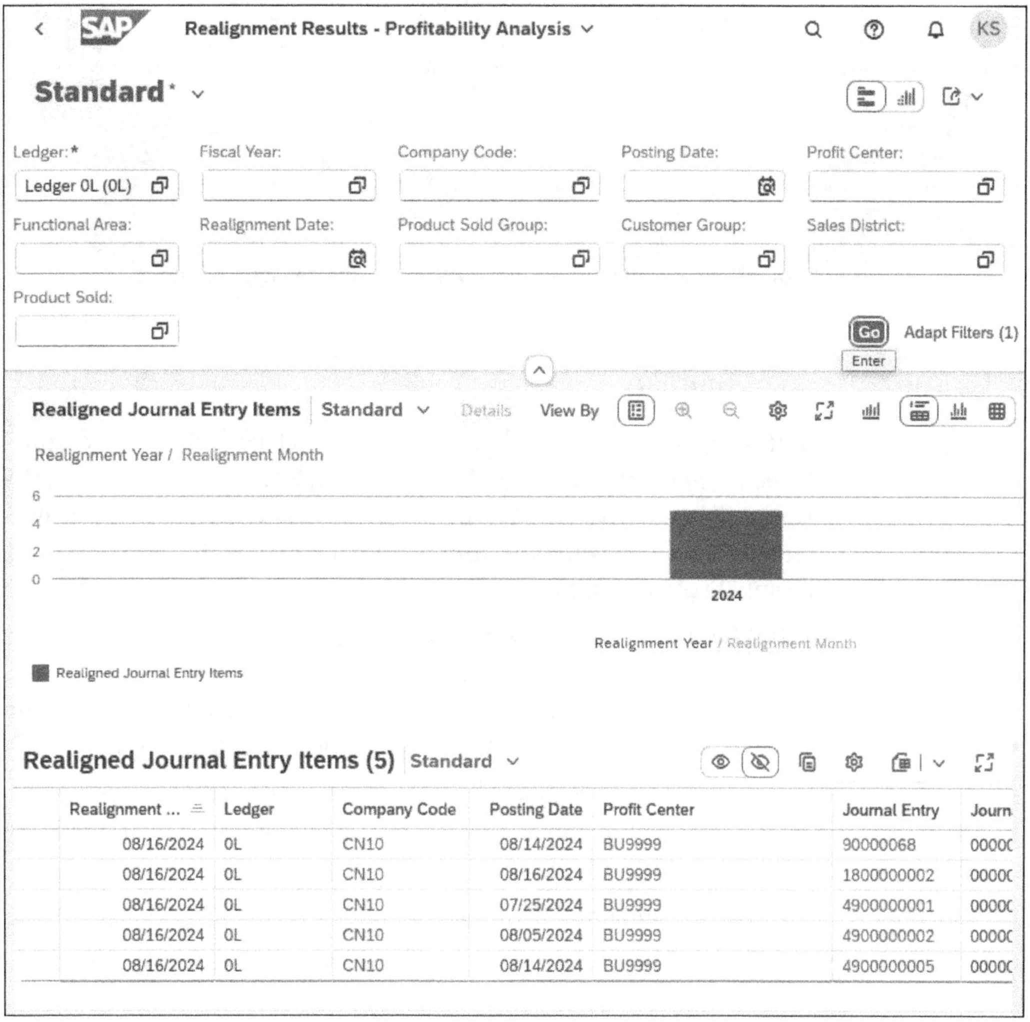

Realignment ...	Ledger	Company Code	Posting Date	Profit Center	Journal Entry	Journ
08/16/2024	0L	CN10	08/14/2024	BU9999	90000068	00000
08/16/2024	0L	CN10	08/16/2024	BU9999	1800000002	00000
08/16/2024	0L	CN10	07/25/2024	BU9999	4900000001	00000
08/16/2024	0L	CN10	08/05/2024	BU9999	4900000002	00000
08/16/2024	0L	CN10	08/14/2024	BU9999	4900000005	00000

Figure 3.74 Review Profitability Segments in Costing-Based Profitability Analysis

By clicking on [⚙] (Settings) in Figure 3.75, you can choose the columns you want to see. For each characteristic, there is a column with the characteristic name + **(Previous)**, displaying the value in the characteristic field before the realignment run. So, let's look at our example. The realignment run re-derived the customer group, which is maintained in the customer master. In the **Customer Group** column, you can see the derivation after the realignment run, and in the **Customer Group (Previous)** column, you can see the field how it was populated when the document was posted. All documents in our example have a new **Customer Group 04** derived. The Realignment Results (Profitability Analysis) app to view realignments is very user friendly and gives you a clear overview of the changes that were processed in the realignment run.

Realignment Date	Ledger	Company Code	Journal Entry	Journal Entry I...	G/L Account	Customer Group	Customer Group (Previous)
08/16/2024	0L	CN10	90000068	000002	600000	04	
08/16/2024	0L	CN10	1800000002	000002	600000	04	01
08/16/2024	0L	CN10	4900000001	000002	700900	04	
08/16/2024	0L	CN10	4900000002	000002	700900	04	
08/16/2024	0L	CN10	4900000005	000002	700900	04	

Figure 3.75 Review Journal Entries Altered in the Realignment Run

3.3 Summary

In this chapter, we showed you how to create characteristics for costing-based profitability analysis and margin analysis. We also showed you the various forms of characteristic derivation and how you can create characteristic derivation with the user-defined characteristics created at the beginning of the chapter.

Finally, we showed you how to analyze your characteristic derivations and how to execute a realignment in case a characteristic was derived incorrectly or you added new characteristics to the operating concern and wanted to derive the characteristic for historical documents as well.

Everything we showed you here is relevant for both costing-based profitability analysis and margin analysis. This chapter is the foundation for the creation and derivation of user-defined characteristics.

In the next chapter, you'll learn about value flows in margin analysis.

Chapter 4
Value Flows of Margin Analysis

Margin analysis is the future of profitability analysis in SAP S/4HANA—it's completely integrated in the Universal Journal and therefore always current and 100% reconciled with financial accounting.

This chapter explains margin analysis, including the value flow of how data is transferred to margin analysis and how the profitability segments are enriched with characteristics. This chapter describes how to create predictive data and how to transfer billing documents, cost of goods manufactured (COGM), and variances to margin analysis. It also describes how to settle or allocate overhead costs, for example, internal orders and cost centers, to profitability analysis. We'll discuss the required configuration settings, introduce the latest innovations in margin analysis, and provide an outlook on further enhancements if known.

The actual value flow determines how data from preliminary processes (e.g., logistics, sales and distribution) is received and processed in profitability analysis. Margin analysis has more functionality than costing-based profitability analysis, so you can expect many more innovations for margin analysis in the near future.

Actual value flow

Margin analysis is by default reconciled with the Universal Journal, as it's part of the Universal Journal. All profitability segments and their characteristics are stored in different columns in the Universal Journal (table ACDOCA). Every characteristic has a separate column in table ACDOCA, and all those characteristics get replicated in the Universal Journal for plan data (table ACDOCP); this also applies for custom-specific characteristics.

Reconciling the general ledger (FI-GL) with margin analysis

This chapter introduces the configuration transactions and settings to configure the creation of profitability segments for all processes and to ensure correct characteristic derivations. At the beginning of each section, we describe the preliminary processes because it's critical that you understand where the data is coming from to facilitate the analysis of the processes and improve the quality of your profitability reports.

4.1 Predictive Accounting

Predictive accounting functionalities

With SAP S/4HANA and margin analysis, you can take advantage of predictive accounting functionalities. Finance departments deliver real-time insights and analyses to manage in a highly competitive market. Predictive accounting goes one step further and delivers predictive insight.

The objective of predictive accounting is to advise and support your strategy through predictive and prescriptive analytics. A large number of period-end tasks are shifting from the end of the period to the actual moment of the transaction. Predictive accounting can be seen as your strategic advisor. You might think that sounds like planning, but there's a difference between predictive data and planning data. *Planning data* defines what you would like to happen based on your company's goals, whereas *predictive data* shows you what current data tells you about the future.

Predictive logic in SAP S/4HANA

There are two different types of predictive logic applied in SAP S/4HANA:

- **Top-down predictions (predictive analysis)**
  This is the application of machine learning for the analysis of large amounts of historical data. It's used to predict future values by considering trends, cycles, and fluctuations. There's no drilldown functionality on predictive data that has been created with top-down predictions because the algorithms that form the basis for this kind of data aren't self-explanatory. Top-down predictions aren't limited to margin analysis or financial accounting at all but can be applied across all functionalities.

- **Bottom-up predictions (predictive accounting)**
  Bottom-up predictions interpret the future document flow of existing documents as part of an individual business process. In other words, they analyze transactions and determine the future outcomes based on the data that is already available in the system. Bottom-up predictions basically enrich and enhance actuals with additional information. In contrast to the top-down predictions, you can drill down on bottom-up predictions and slice and dice the predictive data in all dimensions.

In the next sections, we'll provide an overview of existing and upcoming predictive accounting functionality in margin analysis.

4.1.1 Predictive Accounting: Incoming Sales Order

Prerequisite for predictive accounting

A prerequisite for predictive accounting is the creation of an extension ledger and its subsequent activation in margin analysis. In Chapter 2, you learned how to create and activate the extension ledger. In this chapter, you'll learn which configuration settings are necessary to generate predictive insight by creating a profitability segment for incoming sales orders and their cost of goods sold (COGS). Predictive insight fulfills the requirements of continuous accounting and allows faster decision-making as you can react to increasing/decreasing demand in real time.

To activate predictive accounting for sales processes, follow the configuration path **Financial Accounting • Predictive Accounting • Activate Predictive Accounting for Sales Processes**. Navigate to your controlling area (**CO Ar**), and select **Active with date of entry** from the **Sales Processes** column. Figure 4.1 shows the **Activate Predictive Accounting** area for **Controlling Area 8000**. Note that as of the SAP S/4HANA 1909 release, not all sales scenarios are supported in predictive accounting. SAP is working on supporting all sales scenarios, such as project sales, in the near future. After you activate predictive accounting in your controlling area, save your changes with **Save** or Ctrl+S.

Predictive accounting for incoming sales orders

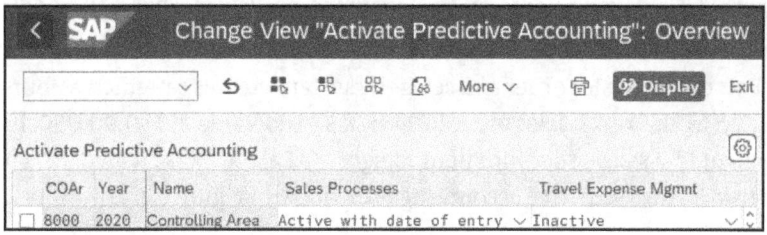

Figure 4.1 Activate Predictive Accounting

You now need to define the sales order item categories for which the predictive journal entries will be created. Only the sales order item categories that are maintained in configuration will be considered in predictive accounting. Follow the configuration path **Financial Accounting • Predictive Accounting • Activate Predictive Accounting for Sales Order Item Categories** to activate sales order item categories for predictive accounting. With **New Entries** or F4, you can add sales order item categories for activation.

Define sales order item categories

In Figure 4.2, the predictive accounting for sales order **Item Category TAN – Standard Item and TAS – Third Party Item** are activated by clicking the **Active** checkbox. If you want to make sure that you capture all sales order item categories in Figure 4.2, you can use 📋 (Get Missing Item Categories). With 🖉 (Activate All) or 🖉 (Deactivate All), you can mass activate or deactivate all item categories at once. It's highly recommended to work together with your colleagues in the sales department on the activation of the sales order item categories to make sure you capture all sales order item categories correctly and completely to have meaningful predictive accounting. After you've maintained all the sales order item categories for predictive accounting, save your changes with **Save** or Ctrl+S.

4   Value Flows of Margin Analysis

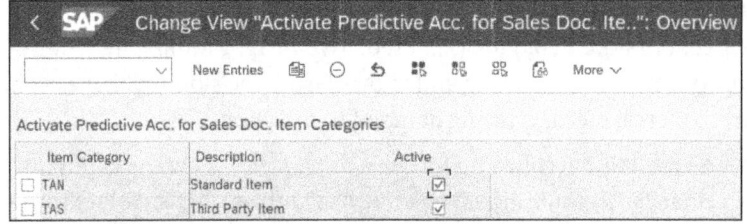

Figure 4.2 Activate Sales Order Item Categories for Predictive Accounting

Predictive journal entry

Now, when a sales order is saved, a program runs in the background to create the predictive journal entry. The program checks the configuration and master data to detect any errors early. If there are any errors, no predictive journal entry will be created. The system also checks for follow-on documents to the sales order and derives predictive journal entries for those. For the sales orders, it's checking whether there's a goods issue following and whether COGS must be derived to the extension ledger. There's no additional configuration necessary for this step.

Sales order creation

Now let's create a sales order and see if the system creates a predictive journal entry. In Figure 4.3, sales order **1011485** has a **Net Value** of **500.00 USD**. In this example, a sales order with item category (**ItCa**) **TAN** was created, which we activated for predictive accounting previously in Figure 4.2.

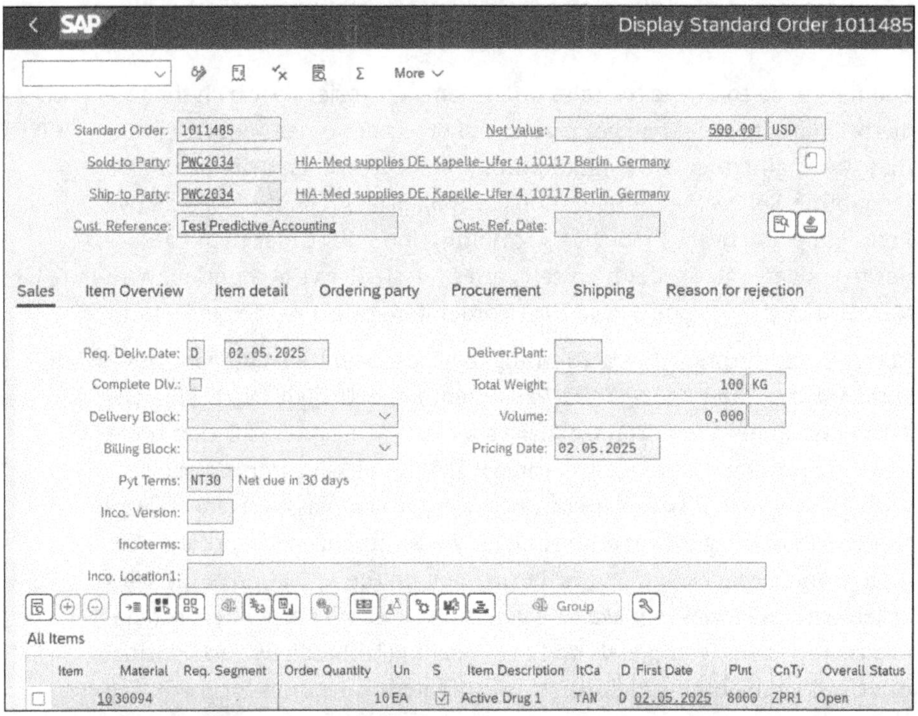

Figure 4.3 Creating a Sales Order

You can also see in Figure 4.3 that the requested delivery date was determined for **02.05.2025**. The posting date (not visible on the screen) of the predictive journal entry is determined by the goods issue date if the sales order is configured for delivery-related billing. Another important characteristic in Figure 4.3 is the **Overall Status**. If the **Overall Status** is blocked, for example, no predictive journal entry is created. The status of the sales order is **Open**, so a predictive journal entry for sales order **1011485** is expected.

Let's double-check the goods issue date in the sales order to be sure about the posting date of the predictive journal entry. Click 🗐 (Schedule Lines for Item) to go to the scheduling details of the item. Double-click on the delivery date, and navigate to the **Shipping** tab. In Figure 4.4, you can see that the **Goods issue date** on the line item is **05.05.2025**. This is the date we expect as the posting date in the predictive journal entry. The requested delivery date shown earlier in Figure 4.3 seems to not have been updated correctly.

Review goods issue date

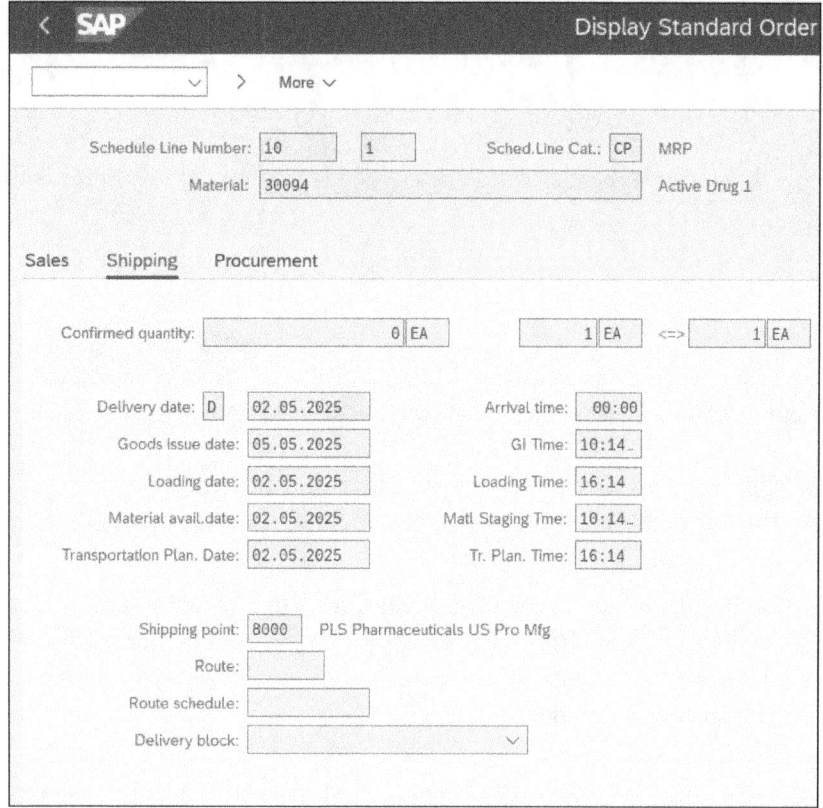

Figure 4.4 Verify Goods Issue Date in the Sales Order

Unfortunately, there are no predictive journal entries in the document flow. To show you some additional characteristics of a predictive journal

Sales order document flow

4   Value Flows of Margin Analysis

Search for predictive journal entry

entry, we'll go straight to table view and display the predictive journal entry in table ACDOCA.

You can search for your predictive journal entry with the sales order number. Figure 4.5 shows a search for the predictive journal entry by entering sales order "11011485" that you created earlier in Figure 4.3 in the **SRC_AWREF** field. All logistic reference information for predictive journal entries is stored in fields starting with SRC_*. There are no entries in table BKPF for the document header for predictive journal entries. All reports and SAP Fiori apps have been adjusted to show predictive data as actuals. For the audit trail, you can drill down to the source of the document or add an additional column for the source ledger to differentiate between actual and predictive journal entries.

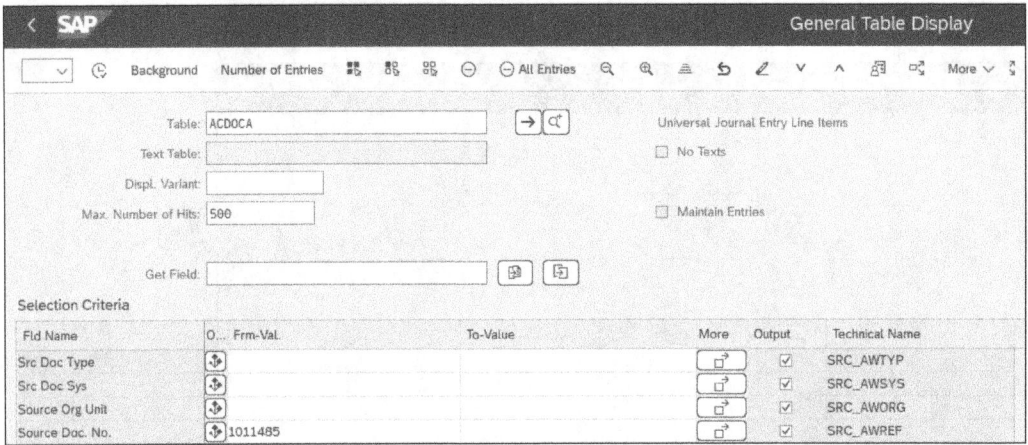

Figure 4.5 Select a Predictive Journal Entry by Sales Order Number

Review predictive journal entries

In Figure 4.6, let's look at the predictive journal entries the system has created via the following columns:

- **DocumentNo**
  The document number of the predictive journal entry is a technical number. You don't have to maintain any number ranges for predictive journal entries because the system gives the predictive journal entries a technical document number.

- **Src***
  All columns starting with **Src*** are for the storage of the link to the logistics reference documents. In the example in Figure 4.6, the **SrcDocTy** (document type), **SrcfDoc No** (source reference document), and **ScItm** (source reference item) columns are filled. The sales order number and the sales order item number of the sales order created earlier in Figure 4.3 are also shown.

4.1 Predictive Accounting

- **ObsR**
  The obsoleteness reason column tells you whether the line item is part of a current active prediction or outdated. The following entries are possible in this column:
  - blank: Active prediction
  - **1**: Outdated prediction
  - **2**: Cancellation of outdated prediction
  - **3**: Reduction posting
- **Posting Date**
  The posting date is for delivery-related billings derived from the goods issue date. For debit and credit memos, the posting date equals the actual date of the creation of the sales order.
- **Doc. Date**
  The document date is the date the sales order was created. All currency conversions will be converted with the date at the time of the order entry date.
- **S**
  If the document status (BSTAT) is **P**, the document is a predictive journal entry.

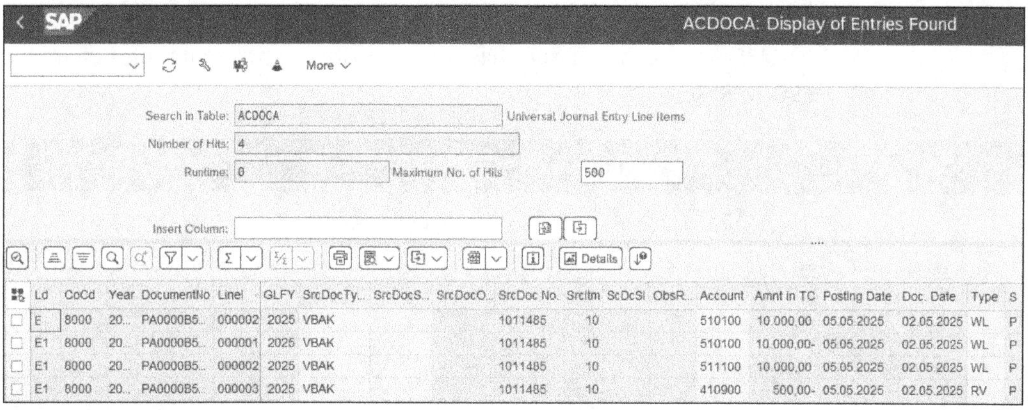

Figure 4.6 Analyzing Predictive Journal Entries in Table ACDOCA

We see all accounts listed in Table 4.1 for which predictive journal entries have been created.

Account	Amt in TC	Explanation
511100 – COGS Direct Materials	10,000.00	This line reflects the reduction in inventory. It's the offsetting posting to COGS.

Table 4.1 Predictive Journal Entry Line Items

4 Value Flows of Margin Analysis

Account	Amt in TC	Explanation
510100 – COGS Standard	-10,000.00	COGS is normally posted with the goods issue. If the COGS splitting in margin analysis is activated, the system will split COGS in the components of the underlying material cost estimate. COGS in the sales order is reflected as the internal price. This is the price the inventory is valuated with. It's the price of the material master. If you don't want to create predictive journal entries for COGS, there is a Business Add-In (BAdI) that allows you to cancel the creation of predictive journal entries for COGS. The total on this G/L account is zero, as the amount is split on the COGS splitting accounts.
410900 – Revenue Domestic - Product	-500.00	The total revenue is the sales price of the material times the sales quantity posted on the revenue account.

Table 4.1 Predictive Journal Entry Line Items (Cont.)

Goods issue posting Let's post the goods issue of the sales order and see what changes in predictive accounting. In Figure 4.7, you can see the document flow of sales order 1011485 The goods issue has been posted with material document 4900019536.

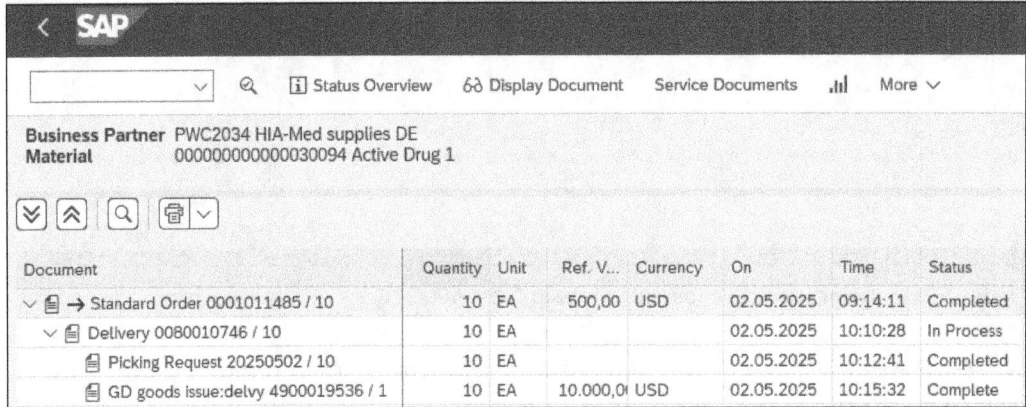

Figure 4.7 Document Flow of Sales Order after Goods Issue posting

Reduction posting What happened in table ACDOCA (Universal Journal) with the predictive journal entries the system created earlier when the sales order was saved? In Figure 4.8, you can see that the initial predictive journal entry for COGS was reversed and that there's an entry of **3** (reduction posting) in column **ObsRsn** (obsoleteness reason).

Ld	CoCd	Year	Document Number	Line Item	GLFY	Trans.Ty...	Ref.org.un	Ref. Doc.	SrcDocTy...	SrcDocS...	SrcDocO...	SrcDoc No	SrcItm	ScDcSI	ObsR...	Account	Amt in BTC
E	8000	2025	PA00000B5R0	000002	2025	RMWL			VBAK			1011485	10			510100	10.000,00
E1	8000	2025	PA00000B5R1	000001	2025	TBCS			VBAK			1011485	10			510100	10.000,00-
E1	8000	2025	PA00000B5R1	000002	2025	TBCS			VBAK			1011485	10			511100	10.000,00
E1	8000	2025	PA00000B5R2	000003	2025	SD00			VBAK			1011485	10			410900	500,00-
E1	8000	2025	PA0000B920	000002	2025	RMWL	2025	4900019536	VBAK			1011485	10		3	510100	10.000,00-
E1	8000	2025	PA0000B921	000001	2025	TBCS	2025	4900019536	VBAK			1011485	10		3	510100	10.000,00
E1	8000	2025	PA0000B921	000002	2025	TBCS	2025	4900019536	VBAK			1011485	10		3	511100	10.000,00-

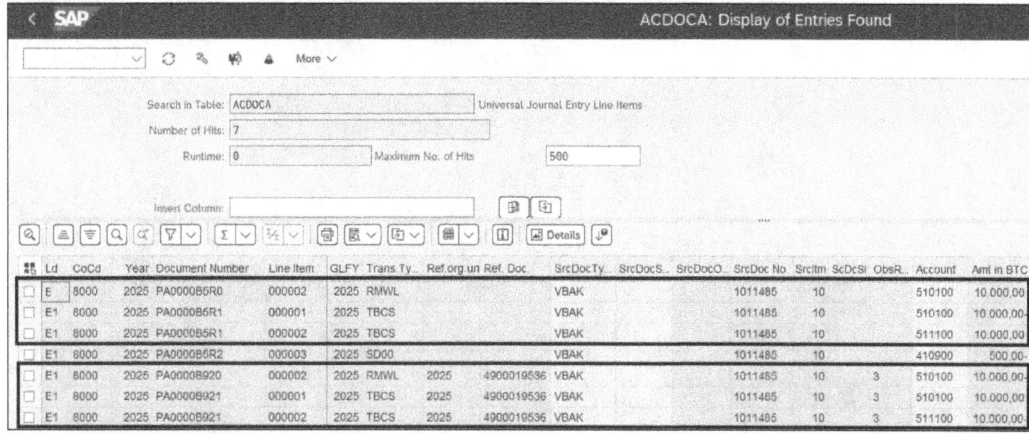

Figure 4.8 Review Predictive Journal Entries after Goods Issue Posting

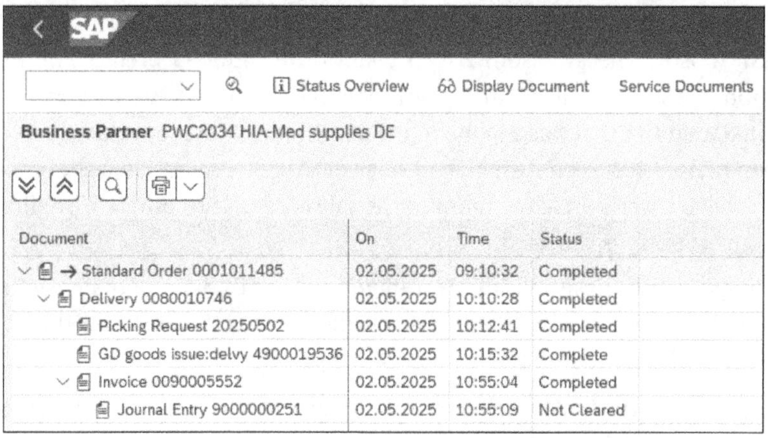

Billing invoice posting

What happens with the predictive journal entry if you're posting the billing invoice? In Figure 4.9, you can see that the billing invoice has been posted with document number **90005552**.

Document	On	Time	Status
→ Standard Order 0001011485	02.05.2025	09:10:32	Completed
Delivery 0080010746	02.05.2025	10:10:28	Completed
Picking Request 20250502	02.05.2025	10:12:41	Completed
GD goods issue:delvy 4900019536	02.05.2025	10:15:32	Complete
Invoice 0090005552	02.05.2025	10:55:04	Completed
Journal Entry 9000000251	02.05.2025	10:55:09	Not Cleared

Figure 4.9 Document Flow of the Sales Order after Billing Invoice Posting

Reversal of predictive journal entry

What changes will we see now in table ACDOCA (Universal Journal)? In Figure 4.10, you can see that the initial predictive journal entry for the billing invoice was reversed and that there's an entry of **3** (reduction posting) in the **ObsRsn** (obsoleteness reason) column. The total of the predictive journal entry for sales order **1011485** is now **0.00**, as the sales order has been completed.

4   Value Flows of Margin Analysis

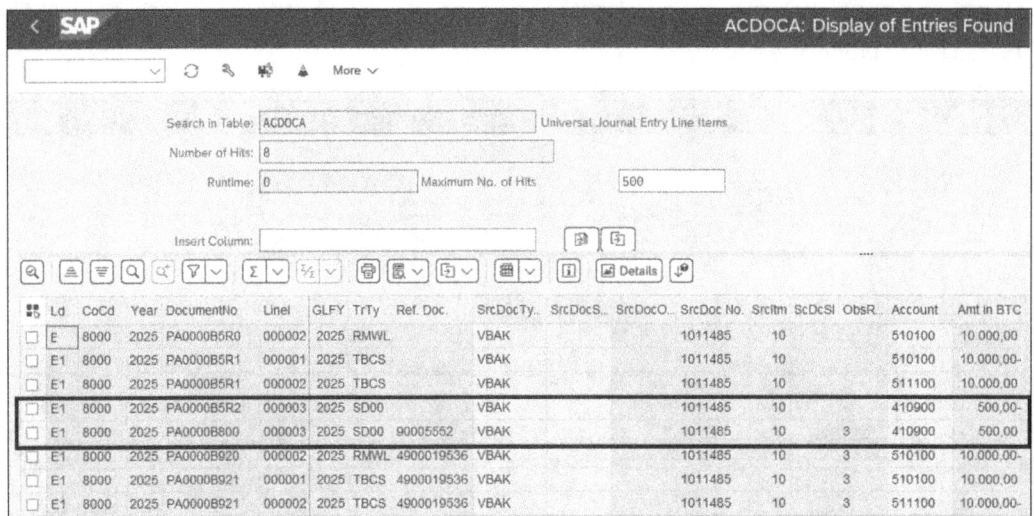

Figure 4.10 Review Predictive Journal Entries after Billing Invoice Posting

Adjustments to the sales order What happens with the predictive journal entry if there are adjustments made to the sales order? If you adjust the sales quantity after the sales order was saved, the system is updating the predictive journal entry. The original document will have **ObsRsn** (obsoleteness reason) **1** (outdated prediction) and there will be a cancellation of the same predictive journal entry with **ObsRsn** (obsoleteness reason) **2** (cancellation of outdated prediction). In addition, the system will create a new predictive journal entry based on the adjusted quantity that has a blank in the **ObsRsn** (obsoleteness reason) column in table ACDOCA. The system isn't able yet to create reversals at the line-item level. If there's an adjustment to the sales order, the complete predictive journal entry is canceled.

Archiving predictive journal entries You might think that there's now a huge number of documents created with predictive accounting. Predictive journal entries can get archived with archiving object FINS_PRED.

In production, there can be a balance between the predictive journal entries, for example, when there has been a rounding difference or a currency difference. With report FINS_PRED_FIN_REDUCTION, those differences are cleaned up, and a zero predictive balance for completed orders is guaranteed. In SAP S/4HANA Cloud, a job for this report is scheduled automatically as soon as predictive accounting is activated. In SAP S/4HANA, on the other hand, you have to manually schedule the job. There's no selection screen for the report; you just need to schedule the execution.

Repost predictive journal entries If you transitioned to SAP S/4HANA without activating predictive accounting, you can activate it later. You can repost previously created sales orders into predictive accounting with report FINS_PRED_REPOST. In Figure 4.11,

4.1 Predictive Accounting

you can see the **Repost Predictive Accounting Data** selection screen. You should at least enter the **Sales Organization** as a selection criterion; otherwise, the system won't find any data to repost. The settings in Figure 4.11 are recommended for a test run to repost predictive accounting data for all sales organizations in the system. You can execute the report with **Execute** or F8.

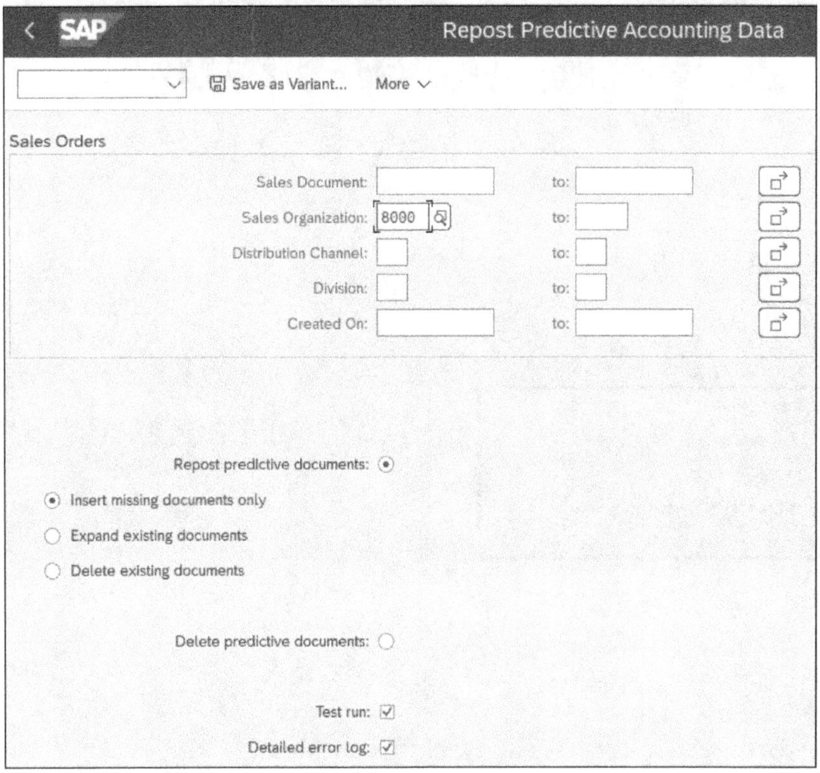

```
< SAP                              Repost Predictive Accounting Data

[        ] ∨   🖫 Save as Variant...   More ∨

Sales Orders
                  Sales Document: [        ]   to: [        ]   ⬚
               Sales Organization: [8000  ]🔍  to: [        ]   ⬚
              Distribution Channel: [  ]       to: [  ]         ⬚
                        Division: [  ]         to: [  ]         ⬚
                      Created On: [        ]   to: [        ]   ⬚


         Repost predictive documents: ◉
    ◉ Insert missing documents only
    ○ Expand existing documents
    ○ Delete existing documents

          Delete predictive documents: ○

                        Test run: ☑
                Detailed error log: ☑
```

Figure 4.11 Report to Repost Predictive Accounting Data

In Figure 4.12, you can see the log of the report to repost predictive accounting documents. The system processed 38 documents of which 4 documents were processed successfully, 4 documents were ignored because they already exist, and 33 documents can't be transferred because they're erroneous, one of which has a fatal error. You can see in the report the reason why the documents couldn't be transferred; usually the reason is master data or configuration related. After you analyze the results and correct all the error messages, you can go back with F3 to the selection screen and execute the report in a real run.

Protocol of reposting predictive journal entries

As of now, the migration of predictive journal entries isn't yet supported, and no history of incoming sales orders can be displayed. It's recommended to also use report FINS_PRED_REPOST if you happen to migrate to a new system.

Migration of predictive accounting

137

4 Value Flows of Margin Analysis

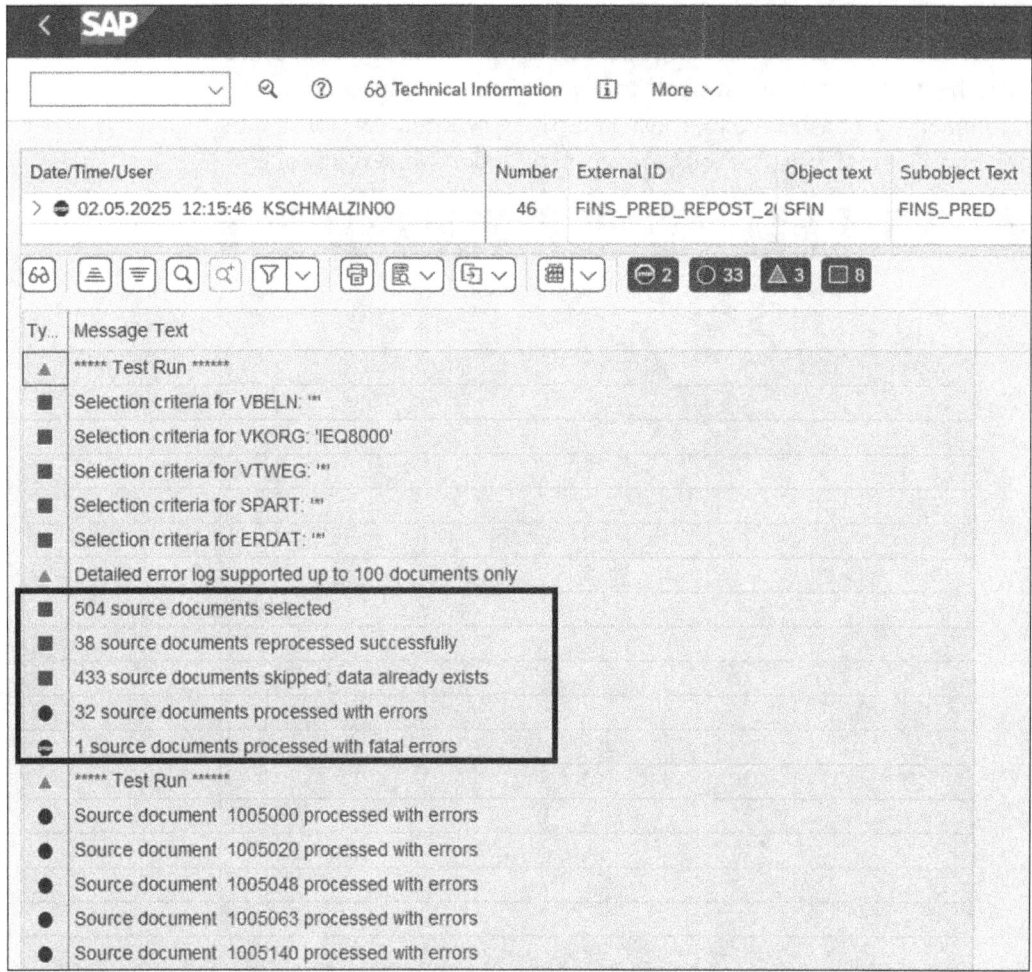

Figure 4.12 Log of Reposting Predictive Accounting Documents

Reporting predictive accounting will be explained and demonstrated in Chapter 7.

4.1.2 Predictive Accounting Commitment Management

Predictive accounting for purchasing isn't yet fully developed and available in SAP S/4HANA. The logic for the predictive journal entries for purchasing documents will be the same as for sales orders. The objective is to be able to simulate goods receipt, invoices, and even purchase requisitions.

Activate extension ledger for availability control

When you created the extension ledger and activated it in your operating concern, you might remember the **Relevant for Commitment Management** checkbox, shown in Figure 4.13.

4.1 Predictive Accounting

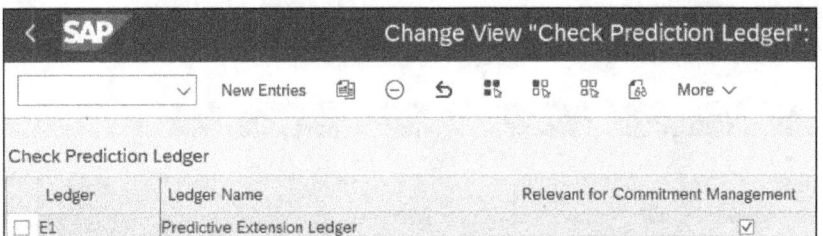

Figure 4.13 Activate the Extension Ledger for Commitment Management

The system is already creating predictive journal entries for commitments on cost center and projects in table ACDOCA (Universal Journal). Database table COOI (Commitments Management Line Items) is still updated with data to be able to display old standard reports in SAP GUI. Let's look at a commitment posting and verify the journal entry in table ACDOCA. In Figure 4.14, you can see a purchase order that has an account assignment to **Cost Center Z10010**. The value of the purchase order is 600.00 USD.

Predictive journal entries for commitments

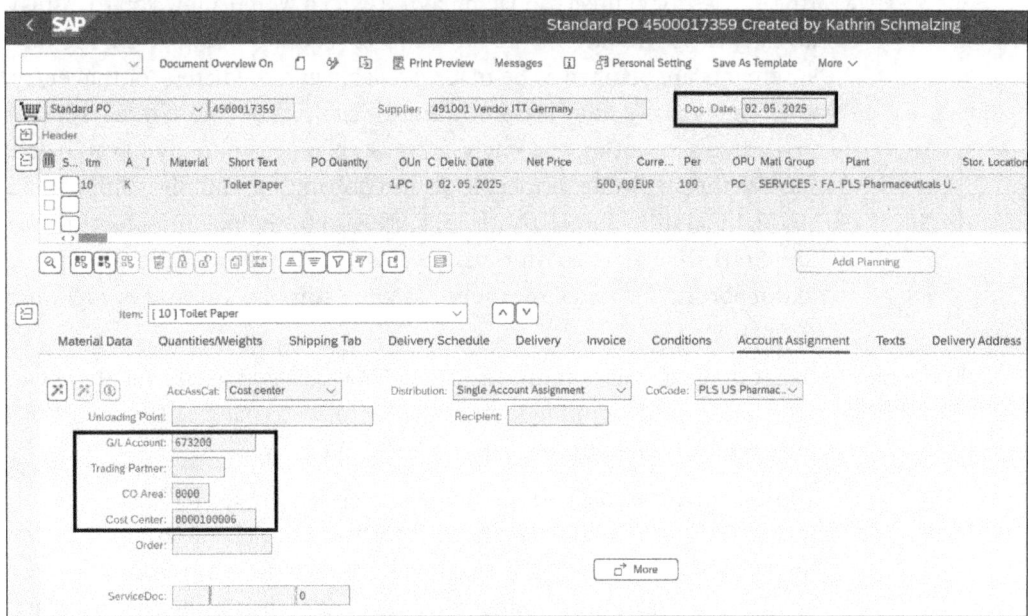

Figure 4.14 Purchase Order with Account Assignment to Cost Center

Let's see what this posting looks like in table ACDOCA (Universal Journal). In Figure 4.15, you see the predictive journal entry for the commitment posting. The **SrcDoc No** (source reference document) has been updated with the purchase order number. There's also a flag in the **Commt** (commitment) column for the expense line item. If there are changes to the purchase order

Review commitment posting in Table ACDOCA

139

4   Value Flows of Margin Analysis

before the goods receipt is posted, the behavior is similar to the sales order predictive journal entry.

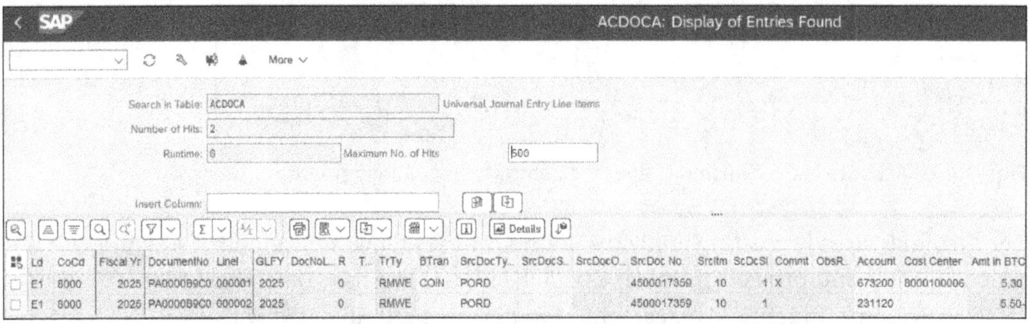

Figure 4.15 Review Commitment Posting in the Universal Journal

Reporting for commitments will be demonstrated in Chapter 7.

Outlook for predictive accounting

In upcoming releases, SAP plans to deliver predictive accounting for recurring entries for depreciation and payroll, which would allow you to display an early close. In addition, the simulation of allocations or currency remeasurements are planned to be made available as a predictive journal entry. The direction is going clearly toward the early close. More integration of predictive accounting and analysis is also planned so that you can better include the predictive insights in your reporting, execute simulations, and so on. In addition, predictive events are also in the planning pipeline for SAP S/4HANA. An event, for example, is an entry in revenue accounting and reporting for actual revenues based on complex revenue recognition regulations.

API for external predictive postings

As the predictive accounting in SAP S/4HANA isn't complete yet, there's an application programming interface (API) for external predictive postings until predictive accounting in SAP S/4HANA matures.

This section gave you a comprehensive overview of the predictive accounting capabilities in margin analysis. Predictive journal entries are expected revenues and costs and not actual revenues and cost, which is why they are posted in the extension ledger to clearly differentiate them from actual costs. We recommend activating predictive accounting because, in the future, there will be many more functionalities that will allow you to make better decisions as you're made aware of developments in your financials faster and earlier than in the past.

4.2 Billing Data Transfer

In margin analysis, the system generates a profitability segment for the billing invoice when the billing invoice is posted together with the financial accounting document. There's no configuration necessary to create a profitability segment for revenue in margin analysis. The G/L accounts for posting the billing invoice are determined in the automatic account determination, which is the interface between financial accounting and sales and distribution in SAP S/4HANA. So, let's take a closer look at how price determination in the sales order works and how G/L accounts are determined when the billing invoice is created.

Billing documents in margin analysis

4.2.1 Price Determination in the Sales Order

Price determination in the sales order uses the condition technique to determine the price of the material. The conditions form the basis for the determination of the price. The system automatically determines the gross price as well as surcharges and discounts that are relevant to specific customers according to pricing rules. The maintenance of conditions is considered master data and executed by the sales department. The conditions are maintained based on key combinations that are maintained in configuration.

Price determination in the sales order

The example shown in Figure 4.16 contains the entries for a pricing condition for **ZPR1**, which is the gross price. When the choices are confirmed by pressing [Enter], a **Key Combination** popup appears that shows the level of detail on which you can maintain your prices. The system always searches for a price from the first key combination to the last in the sales order. For this example, choose **Customer/Material**, and press [Enter].

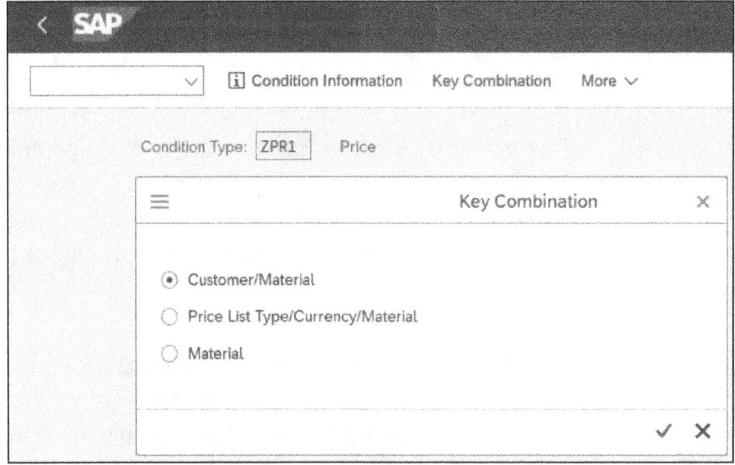

Figure 4.16 Choose the Key Combination for Pricing Condition Maintenance

4 Value Flows of Margin Analysis

Maintain sales price in pricing condition

In the next screen, shown in Figure 4.17, maintain the price of **50,000.00 USD** for **Material 30094**. The price is maintained for the **Sales Organization 8000**, the **Distribution Channel 80**, and the **Customer PWC2034**. Every pricing condition has a validity date. After maintaining the pricing condition, save it with **Save** or [Ctrl]+[S].

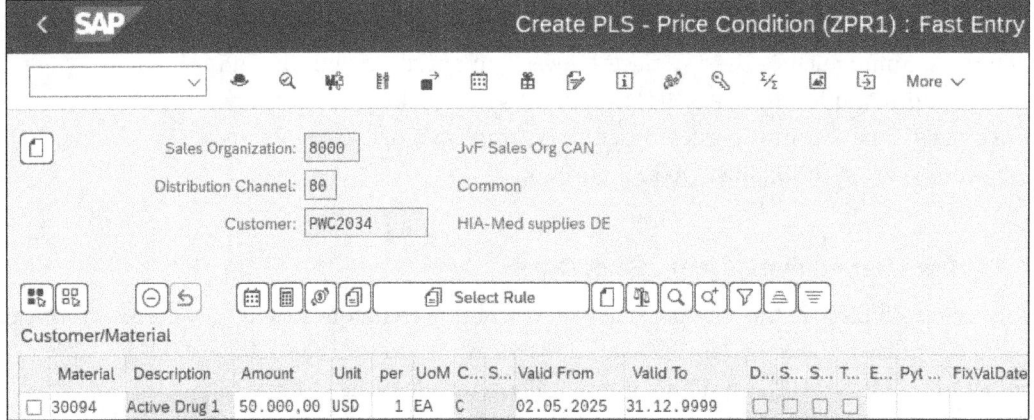

Figure 4.17 Maintain the Pricing Condition

Create a sales order

4.2.2 Create Sales Order

Let's create a sales order with the combination of customer, material, and sales organization for which you just maintained the condition type PPRO and see if the condition gets automatically populated in the sales order. In Figure 4.18, you can see the pricing elements of the sales order you're about to create. There are three conditions to consider:

- **ZPR1**

  You can see pricing condition ZPR1 with the price you maintained in Figure 4.17, so that worked and the system was able to automatically derive the price.

- **SKTO**

  Condition **SKTO** for **Cash Discount** is determined based on the payment term the system derives from the customer master data or the payment term you manually maintain in the sales order. This is a statistical condition, meaning that there's no journal entry posted with the billing invoice because you don't know if the customer pays in time to take advantage of the cash discount or not.

- **VPRS**

  Condition **VPRS** for the **Internal price** is the price from the material master used to valuate the inventory. In margin analysis, COGS entries are posted with the goods issue, so condition type VPRS isn't used in margin analysis. In costing-based profitability analysis, COGS is determined based on that condition type when the billing invoice is posted.

4.2 Billing Data Transfer

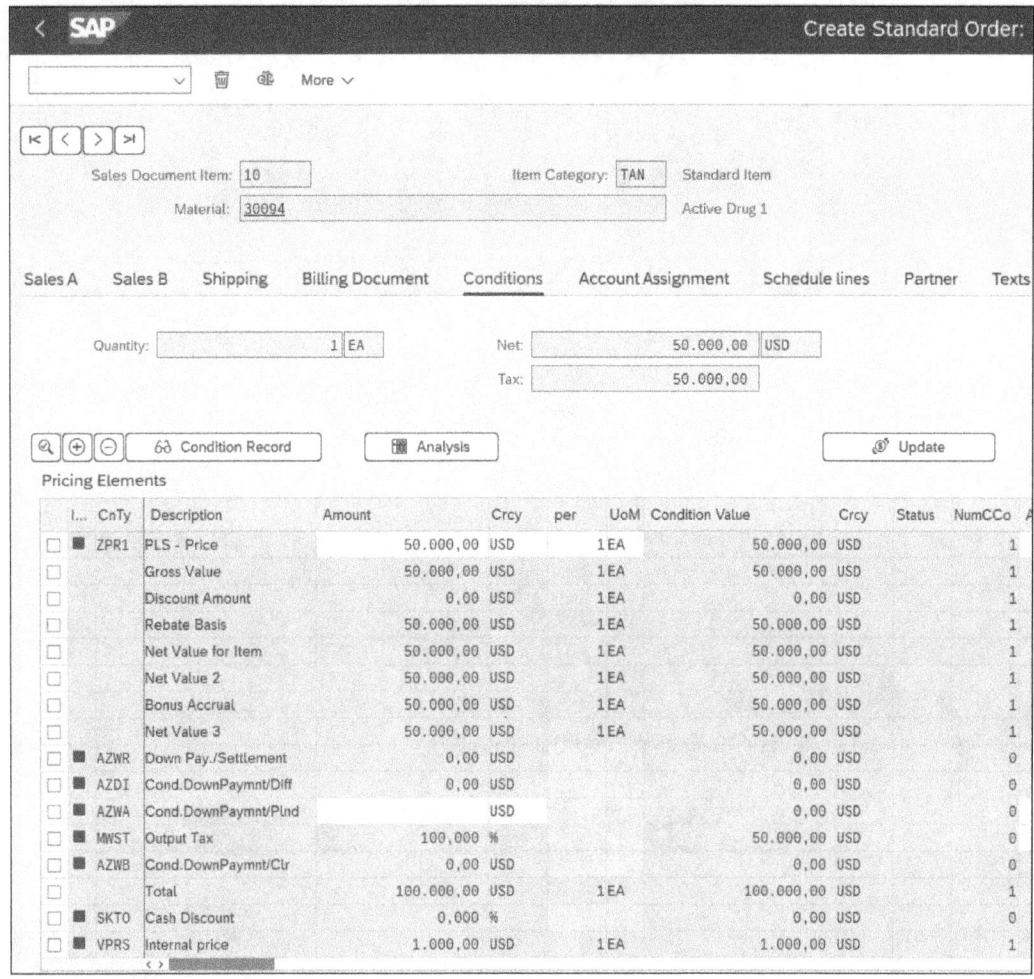

Figure 4.18 Sales Order Pricing

Double-click on condition **ZPR1**. In Figure 4.19, you can see the details of condition **ZPR1 (PLS - Price)**. The condition details are maintained in configuration. The **Account Key** field in the lower **Account determination** section plays a major role in the account determination. Account key **ERL** (revenue) is assigned to condition **ZPR1**. This account key is used in automatic account determination.

Details of pricing condition

Now let's look at the statistical condition to see the difference. Double-click on condition **SKTO – Cash Discount**. In Figure 4.20, you can see the details of condition **SKTO**. The condition details are maintained in configuration. Condition category **E** and the **Statistical** checkbox determine that the condition is statistical. New to this screen is the **Acct Detn Relevant** checkbox, which means that the condition is relevant for account determination. When this checkbox is selected, the **Account Key** field in the **Account Determination** section opens.

Review details of statistical condition

143

4   Value Flows of Margin Analysis

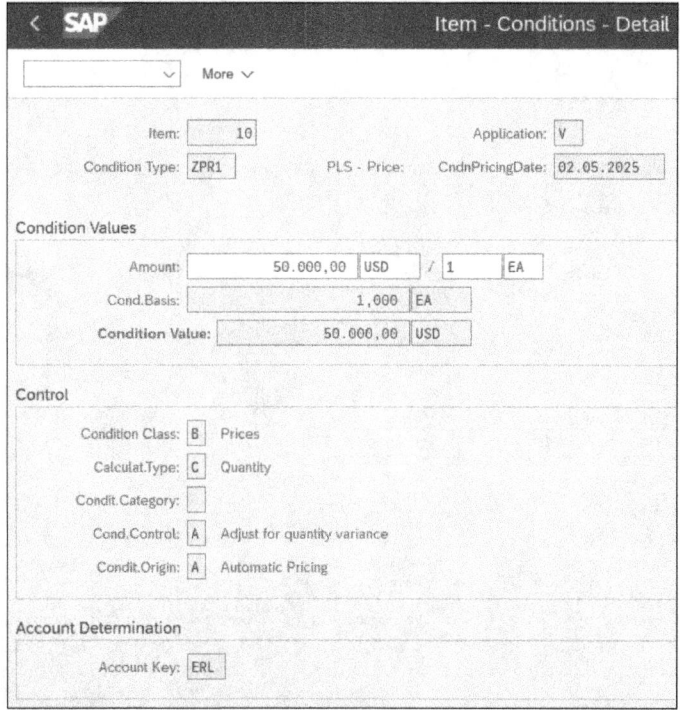

Figure 4.19 Pricing Condition Details

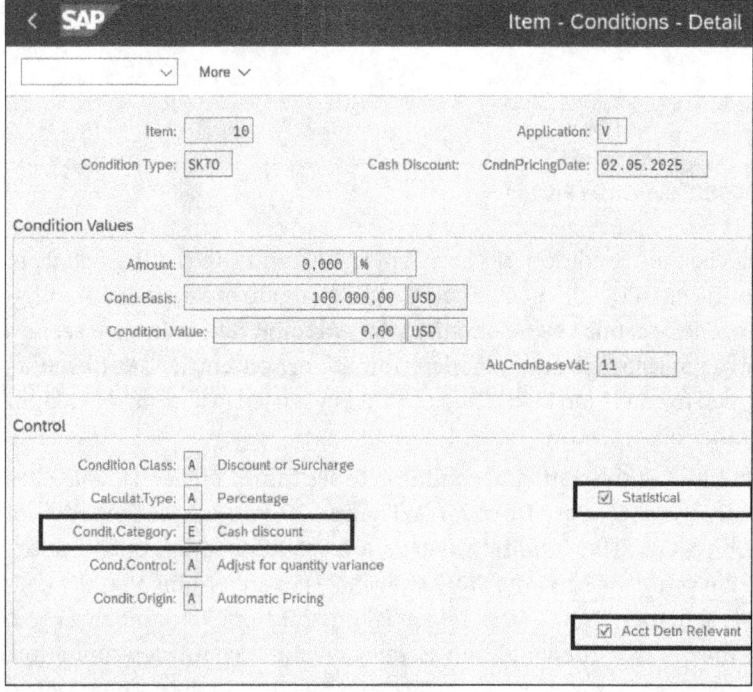

Figure 4.20 Statistical Condition Details

4.2.3 Account Determination

Next let's look at the account determination via Transaction VKOA or the following configuration path: **Sales and Distribution • Basic Functions • Account Assignment/Costing • Revenue Account Determination • Assign G/L Accounts**. Figure 4.21 provides an overview of the tables that determine the characteristics based on which you can maintain G/L accounts. You can create additional tables in configuration if the level of detail of the default tables isn't sufficient. You can double-click table **005 - Acct Key** (account key) to verify the account determination for the account key ERS – Sales Discounts.

Account determination

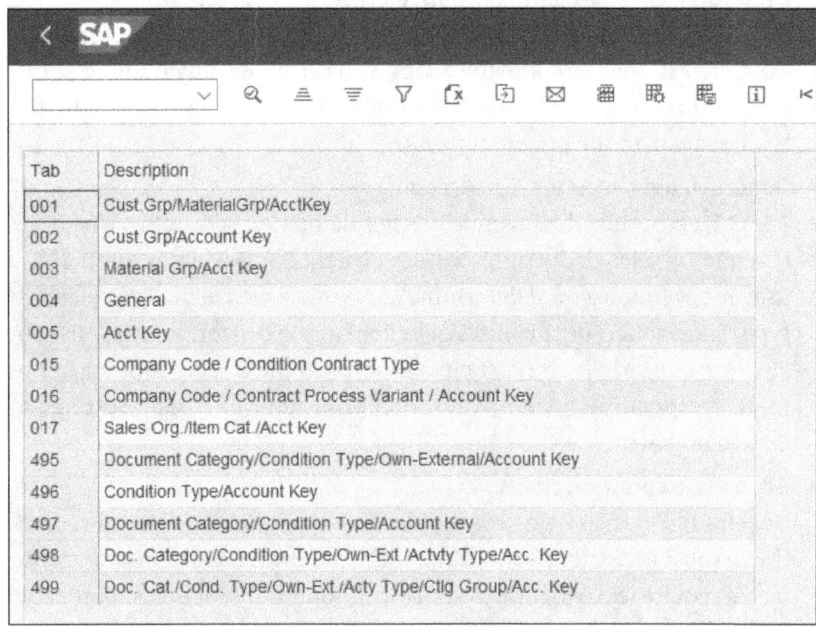

Figure 4.21 Review Sales Account Determination

Figure 4.22 shows the detail maintenance of table **004** (**General**). The table has the following columns:

Maintenance of account determination

- **Application (App)**
  The application determines how the condition is used. **V** is for sales and distribution.
- **Condition type for account determination (CndTy.)**
  There are two types for the determination of accounts:
  – **KOFI** for billing documents without an account assignment object
  – **KOFK** for billing documents with an account assignment object

  KOFI is used when a profitability segment is generated for billing documents. KOFK is used when billing documents are assigned to internal orders, for example, if you use the Customer Service (CS) component in SAP ERP. Consequently, to transfer billing documents to profitability analysis, you have to maintain condition type KOFI.

- **Chart of accounts (Chrt/Accts)**
  The chart of accounts determines where the G/L account is defined. In the example, this is the chart of accounts **8100**, which is assigned to company code 8000.

- **Sales organization (SOrg.)**
  The sales organization is an organizational element from Sales and Distribution (SD) in SAP ERP that classifies the area of responsibility for the sales and distribution of specific products or services. Sales organizations are assigned to company codes. You can assign any number of distribution channels and divisions to sales organizations to add a higher level of detail to the pricing process.

- **General ledger account (G/L Account)**
  The G/L account is the revenue or sales deduction account to which the revenue or sales deductions will be posted. For G/L accounts that are used in revenue account determination, you have to define cost element type 11 (**Revenues**) or 12 (**Sales Deduction**) in the master. In SAP S/4HANA Finance, you specify the cost element type in the G/L account master of the corresponding account on the **Control** tab in **Account Settings in Controlling Area**.

- **Accruals account (Accruals Acc.)**
  Depending on the settings of the condition, you can also use the condition to post accruals. This is often done for bonus conditions, for example. If you're working with statistical conditions, it's mandatory to maintain an accruals account so that the system can create a balanced statistical posting for margin analysis.

Cash discount account determination
In Figure 4.22, you'll see the account determination maintained for the cash discount: **G/L Account 422000** for cash discount and **121130** for **Accruals Acc.**. The entries are saved with **Save** or Ctrl + S.

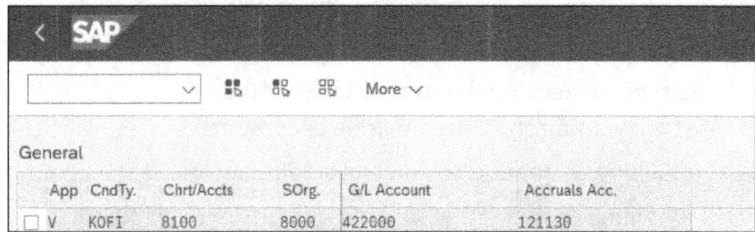

Figure 4.22 Maintain Account Determination for Account Key

4.2 Billing Data Transfer

4.2.4 Activate Transfer of Statistical Sales Conditions

Before you can post the sales order and review the document, you have to activate the transfer of statistical conditions in the ledger group. Therefore, go to Transaction SM30, enter "FCOV_STAT_ACT" in the **Table/View** field (see Figure 4.23), and choose ✎ **Edit**.

Activate statistical conditions transfer

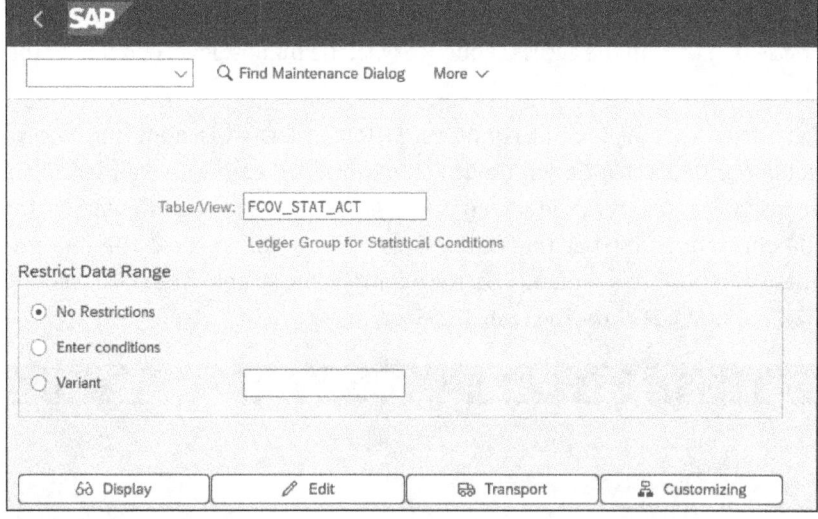

Figure 4.23 Activate Statistical Conditions in the Ledger Group

In the next screen, shown in Figure 4.24, assign the **Ledger Group E1** of the extension ledger to the view with **New Entries** or F5.

Assign the ledger group

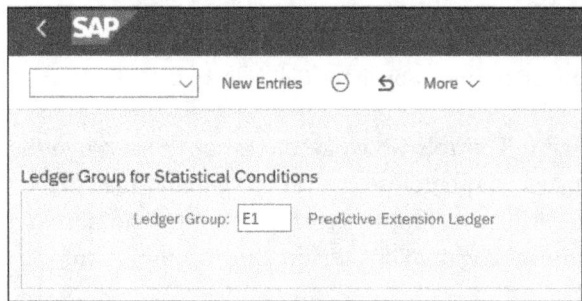

Figure 4.24 Assign the Ledger Group to the Transfer of Statistical Conditions

4.2.5 Create Billing Document

Let's post the sales order and create a billing document to verify the account determination and see if a profitability segment was created. After the billing invoice has been created, you can follow **Environment • Analysis Account Assignment • Revenue Accounts** in Transaction VF03 to display the billing document to analyze the account determination. The left part of the

Create billing documents

147

screen, **Procedure**, lists the pricing procedure that is assigned to the billing invoice. In the pricing procedure, you can see all the pricing conditions that were determined in the billing document. In the example in Figure 4.25, this is condition **ZPR1 (PLS - Price)** and condition **SKTO (Cash Discount)**. Under the folders of the single condition, you can see the various access sequences; based on these, the level of detail was maintained in the G/L account in account determination. The order in which the system performs the search is from the access sequence with the highest level of detail to the bottom access sequence.

Let's take a closer look at condition **SKTO** for **Cash Discount** and **Access sequence 050(KOFI) General** that you maintained earlier in Figure 4.22. In the right-hand part of the screen under **Access Details**, you can see the single characteristics that the system was looking for to meet the requirements of the access sequence. In the example, the system finds G/L account 422000 and 121130 for the cash discount.

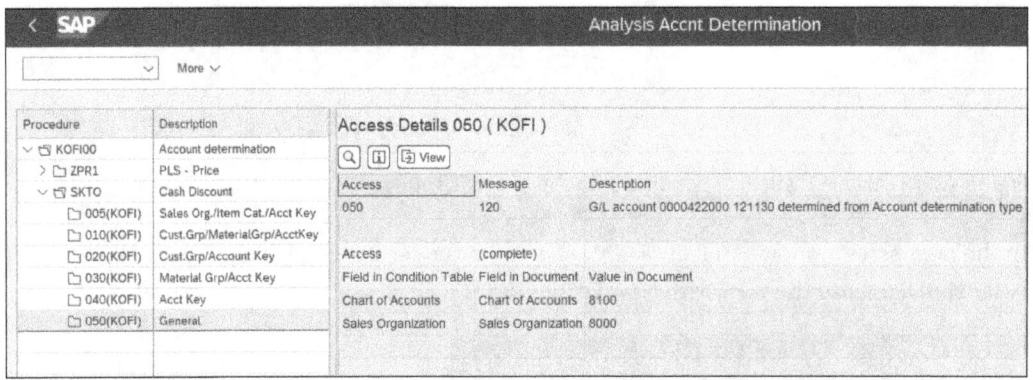

Figure 4.25 Review Account Determination in the Billing Invoice

Account determination for billing documents

The account determination controls which accounts are found when the billing document is posted. It's therefore critical to maintain master data properly to enable the system to find the correct pricing condition and reduce efforts for postprocessing of billing documents to a minimum.

4.2.6 Review Accounting Documents

Review accounting document

Now that you've made sure the account determination worked as designed, let's look at the accounting document, which is also the margin analysis document, as shown in Figure 4.26. There's no separate document for margin analysis because the margin analysis is part of the Universal Journal and hence covered in the accounting document. In the overview of documents to the billing invoice, you see the accounting document, which is the financial accounting document as well as the document for margin analy-

4.2 Billing Data Transfer

sis. You also can see the accounting document for **Ledger Grp E1**, which is the document for the statistical conditions.

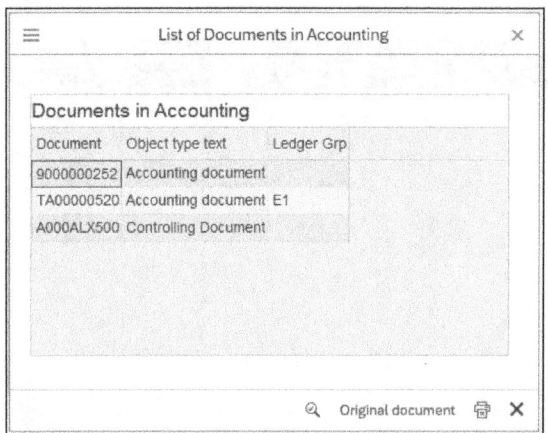

Figure 4.26 Review Documents of the Billing Invoice

Let's look at the document in table ACDOCA (Universal Journal). In Figure 4.27, you can see that the system created a **Prof. Seg.** (profitability segment) for the revenue line item and derived several characteristics (not all characteristics that have been derived are shown in the screenshot). The posting was created in ledger **OL**, the leading ledger.

Review the document in table ACDOCA

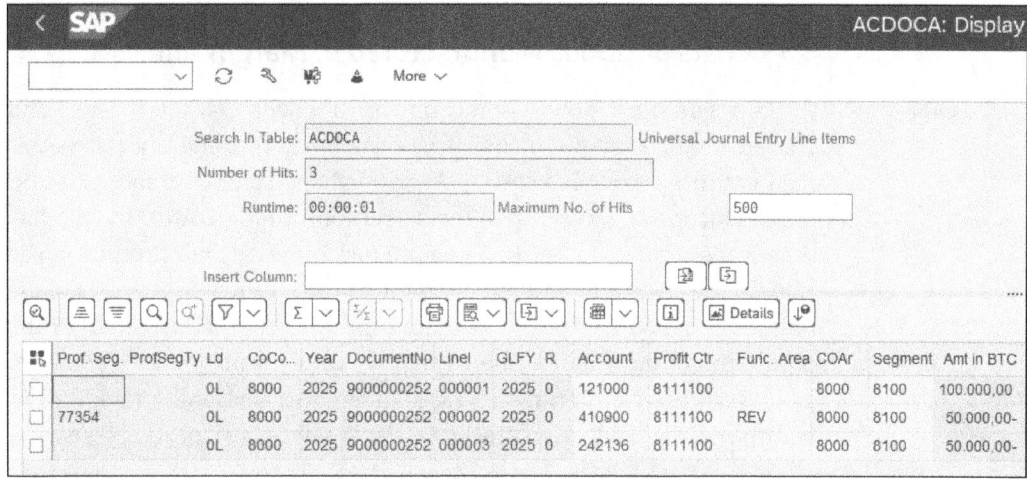

Figure 4.27 Review the Financial Document of the Billing Invoice

Next, let's look at the document in extension ledger **E1** for the statistical sales condition document. In Figure 4.28, you can see that the system created a document in extension ledger **E1** for the cash discount. The offsetting account is the accrual account you maintained earlier in Figure 4.22. The

Review the document in the extension ledger

149

statistical value is visible if you execute a profit and loss (P&L) and balance sheet for the extension ledger. (We show different reporting functionalities in Chapter 7.)

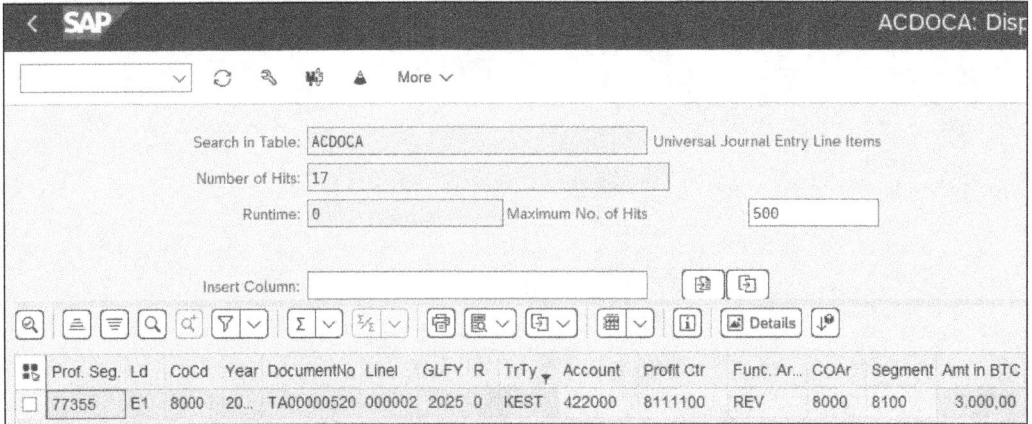

Figure 4.28 Review the Financial Document of the Billing Invoice in the Extension Ledger

Now, you've learned about the transfer of billing data and statistical sales conditions. In the next section, we'll look at how to transfer cost of goods manufactured (COGM) to margin analysis.

4.3 Costs of Goods Manufactured in Margin Analysis

Definition of COGM

COGM refers to the total production cost of goods that are finished and delivered to inventory. It includes the cost of materials and labor, as well as manufacturing overhead costs. The material and labor cost are posted on the production order based on the actual consumption. After the production of the product is finished, the product is posted into inventory. The value of the posting is determined by the price in the material master of the finished product. This price can differ from the cost that occurred for the production of the product. There can be various reasons for a difference in price, such as scrap, price variances, differences in labor, and so on. At month end, the production order is settled. There are two posting scenarios:

- **The production order isn't yet finished, but there are costs on the production order.**
  Before the settlement of the production order, you need to determine work in process (WIP). During the settlement, the system will create a document for the WIP that doesn't affect the P&L. The value of the production order is capitalized in the balance sheet. If the production order is completed in the next month, the WIP will be reversed.

- **The production order is completed.**
  The variance calculation, which is described in detail in Section 4.5, will determine the difference between actual cost and planned cost on the production order, as well as classify and settle the variances. This posting in the P&L won't affect the P&L. It's adjusting the cost for the delivery of the finished product to the inventory. The offsetting account is the price difference account or the different accounts for the variances.

In Figure 4.29, you see a simplified sample posting of the COGM in margin analysis. It shows the value flow from purchasing to sales. The left side of the figure illustrates postings in financial accounting, mapped in T-accounts, while the right side of the figure shows how postings are mapped in the P&L statement in margin analysis in comparison to costing-based profitability analysis. If the configuration settings are correct, the results both in margin analysis and costing-based profitability analysis should be the same.

Sample posting of COGM

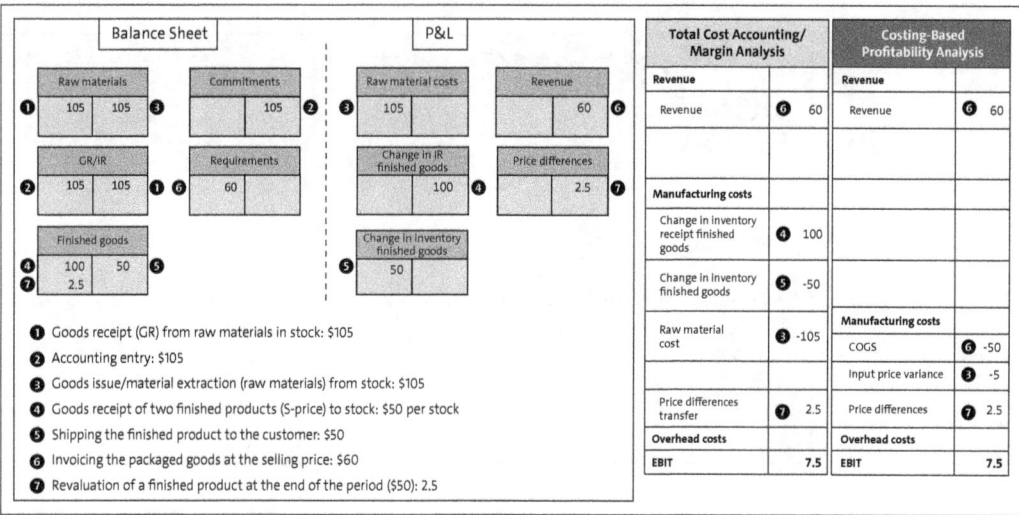

Figure 4.29 Simplified Postings of COGM in Margin Analysis

Only the COGS and the variances will show up in margin analysis unless you have the attributed profitability segments activated; in that case, you'll see statistical profitability segments throughout the month for COGM.

Statistical profitability segments

4.4 Splitting the Cost of Goods Sold

SAP S/4HANA Finance enables you to split the COGS into the individual cost components according to the cost component structure of the material cost estimates. COGS are posted in financial accounting and margin analysis

Cost component split

when the goods issue to a delivery is posted. We know from the past that, initially, the goods issue in margin analysis is posted to one G/L account according to the valuation class of the material. This account is defined in account determination of materials management with Transaction GBB-VAY (Offsetting Entry for Inventory Posting-Goods Issue for Sales Order). The posting for the goods issue still happens on the account that is maintained in the material account determination. In addition, SAP S/4HANA Finance enables you to split these costs according to the components of the cost component structure of the material cost estimate that has been used for inventory valuation.

Create a G/L account for splitting COGS

Before you start configuring the split of the COGS, you need to create a G/L account with cost element category 1 for every cost component of the cost component structure. In Figure 4.30, G/L account **510100** for **COGS - Standard** is being created. Direct material is a cost component in the cost component structure. Cost element category **1** has been chosen (**Column Cost Ele…**) for **Primary costs/cost-reducing revenues**.

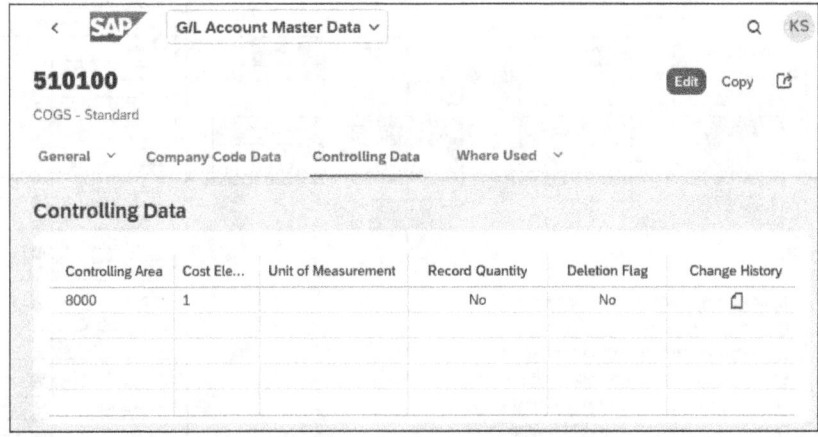

Figure 4.30 Create the G/L Account for the COGS Split

After you've created all the G/L accounts necessary for the COGS split, you can start with the configuration. To configure the COGS split in margin analysis, follow the configuration path **Financial Accounting** • **General Ledger Accounting** • **Periodic Processing** • **Integration** • **Materials Management** • **Define Accounts for Splitting the Cost of Goods Sold**.

Create a splitting profile

With **New Entries** or F5, you can create a new splitting profile. For this example, you're creating **Cost Splitting Profile Z8000** in Figure 4.31 and assigning it to **CO Area 8000**. The chart of accounts in the **Chrt/Accts** column will be derived automatically from the controlling area. If you activate the checkbox in the **Acc Based Split** (account-based split) column, the system will always split the cost into the cost components when there's a post-

4.4 Splitting the Cost of Goods Sold

ing on the **Source Accounts** you'll determine in the next section. For this example, don't select the checkbox because you only want the system to split the COGS based on movements of the sales order. Mark your **Cost Splitting Profile** by selecting the checkbox at the beginning of the line, and navigate in the left section of the screen to the **Source Accounts** folder.

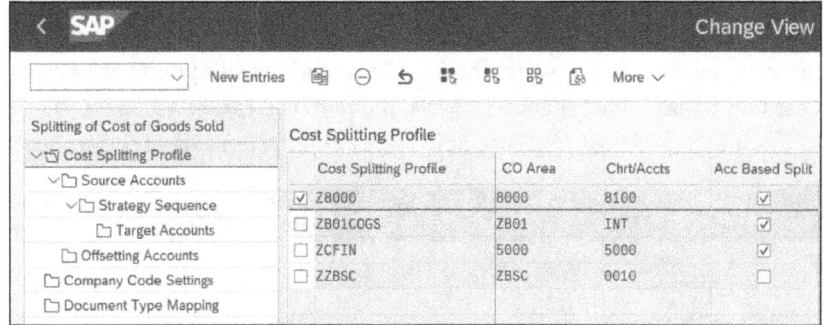

Figure 4.31 Create the Cost Splitting Profile

Use **New Entries** or F5 to maintain the **Source Accounts** (Figure 4.32), which are the G/L accounts that get hit when the goods issue for the delivery is posted. The accounts for the goods issue are maintained in the materials management account determination with Transaction GBB-VAY. In the example, you're maintaining G/L account **510100** for inventory change COGS as the **Source Account**. In the **Valuation View** column, you can maintain a specific valuation, such as a group valuation. If you don't maintain a valuation, the splitting will happen for all valuations that exist. Mark your **Source Account** by selecting the checkbox at the beginning of the line, and navigate in the left section of the screen to the **Strategy Sequence** folder.

Maintain source accounts

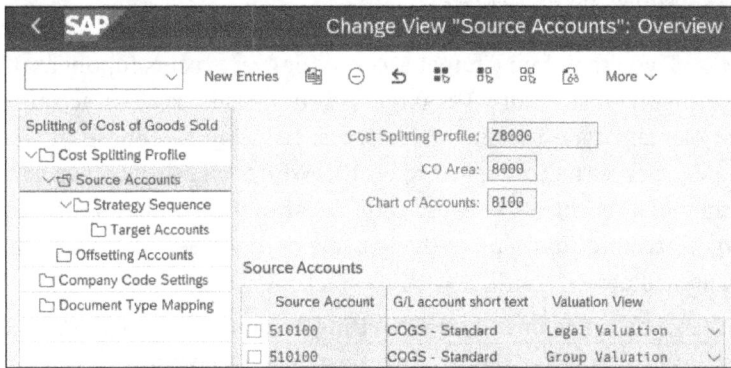

Figure 4.32 Maintain Source Accounts

In the **Strategy Sequence**, you define a strategy for every source account after which the system is trying to split the cost. If the system doesn't find

Define the strategy sequence

153

a material cost estimate in the first sequence, it goes on to the next sequence, and so on.

The example in Figure 4.33 is creating **Strategy 001** with **Sequence Number 1**, and the system will split the cost with the **Current Standard Cost Estimate**. This is the cost estimate that is active in the material master at the time of goods issue posting. You can also split your COGS with the actual cost component split of the material ledger. If you're using actual costing in the material ledger, you can select the **Split Revalued Consumption with Actual Cost Component Split** checkbox. Next, move to the **Target Accounts** folder in the **Splitting of Cost of Goods Sold** section on the left side of the screen.

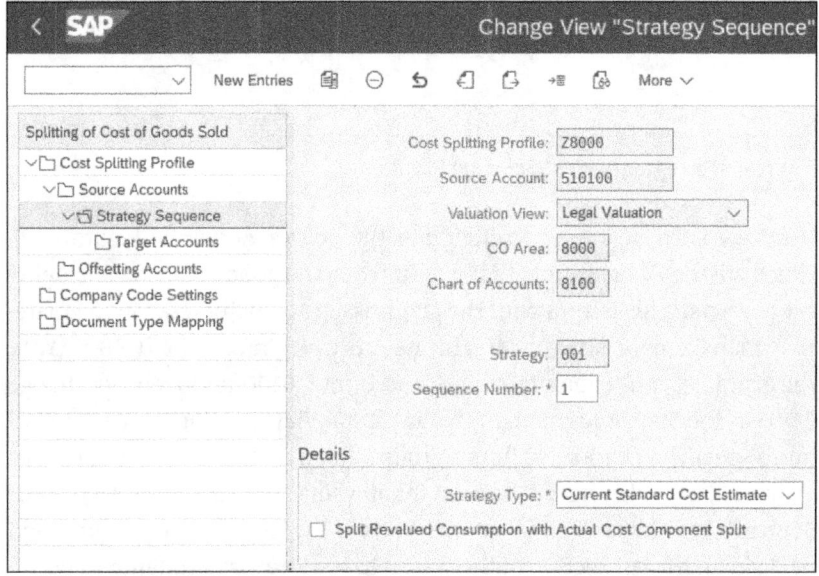

Figure 4.33 Maintain the Strategy Sequence

Maintain target accounts

In Figure 4.34, you maintain a target account for every cost component of the cost component structure. The target account has to be a G/L account with cost element type 1 (primary costs/cost-reducing revenues); otherwise, you can't save the cost splitting profile. With the F4 Help, you can display the possible entries in every column. After you've maintained all your **Target Accounts**, navigate to the left side of the screen to the **Offsetting Accounts** folder.

Maintain offsetting accounts

In Figure 4.35, you can maintain the **Offsetting Accounts**. Usually, the COGS account is used as an offsetting account for the COGS component split. If you're using the material ledger, you can compare the originally posted costs with the actual costs according to the material ledger. This isn't the case in this example, so there's no need to maintain an offsetting account. Move to the **Company Code Settings** folder on the left side of the screen.

4.4 Splitting the Cost of Goods Sold

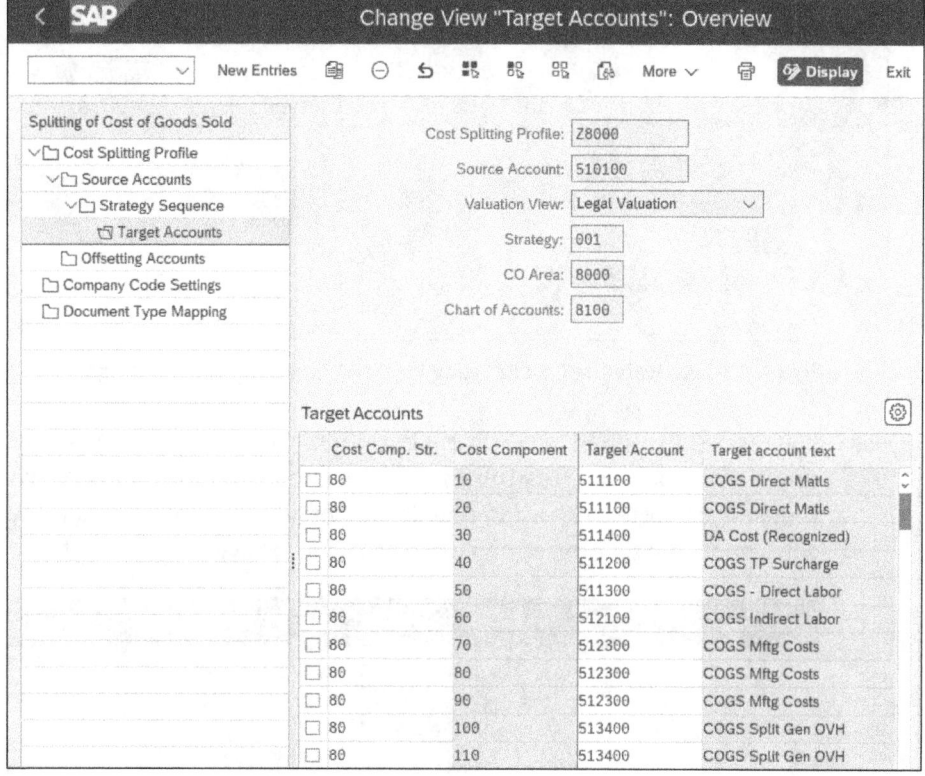

Figure 4.34 Maintain Target Accounts

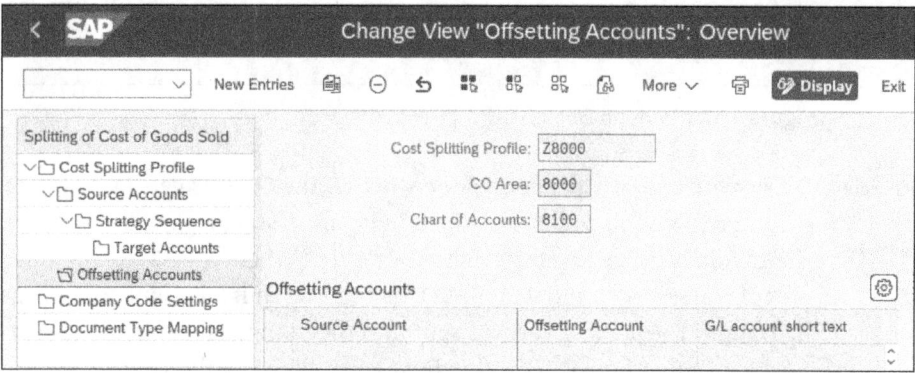

Figure 4.35 Maintain Offsetting Accounts

Before you can save the cost splitting profile, you have to activate it for the company code. In Figure 4.36, add with **New Entries** or [F5] a new entry for **Company Code 8000**. Assign **Cost Splitting Profile Z8000** with a **Valid From** date of **01.01.2020** to **Company Code 8000**. Mark your **Company Code** by selecting the checkbox at the beginning of the line, and navigate in the left section of the screen to the **Document Type Mapping** folder.

Activate the splitting profile in the company code

4 Value Flows of Margin Analysis

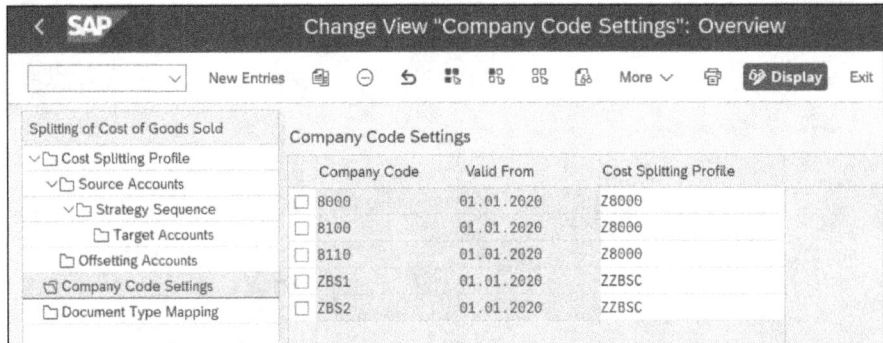

Figure 4.36 Activate the Cost Splitting Profile in Company Code Settings

Assign the document type

You can assign a different document type for the posting of the COGS splitting in Figure 4.37 with **New Entries** or [F5]. For this example, the COGS split shouldn't be posted with a different document type, so don't maintain any entries here but save the cost splitting profile with **Save** or [Ctrl]+[S].

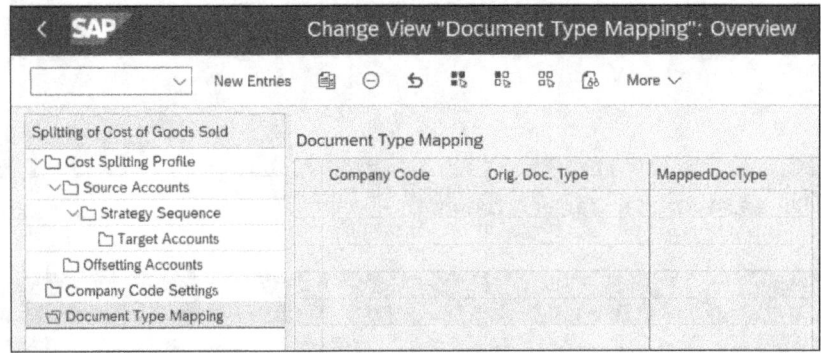

Figure 4.37 Maintain Document Type Mapping

Review COGS split in the predictive journal entry

Let's create a sales order and review whether the COGS is getting split into the target accounts you just defined in Figure 4.34. In Figure 4.38, you see the predictive journal entry that was created when you saved the sales order in table ACDOCA and that the COGS is getting split according to the configuration. The goods issue was split into the individual items of the costing. The value of the COGS is credited and debited to the account for the goods issue posting that is defined for record GBB-VAY in materials management account determination. The balance on change to stock account GBB-VAY is zero.

Review the accounting documents

After posting the goods issue to the sales order, you can see in Figure 4.39 that the system created two accounting documents. In Figure 4.38, the system also posted two predictive journal entries. Let's take a more detailed look at the two accounting documents.

4.4 Splitting the Cost of Goods Sold

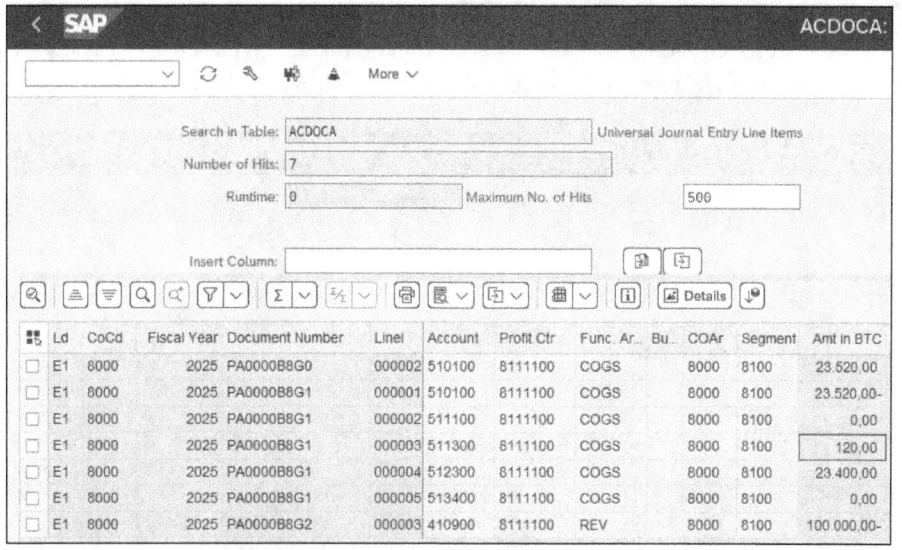

Figure 4.38 Review the COGS Split in the Predictive Journal Entry

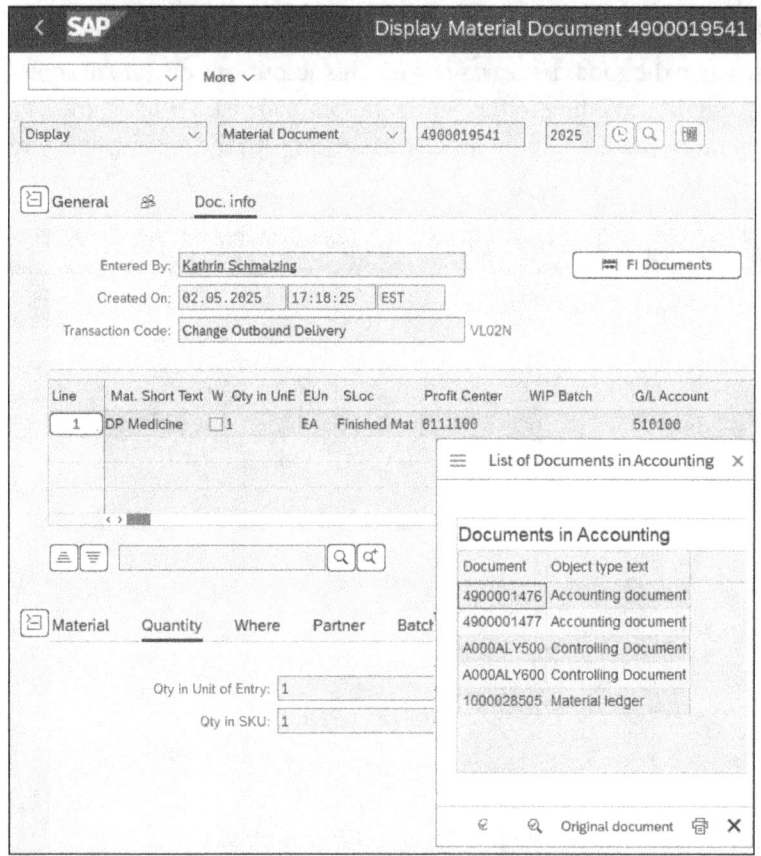

Figure 4.39 Review Documents in Accounting of Goods Issue

4   Value Flows of Margin Analysis

Figure 4.40 shows document **4900001476** in which you can see that the system reduced the inventory account and posted an expense to the inventory change account for COGS. There is no COGS split here.

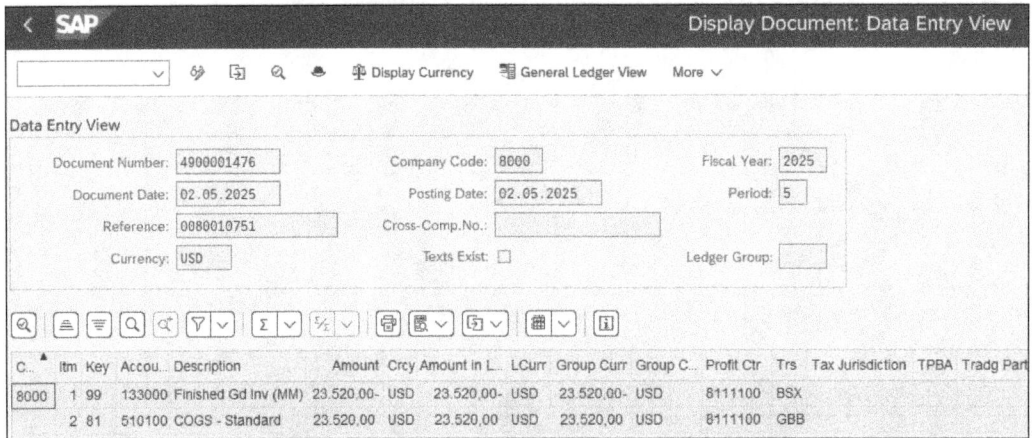

Figure 4.40 Review Accounting Document of Goods Issue

Review the COGS split

Figure 4.41 shows the second accounting document **4900001477** that was created when the goods issue was posted. This accounting document shows the COGS split according to the configuration made earlier in Figure 4.34. The inventory change account for COGS is offset with the posting in Figure 4.40.

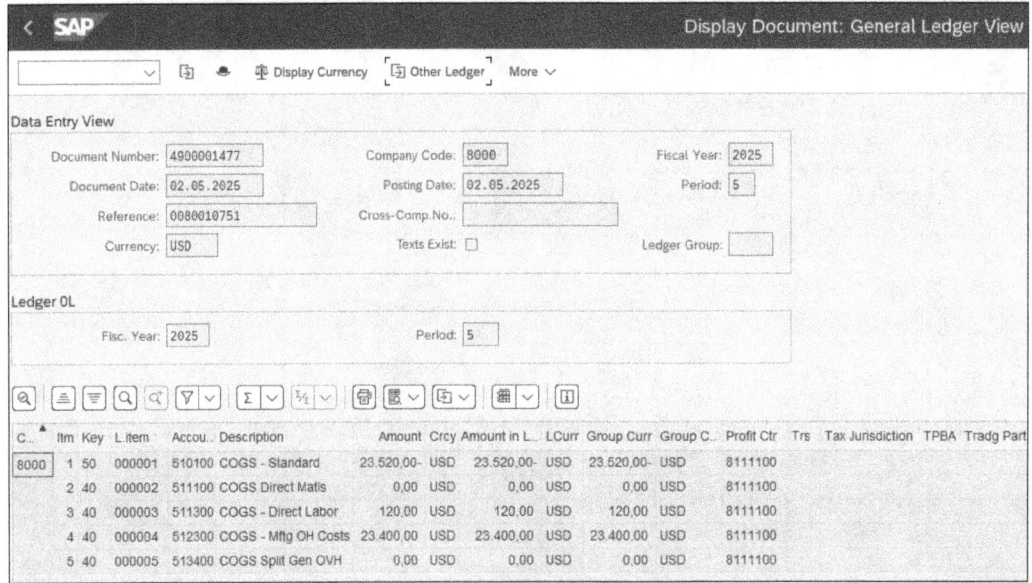

Figure 4.41 Review the Accounting Document for the COGS Split

4.5 Variance Calculation

Figure 4.42 shows the material cost estimate of the material that the goods issue was posted for to verify that the correct amounts have been split during the COGS split posting. The numbers do match exactly with the documents in Figure 4.41 and Figure 4.38.

Review the material cost estimate

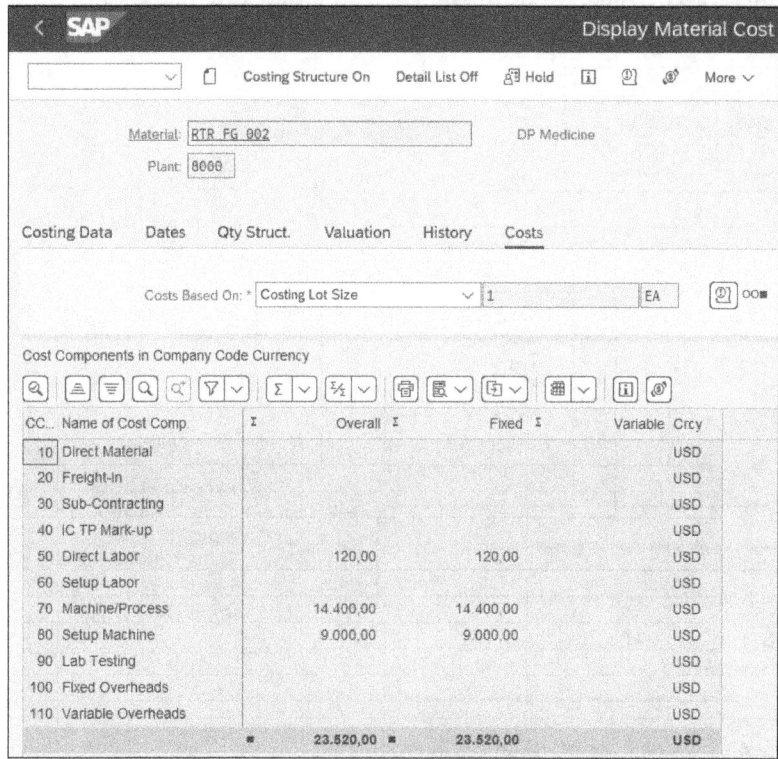

Figure 4.42 Review Material Cost Estimate

The COGS split allows for a detailed analysis of the COGS in financial accounting and margin analysis. You can also activate the COGS split without using margin analysis, but you have to use the controlling functionality to be able to take advantage of the COGS split.

4.5 Variance Calculation

In Section 4.3, we quickly touched on variances in the production process and that balances on production orders get settled to margin analysis at month end. In this section, we'll explain how variances occur and what configuration settings are required to settle them correctly. The variance calculation focuses on the variances on production orders, but there are also variances on cost centers that have to get transferred to margin analysis to see the complete COGM in margin analysis.

Value flow production variances

159

Let's look at the value flow of the production variances in Figure 4.43. You can see different cost center types that have an impact on the variances besides the production orders. We'll explain the assessment of cost center cycles in Section 4.7, but to understand the value flow of the production variances, we'll explain the different cost center types in this section.

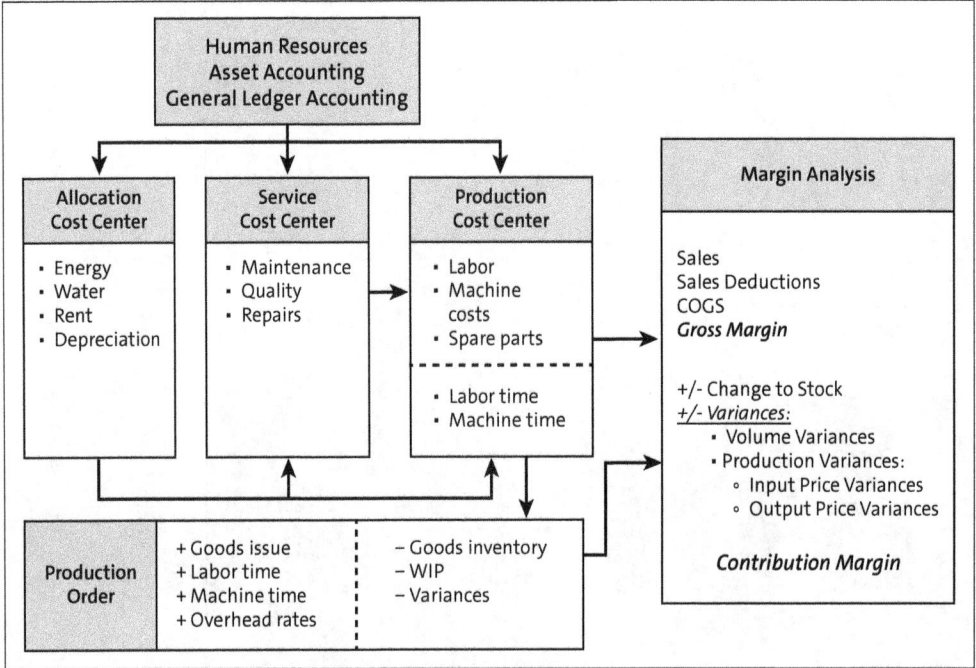

Figure 4.43 Value Flow of Variances in Production

Different cost center types

The different cost center types all receive their costs from different functionalities such as financial accounting, asset accounting (AA), human resources, accounts payable, and so on. Now let's take a more detailed look into the cost center types and what kinds of cost they are collecting:

Allocation cost center

- **Allocation cost center**
  Costs are collected that can't be assigned directly to a functional cost center, for example, energy and rent. You receive one bill for the whole company, and it's too complicated to split that bill when you enter it in financial accounting and assign it to the different cost centers. Instead, you collect all costs on one cost center and allocate them at month end with a cost center assessment to other cost centers based on various criteria, for example, the size of the site.

Service cost center

- **Service cost center**
  For companies with manufacturing processes, there are usually several different service cost centers, but even nonproducing companies have

service cost centers, which are for maintenance, IT, HR, production support, and so on. All those cost centers provide services to other functions. Service cost centers get allocated to cost centers they are providing services for, similar to the allocation cost centers, for example. The allocation takes place based on the number of employees assigned to a cost center or the number of hours the employees provided service for another cost center.

- **Production cost center**
All costs associated with production are collected, such as labor cost, depreciation of machines, energy, and so on. The objective is to assign the total production costs to the goods manufactured so that at month end, the balance of the cost center equals zero. This requires a lot of detailed planning and rarely happens as there are usually variances. The costs in the production cost centers are used to determine the price for personnel hours and machine hours, as well as any other prices that are required for production.

Production cost centers

In the planning process, the expected total production costs are determined for each cost center. You can determine the price based on the total costs, output quantity, and personnel hours. Here, you assume that the cost center is credited to 100%. You can determine your prices manually, but the SAP system also provides an automatic price calculation option. These prices are then used in the material cost estimate of the goods manufactured to cost the personnel hours and machine hours that are required to produce the product.

Price determination in production cost centers

The routing defines how many machine hours or personnel hours are used to manufacture the product. The COGM for the finished goods are determined from the material costs and the personnel hours/machine hours costs, as well as potential overhead rates. The COGM is the standard price of a product and is updated in the material master. This data is the basis for the inventory valuation and therefore is used for any valuation of material movements in the system. Any balance at month end on a production cost center has to be allocated to margin analysis with a cost center assessment cycle. You'll learn how to create cost center assessments in Section 4.7.

4.5.1 Production Order

When you create a production order, the system resolves the bill of materials (BOM) and routing to determine the target costs of the order. With goods issued on the order and confirmations of hours for the production order, costs are debited to the order. If the manufacturing process for the order has been completed, the receipt of the manufactured product is

Production order

posted to the warehouse/inventory. The goods movement for the delivery of the finished goods to the warehouse is valuated using the standard price (or the price according to the price control in the material master), which reflects the price determined in the material cost estimate.

Variance on production orders

After the production is complete and the order shows a balance, this is referred to as a variance. The total of the actual costs doesn't correspond to the calculated costs for several possible reasons: more or fewer material components were consumed, more or fewer working hours were necessary, or price fluctuations occurred for material components. These variances are determined at month end and are settled to margin analysis. The settlement of the production order also creates a financial posting. The change of stock is corrected by the variance. We'll explain this in more detail with an example later in this section.

Work in process

Let's talk about production orders in more detail. At the end of the month, WIP is determined on the basis of the production orders/process orders that have neither status **TECO (Technically Completed)** nor status **DLV (Delivered)**. When the orders are settled, the system generates a posting that doesn't affect the net income and capitalizes the actual costs of the order as WIP in the balance sheet.

Variance calculation

For orders that have already been delivered or completed, the system determines variances and the type of variance. For this purpose, the SAP system uses various default variance categories that distinguish variances as either on the output side or the input side. When the orders are settled, the system posts a price difference in financial accounting and corrects the change in the stock account. This posting doesn't affect the net income.

In traditional SAP ERP systems, the system uses the account that is defined in the Materials Management (MM) account determination of account grouping PRD-PRF to post variances in Financial Accounting (FI). With SAP S/4HANA, you're able to split those variances by variance categories on separate G/L accounts.

Splitting price differences

Now let's discuss how to configure the variance calculation, set up the settlement process for margin analysis, and split the variance calculation according to the variance categories.

Create a variance key: Transaction OKV1

To determine variances for production orders, the orders have to be assigned to variance keys. To create variance keys, go to Transaction OKV1 or use the configuration path **Controlling** • **Product Cost Controlling** • **Cost Object Controlling** • **Product Cost by Order** • **Period-End Closing** • **Variance Calculation** • **Define Variance Keys**.

You can create a new variance key with **New Entries** or F5. This example creates variance key **000001 – Variance Calculation for Orders** (Figure 4.44). You can set two checkmarks in the variance keys (not shown):

4.5 Variance Calculation

- **Scrap**
  In the **Variance Calculation** section, selecting the **Scrap** checkbox makes the variance calculation determine the value for scrap and subtract it from the total variance. A prerequisite is that scrap is properly confirmed in production.

- **Write Line Items**
  In the **Update** column, selecting the **Write Line Items** checkbox creates an additional document when the variance is determined. This document provides information on who determined the variances when and which variances have changed.

For this example, don't set any of the checkboxes, but save the entries with `Save` or `Ctrl`+`S`.

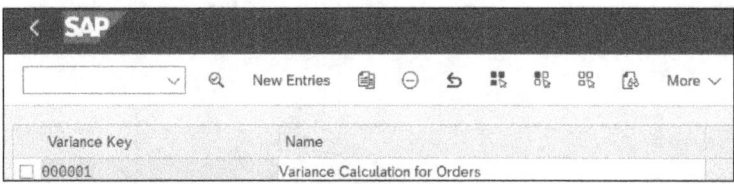

Figure 4.44 Create the Variance Key

The next step is to assign the variance key to a plant as a default value. Go to Transaction OKVW or follow the configuration path **Controlling • Product Cost Controlling • Cost Object Controlling • Product Cost by Order • Period-End Closing • Variance Calculation • Define Default Variance Keys for Plants**.

Assign the variance key to a plant: Transaction OKVW

Figure 4.45 shows variance key **000001** assigned to plant **8000** for this example. From now on, when a material master record or production order is created, this variance key will be used as a default value, but it can be overwritten. Maintaining the default value ensures that all orders are assigned to a variance key and that variances can be determined at the end of the month. Save your entries with **Save** or `Ctrl`+`S`.

Plnt	Variance Keys	Name
8000	000001	Variance Calculation for Orders
8001	000001	Variance Calculation for Orders
8002	000001	Variance Calculation for Orders
8003	000001	Variance Calculation for Orders
8100	000001	Variance Calculation for Orders
8101	000001	Variance Calculation for Orders
8110	000001	Variance Calculation for Orders

Figure 4.45 Assign the Variance Key to the Plant

4 Value Flows of Margin Analysis

4.5.2 Define Variance Categories

Define variance categories: Transaction OKVG

Now let's define the variance categories that the system will use. Go to Transaction OKVG or use the configuration path **Controlling • Product Cost Controlling • Cost Object Controlling • Product Cost by Order • Period-End Closing • Variance Calculation • Check Variance Variants**. This example is activating all variance categories in Figure 4.46 for **Variance Variant 001**. If variances occur in a variance category that isn't activated, these variances are assigned to the remaining variance category. As mentioned earlier, the system divides the variance categories into two areas: input side variances and output side variances.

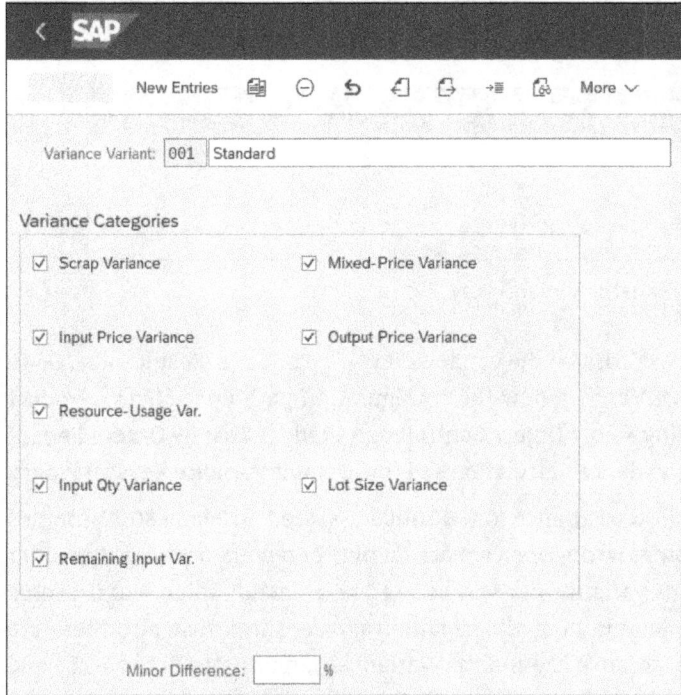

Figure 4.46 Activate Variance Categories

Variances on the input side

Variances on the input side include the following:

- **Scrap variance**
  The scrap variance determines the value of unplanned scrap. Scrap has to be reported separately in the order confirmation to enable the system to identify scrap.

- **Input price variance**
  The input price variance determines the variance of prices of material components and services. The variance is derived from the difference between target costs and actual costs.

- **Resource-usage variance**
  The resource-usage variance is derived from the difference between target costs and actual costs if a material component that wasn't listed on the BOM was used.

- **Input quantity variance**
  The input quantity variance is the difference between target costs and actual costs if there's a difference between planned and actually used quantities.

- **Remaining input variance**
  All variances on the input side that can't be assigned to any of the input variances mentioned previously are assigned to the remaining input variance.

The variances on the output side include the following:

Variances on the output side

- **Mixed-price variance**
  If you use mixed price calculations, the mixed-price variance determines the variance between the standard price and the material cost estimate of the procurement alternative selected in production.

- **Output price variance**
  Output price variances only occur if finished goods are delivered to inventory with a price that deviates from the standard price (e.g., moving average price).

- **Lot size variance**
  Lot size variances occur if the actual quantity differs from the planned quantity. They result from the fact that some of the total costs don't change even if the output quantity changes.

It's necessary to define a target cost version for the variance calculation. This version specifies how variances will be determined. To create target cost versions, go to Transaction OKV6 or use the configuration path **Controlling • Product Cost Controlling • Cost Object Controlling • Product Cost by Order • Period-End Closing • Variance Calculation • Define Target Cost Versions**.

Define the target cost version: Transaction OKV6

Create a new target cost version with **New Entries** or F5. Figure 4.47 shows the creation of target cost version **0** for controlling area **8000**. Target cost versions determine which cost estimate is used as the basis for variance calculation.

In the target cost version, you specify in detail which variance variant will be used for variance calculation. For this example, assign the **Variance Variant 001**, which was created earlier in Figure 4.46.

Assign the variance variant to the target cost version

In the **Control Costs** section, select **Actual Costs** to analyze the actual costs. In the **Target Costs** section, define the comparison value for the actual costs, which is usually the current standard cost estimate (**Current Std Cost Est**). These settings are the recommended settings for the determining variances that are later transferred to financial accounting during the settlement process because these settings guarantee the balance of the production order will be zero after settlement. Now, save your entries with **Save** or [Ctrl]+[S].

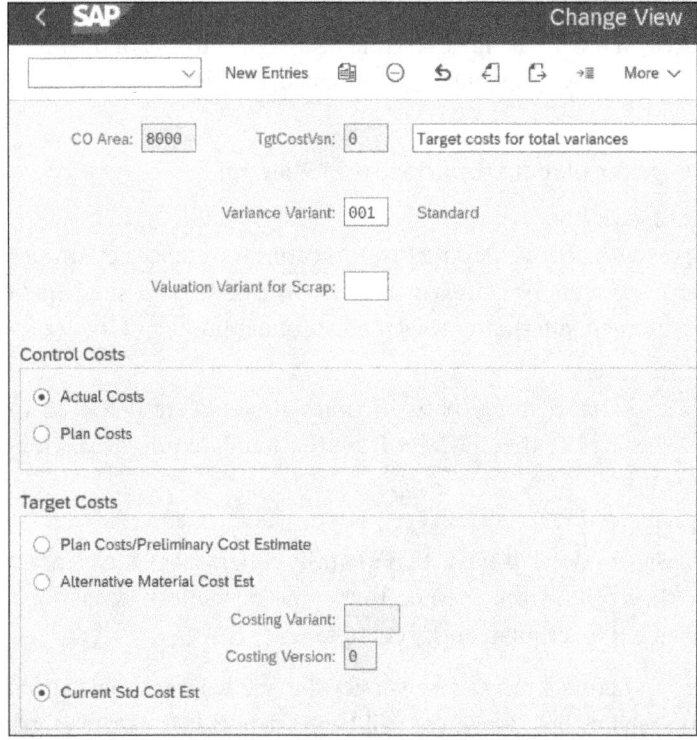

Figure 4.47 Create the Target Cost Version

Additional target cost versions

You can create further target cost versions and compare them to alternative material cost estimates. For this purpose, select **Alternative Material Cost Est** in the **Target Costs** section of the target cost version, and specify the costing variant that you want to compare your values to.

Defining number ranges: Transaction KANK

Before you can perform variance calculations, you have to assign a number range to the variance documents. To do so, go to Transaction KANK or follow the configuration path **Controlling** • **Product Cost Controlling** • **Cost Object Controlling** • **Product Cost by Order** • **Period-End Closing** • **Variance Calculation** • **Define Number Ranges for Variance Documents**.

4.5 Variance Calculation

Enter the controlling area in the selection screen, which, for this example, is "8000" (Figure 4.48).

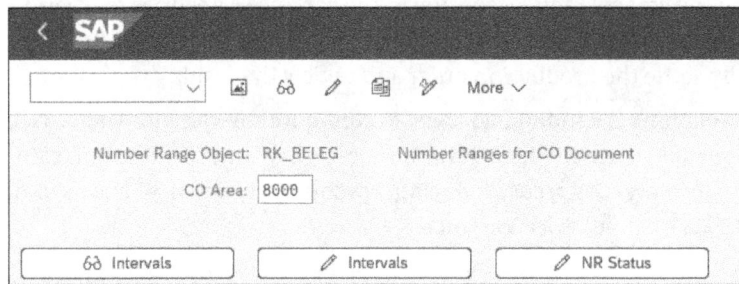

Figure 4.48 Maintain Number Ranges for Variance Documents

Next, navigate to **More • Goto • Groups • Change**. Search for "KVAR (Variance Calculation)" with (Find) or press Ctrl+F (Figure 4.49). Click on **KVAR**, and go to (Assign Element to Group) or press F7.

Assign subobjects to a group

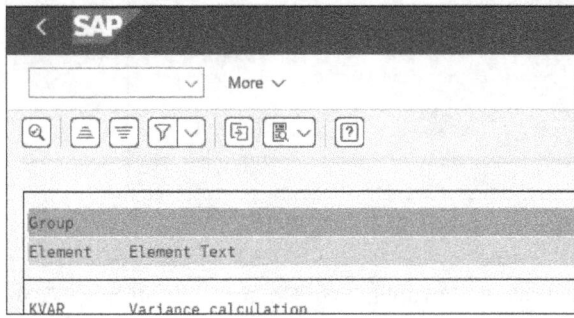

Figure 4.49 Search Element for Variance Calculation

In Figure 4.50, double-click on **Subobject 8000** that has a **Number range number** in the adjacent column. Confirm your entries with Enter, and save them with **Save** or Ctrl+S.

Select subobjects

Subobject	Number range number	Number range number	Group text
8000	01		
8000	02		
8000	03		
8000	04		
8000			Non-Assigned Elements

Figure 4.50 Assign an Element to the Subobject

167

4 Value Flows of Margin Analysis

Create G/L accounts for variance split

4.5.3 Create General Ledger Accounts for Variance Split

Now that you've maintained the variance calculation settings, you need to configure the variance split according to the variance categories for margin analysis and then the settlement profile of the production order before you can finally settle the production order and review the result.

Before explaining the individual steps to configure the variance split, a G/L account must be created for each variance category with cost element category **1 – Primary costs/cost-reducing revenues**. Figure 4.51 shows **G/L Account 522511 – Input Price Variance**.

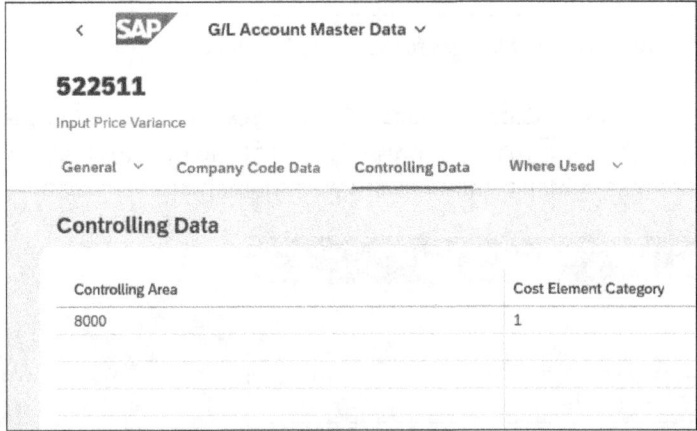

Figure 4.51 Create the G/L Account for Variance Split

Create the splitting profile for the variance split

To configure the variance calculation split, go to **Financial Accounting • General Ledger Accounting • Periodic Processing • Integration • Materials Management • Define Accounts for Splitting Price Differences** in configuration.

With **New Entries** or F4, create the splitting profile **8000SPLIT**, and assign it to the controlling area **8000** (Figure 4.52). The chart of accounts will be derived automatically from the controlling area. Enter a name for the splitting profile (in this case, "PLS Variance Split") in the **Price Diff. Splitting Profile Name** column, and then select the profile via the checkbox at the beginning of the line. Navigate to the **Detailed Price Difference Accounts** folder on the left side of the screen.

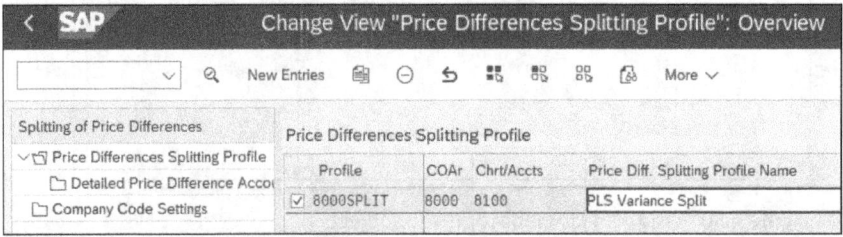

Figure 4.52 Create a Price Difference Profile

4.5 Variance Calculation

With **New Entries** or F5, assign one G/L account to each variance category. In Figure 4.53, maintain the following data:

Assign G/L accounts to variance categories

- **Description**
  The description describes the price difference. We recommend using the name of the variance category, for example, "Input Price Variance".

- **Cost Elem. From**
  You can maintain a single cost element or cost element interval as the source cost elements for the variance calculation. You can't enter a "0" here because it has to be a cost element that actually exists in the controlling area that the price splitting profile is assigned to.

- **Cost Elem. To**
  You can add a cost element if you want to maintain an interval.

- **CElem Group**
  For the cost element group in this example, select **ALL** as the source cost element.

- **Category**
  This refers to the variance category that is split with the target account in the next field. With the F4 Help, you can display all existing variance categories in your controlling area.

- **Target Account**
  This is the G/L account that the variance category will be posted to when the production order is settled.

You can maintain another price difference account with ⊞ (Next Entry) or F8.

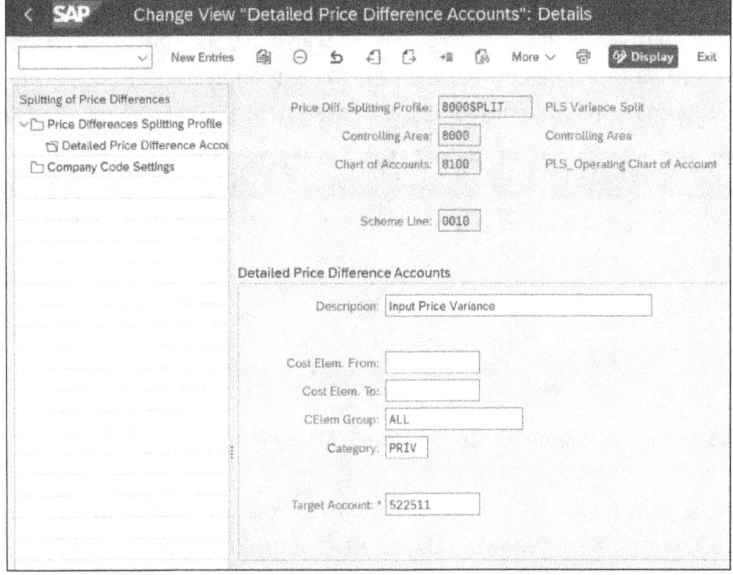

Figure 4.53 Add a Price Difference Account to the Splitting Profile

4   Value Flows of Margin Analysis

Overview of variance categories

After you've maintained price difference accounts for every variance category, you can see the overview of all price difference accounts assigned to the splitting profile by pressing [F3] to go back (Figure 4.54). Save your data with **Save** or [Ctrl]+[S]. Now, you can assign the splitting profile to the company code by navigating to the **Company Code Settings** folder on the left side of the screen.

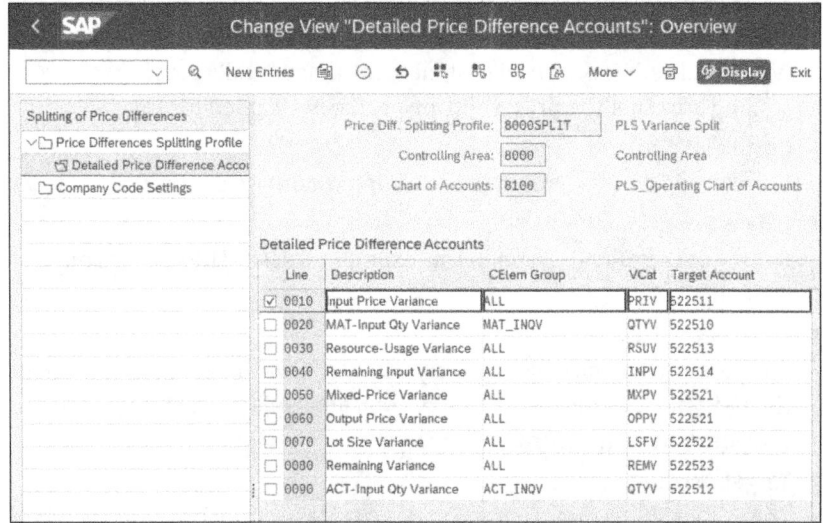

Figure 4.54 Overview of Price Difference Accounts Assigned to the Splitting Profile

Maintaining variance categories

With **New Entries** or [F5], assign the splitting **Profile 8000SPLIT** to the **CoCd** (company code) **8000** with a **Valid From** date of **01.01.2020** (Figure 4.55), and save the data with **Save** or [Ctrl]+[S]. You can also use the splitting of price differences in variance categories without margin analysis, but your company code needs to have a controlling area assigned.

Assigning splitting profiles

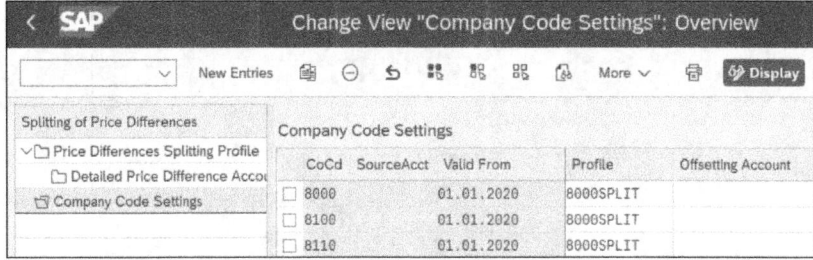

Figure 4.55 Assign the Company Code to the Splitting Profile

Create the PA transfer structure

Before we determine the variances, we need to look at the settlement profile of the production order. We won't explain the setup of the settlement profile in detail because we'll describe the detailed settlement in Section 4.5.4.

4.5 Variance Calculation

To make sure that the settlement profile is set up correctly, follow the configuration path **Controlling • Profitability Analysis • Flows of Actual Values • Settlement of Production Variances • PA Transfer Structure • Assign Settlement Profiles** to review the settlement profile. Select the checkbox at the beginning of line **Z800 – Process orders**, and look at the details with 🔍 (Details).

Assign PA transfer structure to settlement profile

Figure 4.56 shows the details of **Settlement Profile Z800** – PLS **Process Order**. In the **Default Values** section, check that an allocation structure is assigned to the settlement profile. In the **Indicators** section, select the **Variances to Costing-Based PA** checkbox; then, variances are settled in costing-based profitability analysis when the orders are settled. If you don't select this, the system doesn't create a profitability analysis document for costing-based profitability analysis. Now, make sure that in the **Valid Receivers** section, **Material** and **Profit. Segment** (profitability segment) are allowed receivers.

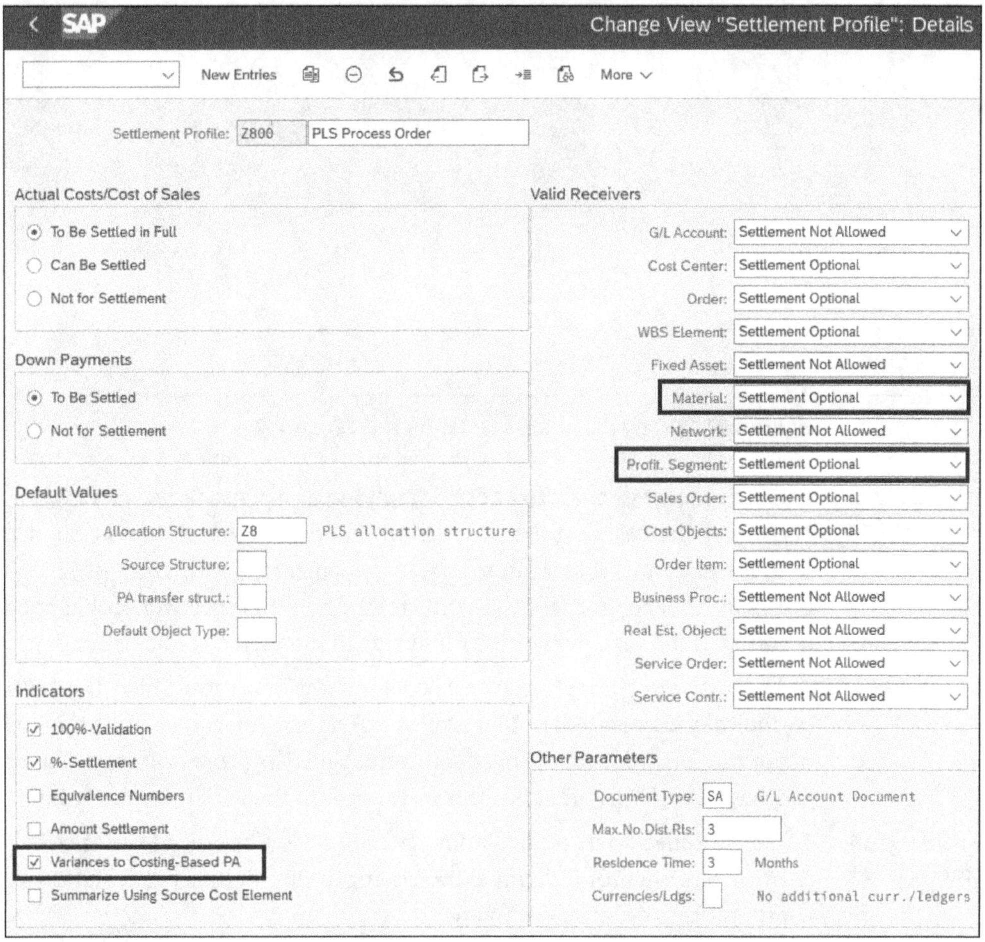

Display settlement profile details

Figure 4.56 Settlement Profile Details

4   Value Flows of Margin Analysis

4.5.4 Check Production Order

Check production order

Now that you've completed all the required settings for variance calculations, you can check the **Control** tab in the production or process order to see whether a variance key is assigned to the order. In Figure 4.57, **Variance Key 000001** that you just created is assigned.

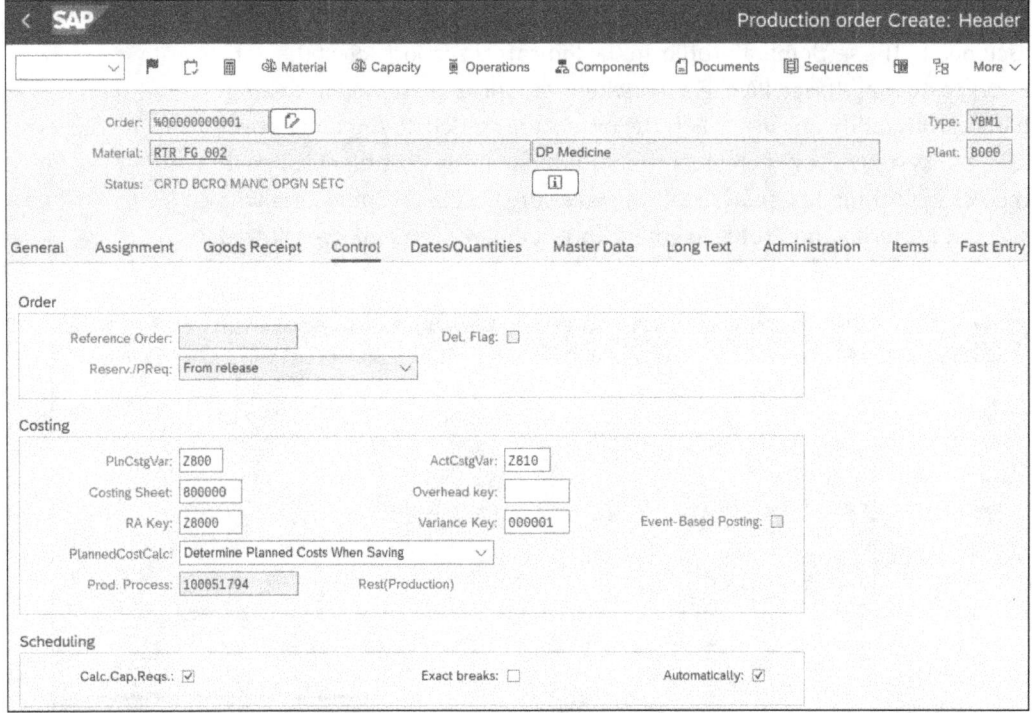

Figure 4.57 Review the Process Order for the Variance Key

Production order costs

You can follow **More · Goto · Costs · Analysis** to display a target/actual comparison of the costs on the production order. In Figure 4.58, you can see that, for example, direct labor activity was planned with costs of 120.00 USD, but no labor activity was confirmed. The difference will probably be reported in the variances of the input side in the variance calculation.

Production order status

To be considered in the variance calculation, the order must have the **TECO** (**Technically Completed**) or **DLV** (**Delivered**) status. You can check the status of production orders in the header information. The process order in Figure 4.59 has **DLV**, so this process order will appear in the variance calculation.

Performing variance calculations

To execute the variance calculation, go to SAP Fiori apps Run Variance Calculation by Order for Individual Processing or Run Variance Calculation by Lot for process orders collectively.

4.5 Variance Calculation

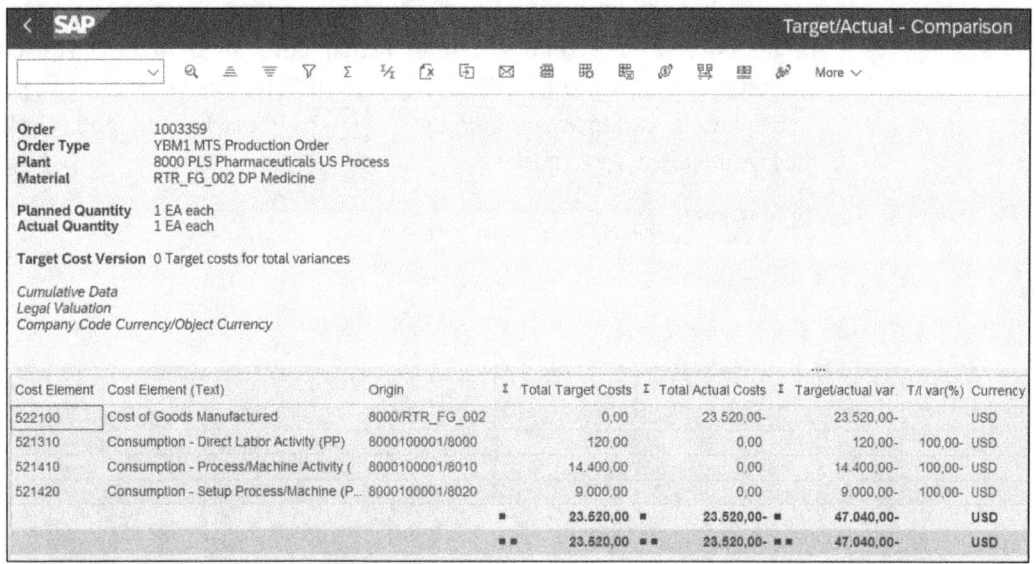

Figure 4.58 Review Costs in the Process Order

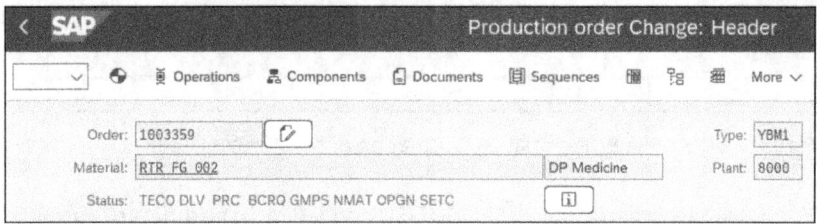

Figure 4.59 Review the Process Order Status

To execute the variance calculation for an individual process order, you enter the process order number as well as the period and year for which you want to execute the variance calculation in the selection screen and execute it with **Execute** or F8.

Execute variance calculation

In Figure 4.60, you see the result of the variance calculation with the following columns:

Analyze the variance calculation result

- **Target Costs**
  Costs that are determined based on the costing variant applied to the produced quantity.
- **Control Costs**
  Costs that are credited on the process order.
- **Allocated Actl Costs**
  Costs posted for deliveries on stock.

4 Value Flows of Margin Analysis

Referring to the numbers shown earlier in Figure 4.58, the variance between **Target Costs** of **23,520.00** USD and **Actual Costs** of **0.00** USD equals 23,520.00 USD, which will be divided into the different variance categories. To display the variance categories, go to 📊 (Choose Layout), and select the layout variance categories.

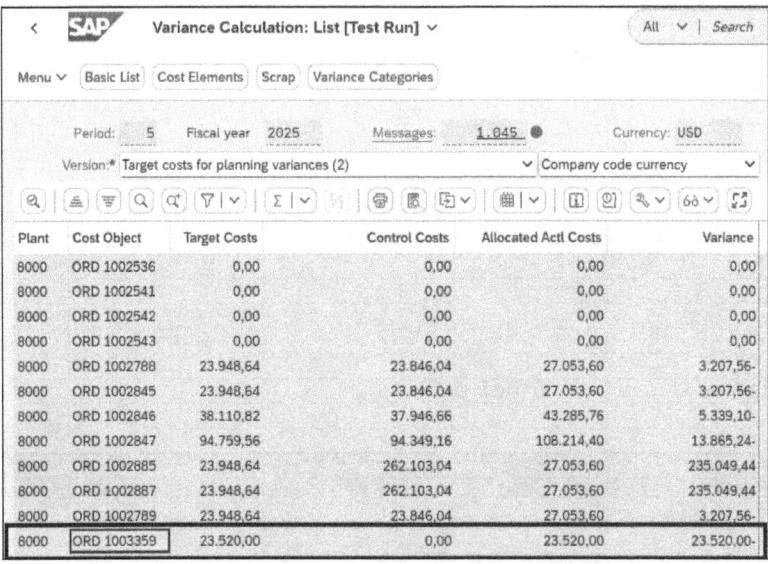

Figure 4.60 Analyze the Variance Calculation of the Process Order

Analyze variance categories
You'll see the variances of 23,520.00 USD divided into the different variance categories in Figure 4.61. The system assigned **23,520.000** USD to **Res-Usage Var.** (resource-usage variance) because no actual costs have been consumed. The system assumes an error with the BOM or routing, whereas no confirmation/goods issue has been posted to the production order, just the goods receipt of the finished good back to stock. You need to execute the variance calculation in the update run before you can settle the variances to margin analysis. The execution of the variance calculation doesn't create any postings in financial accounting nor margin analysis. Postings are created with the order settlement.

Order settlement
To settle the process order, go to SAP Fiori app Run Actual Settlement, or follow the menu path **Order • Single, Run Actual Settlement • Product Cost Collectors, Run Settlement • Maintenance Orders • Actual**.

Of course, you can also settle orders via collective processing with the Settle Orders - Optimized app.

4.5 Variance Calculation

Orders are settled on a monthly basis. In this example, the process order for which the variances were calculated is settled with the Run Actual Settlement app, or by following the menu path **Order • Single, Run Actual Settlement • Product Cost Collectors, Run Settlement • Maintenance Orders • Actual**. In the selection screen, enter the process order, settlement period, and year in which to settle the order. Execute the settlement with **Execute** or F8.

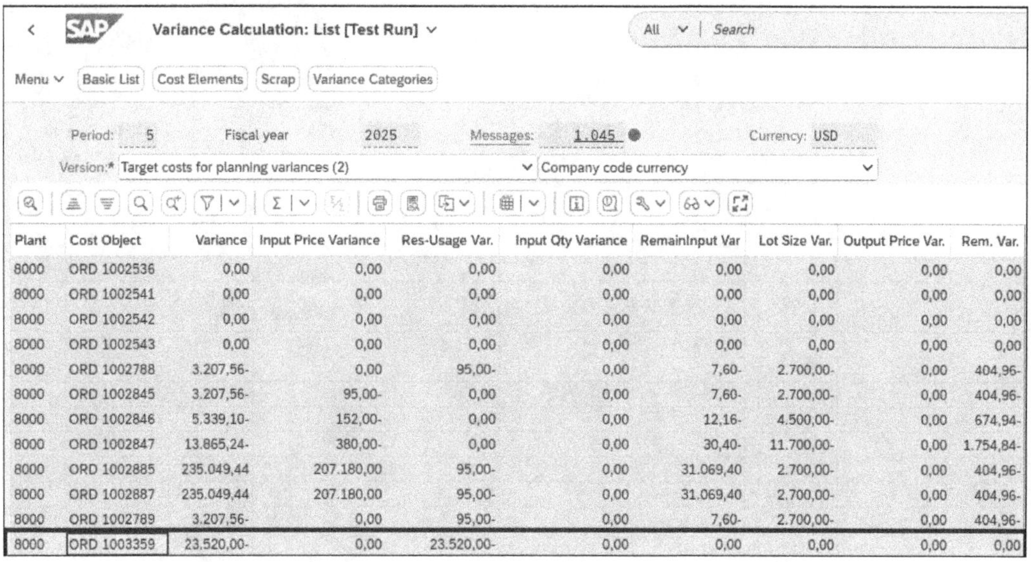

Plant	Cost Object	Variance	Input Price Variance	Res-Usage Var.	Input Qty Variance	RemainInput Var	Lot Size Var.	Output Price Var.	Rem. Var.
8000	ORD 1002536	0,00	0,00	0,00	0,00	0,00	0,00	0,00	0,00
8000	ORD 1002541	0,00	0,00	0,00	0,00	0,00	0,00	0,00	0,00
8000	ORD 1002542	0,00	0,00	0,00	0,00	0,00	0,00	0,00	0,00
8000	ORD 1002543	0,00	0,00	0,00	0,00	0,00	0,00	0,00	0,00
8000	ORD 1002788	3.207,56-	0,00	95,00-	0,00	7,60-	2.700,00-	0,00	404,96-
8000	ORD 1002845	3.207,56-	95,00-	0,00	0,00	7,60-	2.700,00-	0,00	404,96-
8000	ORD 1002846	5.339,10-	152,00-	0,00	0,00	12,16-	4.500,00-	0,00	674,94-
8000	ORD 1002847	13.865,24-	380,00-	0,00	0,00	30,40-	11.700,00-	0,00	1.754,84-
8000	ORD 1002885	235.049,44	207.180,00	95,00-	0,00	31.069,40	2.700,00-	0,00	404,96-
8000	ORD 1002887	235.049,44	207.180,00	95,00-	0,00	31.069,40	2.700,00-	0,00	404,96-
8000	ORD 1002789	3.207,56-	0,00	95,00-	0,00	7,60-	2.700,00-	0,00	404,96-
8000	ORD 1003359	23.520,00-	0,00	23.520,00-	0,00	0,00	0,00	0,00	0,00

Figure 4.61 Analyze the Variance Categories of the Process Order's Variance Calculation

In Figure 4.62, you see the basic list of the actual settlement that repeats the **Selection Parameters** and gives you information in **Processing Options** about whether you executed the settlement in an update or a test run. In **Statistics**, you see how many orders were included in the settlement and if there were any errors. The settlement for order **1003359** was executed in an update run without any errors. Click **Next list level** Next list level on the bottom right side of the screen or press Enter to be redirected to a more detailed view.

Analyze the actual settlement: order basic list

In the detail list you see in Figure 4.63, the settlement has two receivers—material (**MAT**) and profitability segment (**PSG**)—of costing-based profitability analysis. You can review the accounting documents by clicking **Accounting documents** at the top of the screen.

Display the order settlement detail list

4 Value Flows of Margin Analysis

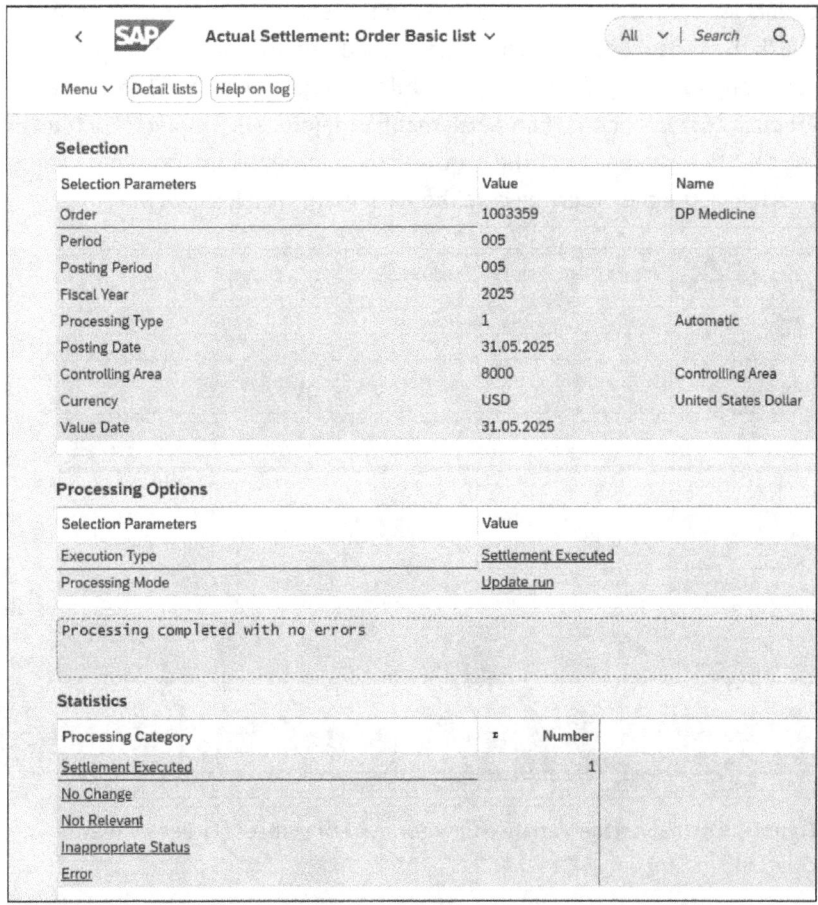

Figure 4.62 Order Basic List: Order Settlement

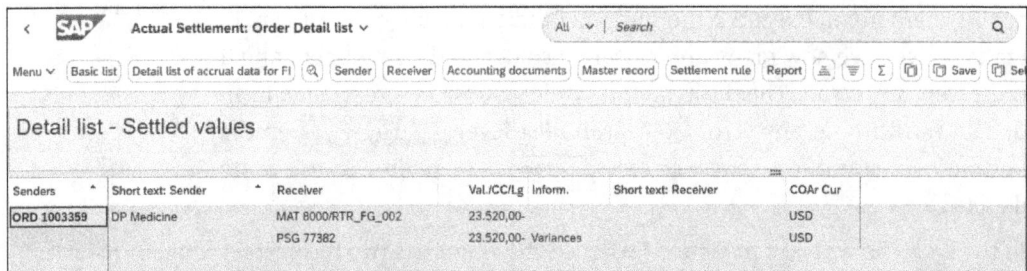

Figure 4.63 Review the Detail List of the Order Settlement

Display order settlement accounting documents In the overview of documents created with the order settlement, double-click on the accounting document to go the financial document, which is also the document for margin analysis. In Figure 4.64, you can see that the system credited the price difference account and debited the production order activity account (as the output costs were higher than the costs of the material used for production).

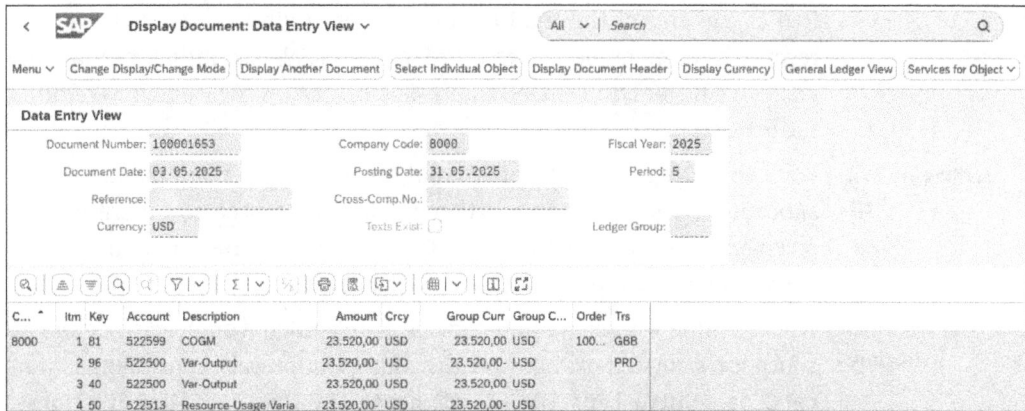

Figure 4.64 Review the Variance Calculation Settlement Accounting Document

You can review the profitability segment the system created for margin analysis by viewing the controlling document. In Figure 4.65, you can see that the system created a profitability segment and derived various characteristics for the expense line items.

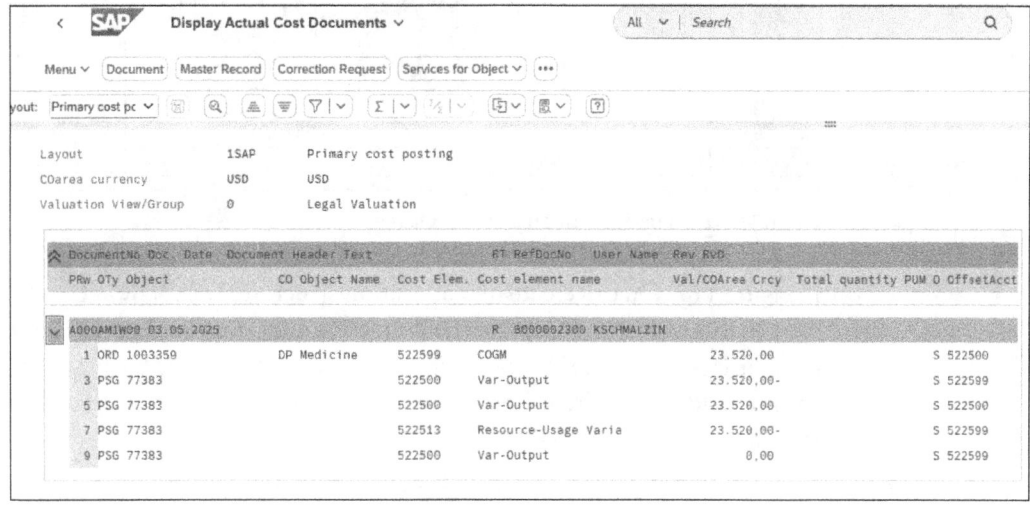

Figure 4.65 Review the Controlling Document

In this section, you learned how to configure the variance calculation, how to calculate variances, and lastly how to transfer them to margin analysis.

4 Value Flows of Margin Analysis

4.6 Order Settlement

Like production and process orders, internal orders always have to be settled at the end of the month. You can settle internal orders to various receivers, for example, internal orders, financial accounting, profitability analysis, and/or cost centers. Figure 4.66 illustrates the value flow of order settlements.

Settling internal orders
You can settle internal orders to margin analysis. In margin analysis, the allocation structure in the settlement profile is used to determine the receiver. You don't have to create a PA transfer structure in margin analysis like you do for costing-based profitability analysis.

Create allocation structure
To determine where and how the costs are settled, you need to define an allocation structure by following the configuration path **Controlling • Internal Orders • Actual Postings • Settlement • Maintain Allocation Structures**.

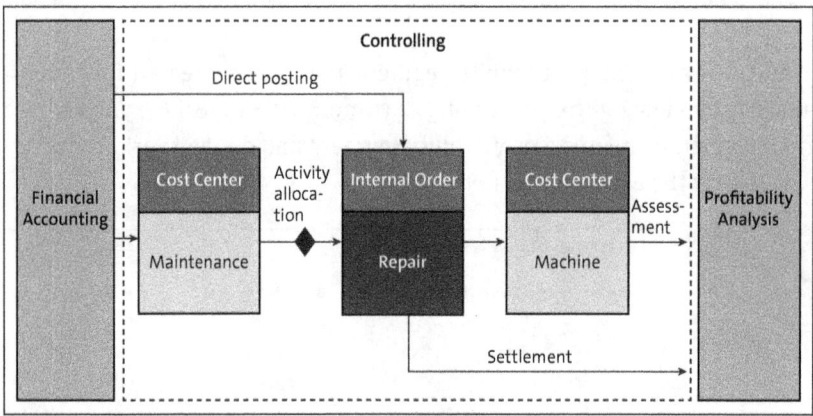

Figure 4.66 Value Flow of an Internal Order

In Figure 4.67, you can create new allocation structures with **New Entries** or F5. In this example, allocation structure **Z9 – Allocation Structure Z200** is created. Now, mark the entry by selecting the checkbox at the beginning of the line, and navigate to the **Assignments** folder in the **Dialog Structure** section on the left side of the screen.

Create an assignment line
In Figure 4.68, you can create new **Assignments** with **New Entries** or F5. In this example, **Assignment 1 = all IC Expense** is created. You have to create a separate assignment line for each settlement cost element that you want to use for the settlement process. After you've created all your assignments, mark the first one by selecting the checkbox at the beginning of the line, and then navigate to the **Source** folder in the **Dialog Structure**. You need to follow the steps we're about to explain for every assignment line: assigning the **Source** and assigning **Settlement cost elements**.

4.6 Order Settlement

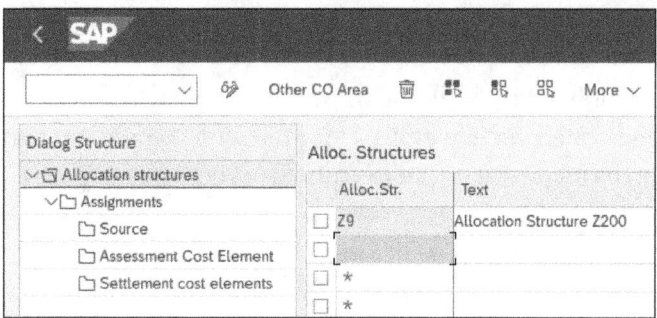

Figure 4.67 Create the Allocation Structure

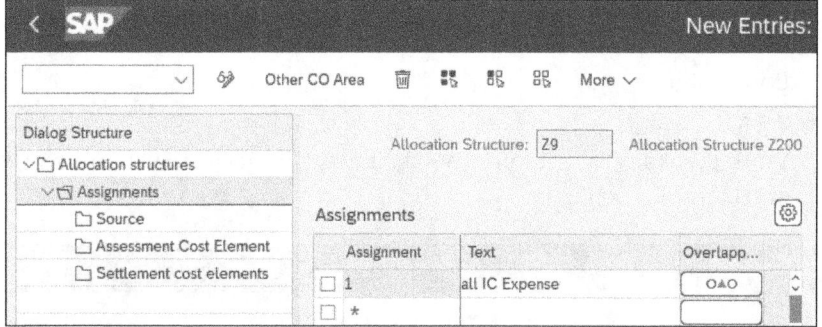

Figure 4.68 Create Assignments in the Allocation Structure

In the **Source**, you'll define which cost elements are considered in the settlement. Therefore, all postings on the internal order with the cost elements defined in the **Source** will be settled. In Figure 4.69, the cost element interval **0–Z** is being assigned to the **Source**, which means all postings on the internal order will be settled with this assignment line. Note that the cost elements in the different assignment lines can't overlap; if they do, the system isn't able to save the allocation structure. Now, navigate to the **Settlement cost elements** folder in the **Dialog Structure**.

Define the source in the allocation structure

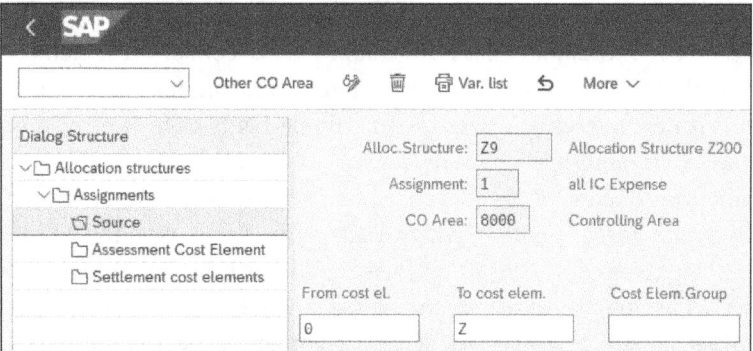

Figure 4.69 Assign Cost Elements to the Source in the Allocation Structure

179

4 Value Flows of Margin Analysis

With **New Entries** or [F5], you can define receiver categories that determine to which objects the internal order can get settled and with which settlement cost element (see Figure 4.70).

Defining the receiver category

Numerous receiver categories are available for the settlement of internal orders; however, the following categories are used most often:

- **FXA**
  Fixed asset (settlement to an asset in AA).

- **ORD**
  Internal order.

- **PSG**
  Profitability segment.

- **CTR**
  Cost center.

- **WBS**
  Work breakdown structure (WBS) element (if Project System in SAP S/4HANA is used).

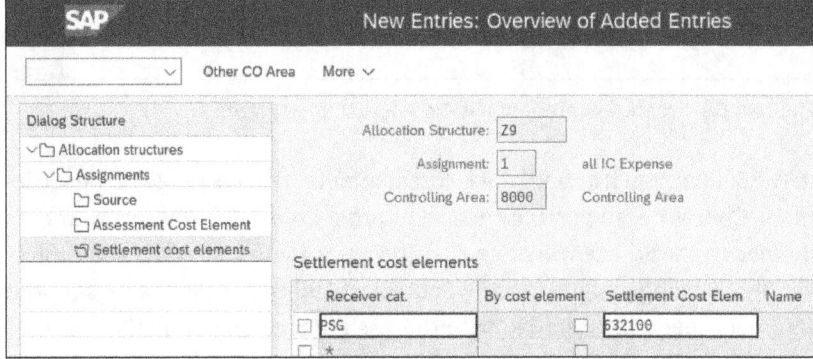

Figure 4.70 Maintain Settlement Cost Elements in the Allocation Structure

In the example in Figure 4.70, one entry is being created for receiver category **PSG** (profitability segment) to settle the internal orders to margin analysis. In the **Settlement Cost Elem** column, the secondary cost element is defined that will be used for the settlement. You can define different cost elements for each receiver category; the cost elements have to be created with the G/L account type for **Secondary Costs G/L** and cost element type **21** (**Settlement**) (not shown) If you don't want to settle your internal orders with a secondary cost element, you can select the checkbox in the **By cost element** column, and the system will use the source cost element to settle the cost. After you repeat the maintenance of the **Source** and the assignment of **Settlement cost elements** for all assignment lines in the allocation structure, you can save your entries with **Save** or [Ctrl]+[S].

180

4.6 Order Settlement

After creating the allocation structure, you need to assign the allocation structure to the settlement profile. The settlement profile is then assigned to the internal order type. To maintain a settlement profile, follow the configuration path **Controlling • Internal Orders • Actual Postings • Settlement • Maintain Settlement Profiles**.

Maintaining settlement profiles

Settlement Profile Z200 is being created in Figure 4.71 with **New Entries** or F5. The settlement profile is divided into various sections:

- **Actual Costs/Cost of Sales**
  Define whether the order will be settled in part or in full. The default setting is **To Be Settled in Full**. If you want to close the order or flag it for deletion, this is only possible if the order has been settled in full.

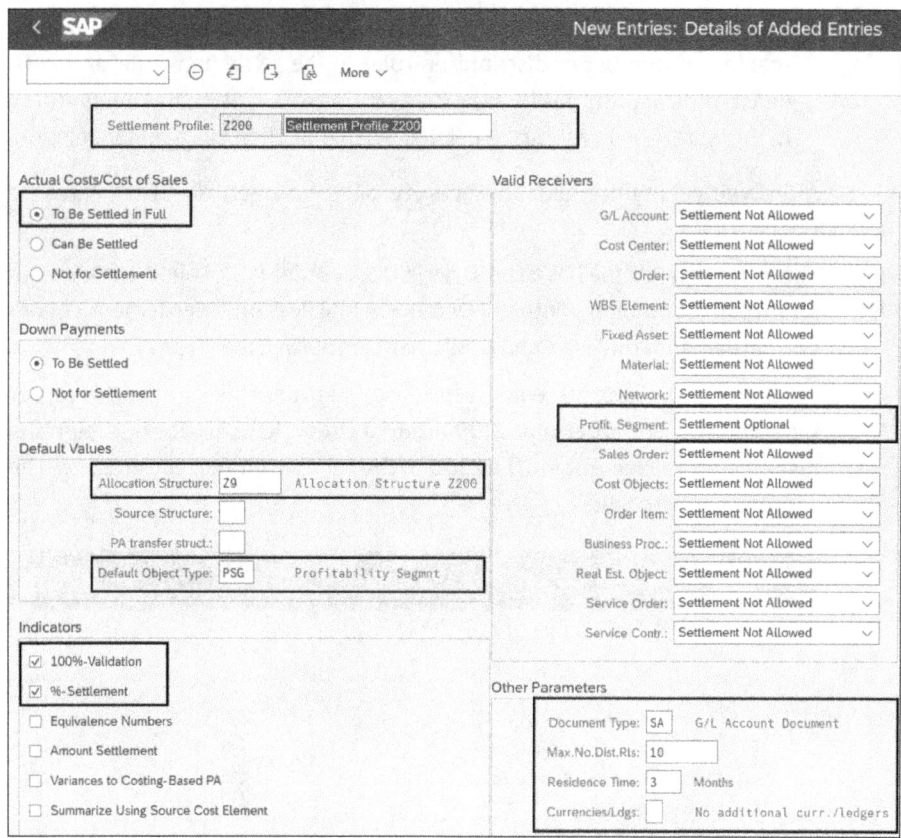

Figure 4.71 Maintain the Settlement Profile

- **Default Values**
  Define **Allocation Structure Z9**, which you've created in this section, and specify **PSG** as the **Default Object Type**. The system always proposes the settlement to margin analysis as the receiver category in the settlement rule.

181

- **Indicators**

  Define that you can use percentages for the settlement (**%-Settlement**); that is, you can settle an order with 50% to a cost center and 50% to the operating profit. In addition, specify that a **100%-Validation** is supposed to be performed in the settlement. This means that the system displays a warning message if orders aren't settled in full.

- **Valid Receivers**

  Define which receivers can or have to be used for the settlement. This is primarily determined by the allocation structure. In the example in Figure 4.71, define the profitability segment as **Settlement Optional**.

- **Other Parameters**

  Enter a specific **Document Type** for the settlement. Otherwise, the system uses the **SA** document type by default. **Max.No.Dist.Rls** determines the maximum number of distribution rules in the settlement rule, and **Residence Time** determines for how many months the internal order is stored in the system before it can be deleted when it has been flagged for deletion.

After you've maintained all sections, save your entries with **Save** or Ctrl+S.

Assigning settlement profiles — The settlement profile has to be assigned to an order type. Follow the configuration path **Controlling • Internal Orders • Actual Postings • Settlement • Maintain Settlement Profiles • Enter Settlement Profile in Order Types** to do so.

In Figure 4.72, you can see an overview of all order types in the **Cat** (order type) column. For order type **Z100 Internal Order: IC Chargeback**, assign the just-created **Settlement Profile Z200** in the corresponding column. Save the assignment with **Save** or Ctrl+S.

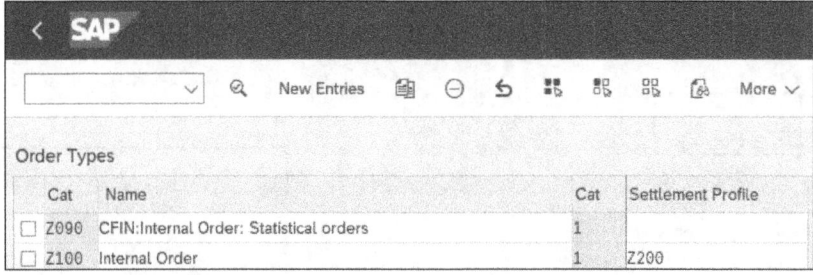

Figure 4.72 Assign the Settlement Profile to the Internal Order Type

Now, let's create an internal order and review the configuration settings. You create an internal order with order type Z100 using the Manage Internal Orders app. To maintain the settlement rule, go to the Manage Settlement Rules – Internal Orders app.

4.6 Order Settlement

Select the internal order you just created to review the settlement rule of the internal order. You can see in Figure 4.73 that a settlement rule with receiver category **PSG** (profitability segment) has been created. With **More** • **Goto** • **Settlement Parameters** or F8, you can review whether the settlement profile from Figure 4.71, shown earlier, has been assigned.

Review settlement rules

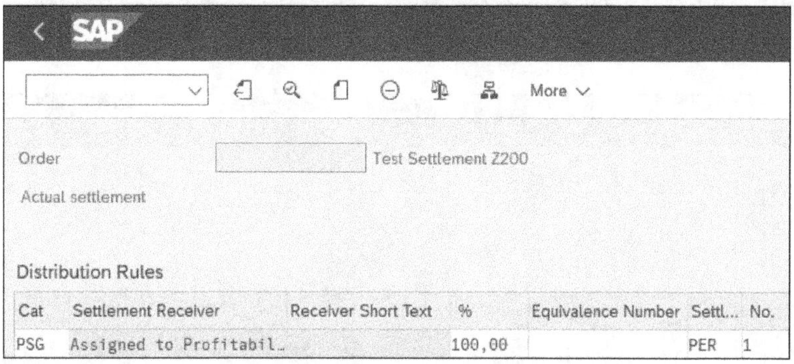

Figure 4.73 Review the Settlement Rule in the Internal Order

In Figure 4.74, you see the settlement **Parameters**, which refer to **Settlement Profile Z200** and **Allocation Structure Z200**, as maintained in Figure 4.72. Let's go back with F3 and release the internal order before saving it with **Save** or Ctrl+S.

Review settlement parameters

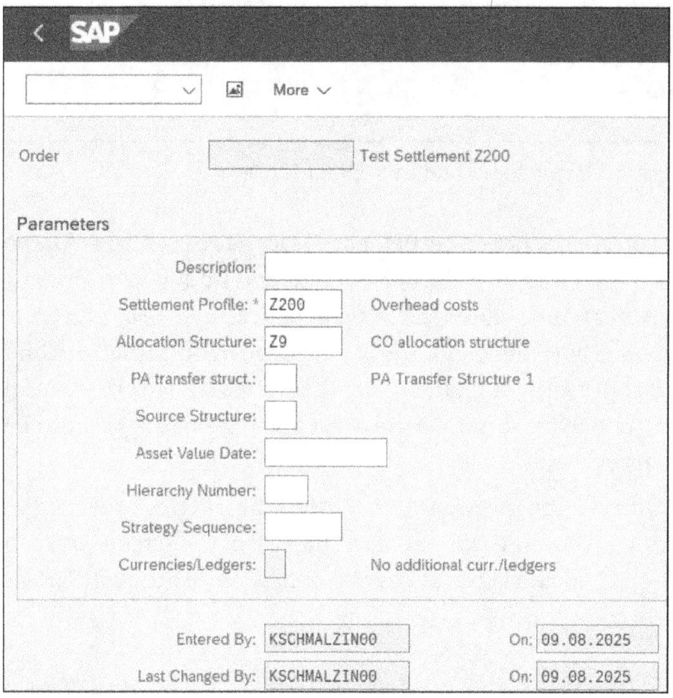

Figure 4.74 Review Settlement Parameters in the Internal Order

183

4 Value Flows of Margin Analysis

Settle the internal order

After creating a journal entry with assignment to the internal order just created, you can settle the internal order. Go to the Individual Processing app. If you want to settle several orders collectively, you can use the Selective Processing app for collective processing.

Orders are settled monthly. The example in Figure 4.75 settles the internal **Order 100081**, created in Figure 4.73 earlier, for **Settlement Period 5** in **Fiscal Year 2020**. Execute the settlement with **Execute** or F8.

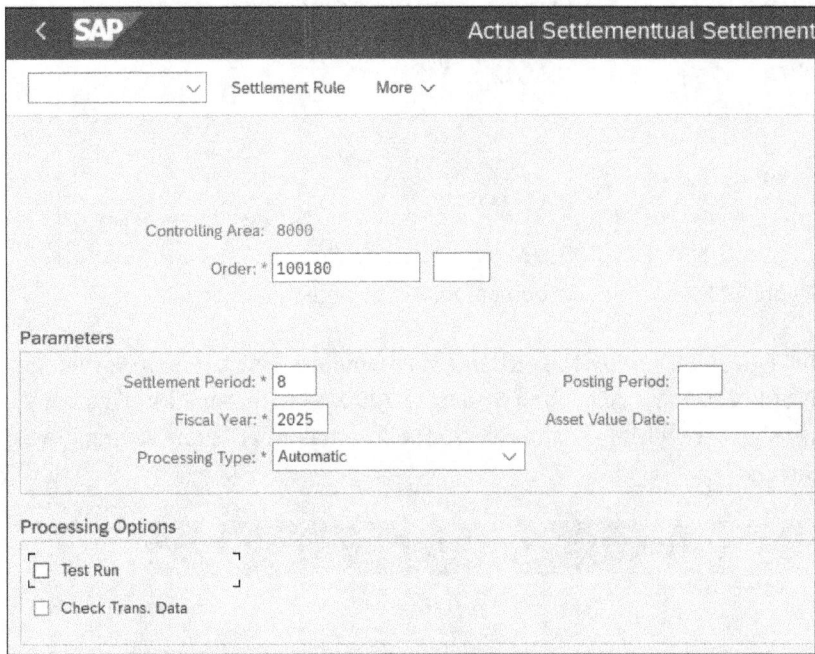

Figure 4.75 Selection Screen for the Internal Order Settlement

Settlement: Order basic list

In Figure 4.76, you see the **Actual Settlement: Order Basic list** screen for the actual settlement that lists the **Selection Parameters** and shows in **Processing Options** whether you executed the settlement in an update or a test run. In **Statistics**, you see how many orders were included in the settlement and if there were any errors. The settlement for order **100081** was executed in an update run without any errors. Click **Next list level** or press Enter to be redirected to a more detailed view.

Settlement: Order detail list

In Figure 4.77, you can see the order **Detail list** of the internal order settlement. By clicking **Accounting documents** at the top of the screen, you can view the accounting documents that have been created with the settlement of the internal order.

4.6 Order Settlement

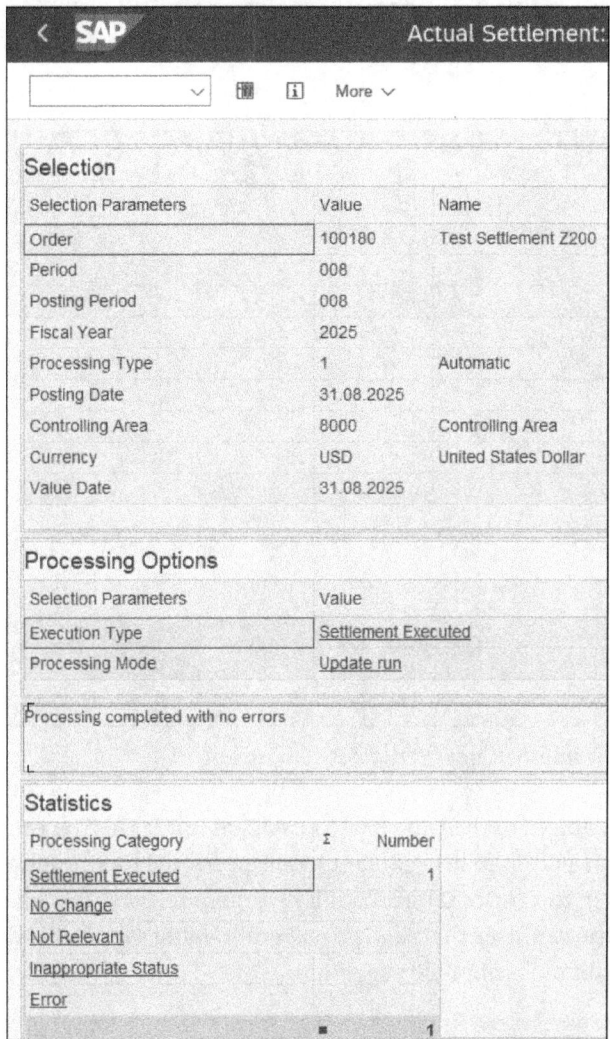

Figure 4.76 Order Basic List of the Internal Order Settlement

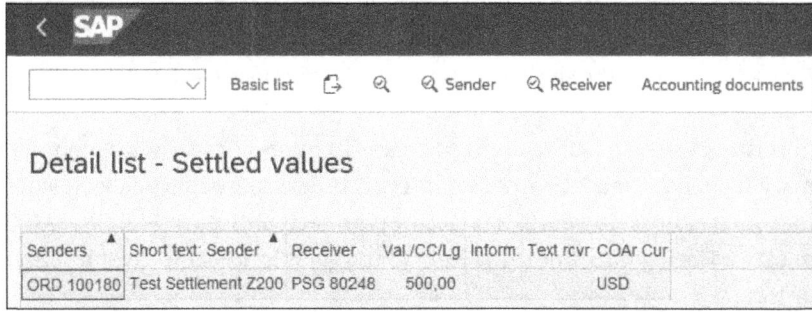

Figure 4.77 Detail List of the Internal Order Settlement

4 Value Flows of Margin Analysis

Review the accounting document The accounting document in Figure 4.78 shows that the internal order has been settled with the settlement G/L **Account 92105000** and the total value of the internal order.

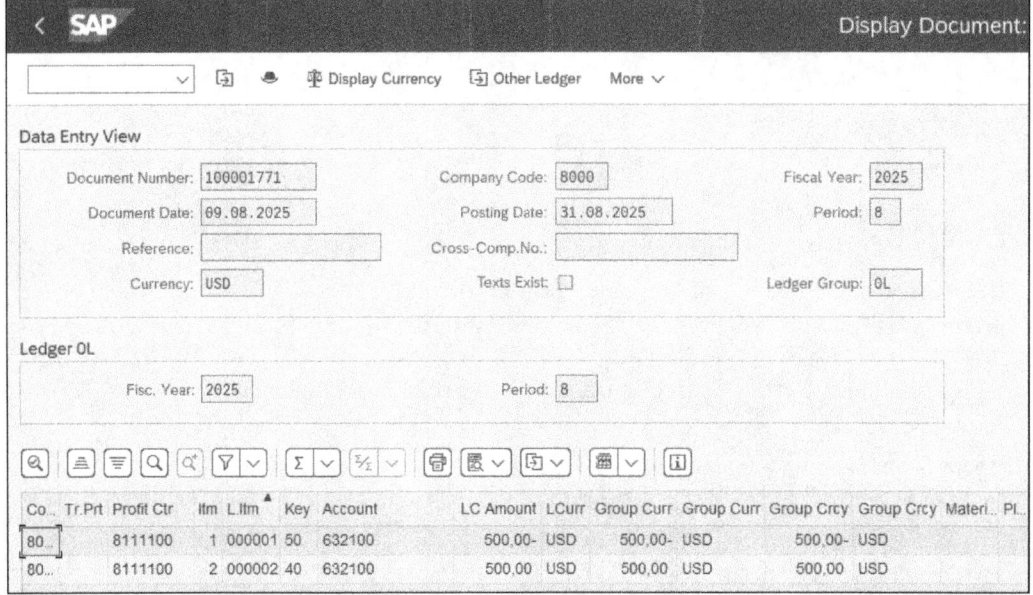

Figure 4.78 Review the Internal Order Settlement Document

Document in the Universal Journal Let's display the document in the Universal Journal (table ACDOCA). You can see in Figure 4.79 that a profitability segment has been created for the debit line item, and the **X** in the **PAObCO** (profitability segment relevant for controlling compatibility) confirms that this is a real profitability segment and not a statistical/attributed profitability segment.

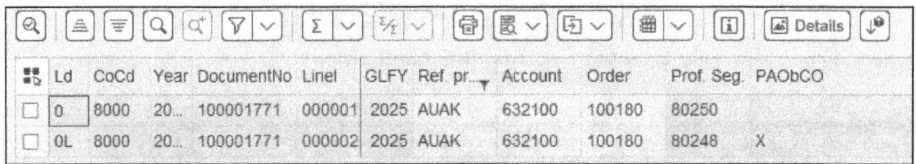

Figure 4.79 Review the Settlement Document in the Universal Journal

In this section, you learned how to create an internal order settlement for margin analysis and how to settle internal orders to margin analysis. Note that internal orders will be replaced by projects in future SAP S/4HANA releases. The logic and the settlement for projects is the same as for internal orders.

4.7 Overhead Allocation

SAP introduced *universal allocations*, which means that all allocations have been summarized and harmonized in two SAP Fiori apps: Manage Allocations and Run Allocations. The vision is to simplify the allocations and combine various functionalities in one place. In SAP GUI, there are currently 120 different transactions for allocations, and there's no overview list available of which cycles are created or even which cycles have been run in the month-end closing. With this in mind, the objective is to have one central SAP Fiori app for all of these transactions.

Universal allocations

4.7.1 Advantages of Universal Allocations

The combination of all allocations in one place is advantageous for the user who creates and runs the allocations because the centralization in one or two SAP Fiori apps simplifies the month-end closing process by giving an overview of all allocations. Other advantages include the following:

Advantages

- Combines cost center and profit center allocations in one SAP Fiori app (financial and management accounting allocations)
- Combines plan and actual allocations in one SAP Fiori app
- Includes predictive data
- Provides simulation capabilities
- Provides the traceability of the value flow
- Provides all reporting currencies and shows a currency breakdown
- Provides guided procedures and validations

The source tables for universal allocations are the Universal Journal for actuals (table ACDOCA) and for planning (table ACDOCP). We recommend using universal allocations for all your cost center and profit center allocations, if available. The transactions in SAP GUI to create cost center and profit center cycles are still available, but they have old tables as sources and will retire in the long term.

Before we create a cost center assessment for margin analysis, we want to explain the business need for allocations and then show you how to create and run them.

In Section 4.5, we already introduced different types of cost centers that need to be allocated to margin analysis to get a complete margin report. In this section, we'll show how overhead costs that can't be directly attributed to products are allocated to the product-specific profit centers. Mostly those overhead costs are posted on overhead cost centers. *Overhead cost centers* collect costs that can't be allocated to the production process or aren't of service for any other cost center but are still required in a company to operate, such as sales, marketing, finance, and so on. Those cost centers get allocated directly to margin analysis at month end. With cost center allocations, the balance of cost centers at the end of the month is transferred to receivers that are specified in the cost center allocation.

4.7.2 Types of Cost Centers

Two types of cost centers are assessed:

Types of cost centers

- **Service cost centers**
  Service cost centers distribute their costs to other receivers via secondary cost distribution. These include, for example, production cost centers, such as machine cost centers or personnel cost centers. Ideally, these costs are credited to 100% if the planned costs correspond to the actual costs, and no production variances occur. If there are variances, the service cost centers show a balance that has to be settled. Alternatively, you can also determine actual prices on the cost centers and subsequently debit the process orders with the delta from actual price and planned price. Then, the costs centers are credited to 100%. They don't show a balance, so the cost centers don't have to be assessed in margin analysis.

- **Overhead cost center**
  Overhead cost centers include administration cost centers or sales cost centers, for example. These cost centers usually don't allocate any costs to other cost centers with secondary cost distribution and are settled to margin analysis to 100%.

The receivers of cost center assessments in margin analysis are a selection of characteristics.

Cost center assessment parameters

Before you create a cost center assessment, you should spend some time setting up a concept for the allocations to margin analysis and give some thought to the following parameters:

- **Sender**
  Determine which cost center or cost center group you want to allocate the costs for.

- **Cost element**
  Determine which cost element or cost element group of the sender cost center you want to allocate. We recommend working with cost element groups in the allocations because they are easier to adjust after you're working in a productive system. In addition, if the group is used several times in the assessment cycle, you only have to make the adjustment afterward in the cost element group without needing to change every single assessment cycle. This also ensures that all assessment cycles are up-to-date and you don't forget to adjust an assessment cycle.

- **Receiver**
  After defining the sender cost center, you have to specify which segment will receive the costs in margin analysis. You have to specify the characteristics and characteristic values of the receiver. The number of receivers is determined by the combination of all values (10 customers and then material groups lead to 100 profitability segments).

- **Assessment cost element**
  To allocate costs to margin analysis, you have to create a G/L account with the G/L account type **Secondary Costs**, which you choose in the **Type/Description** tab of the G/L account. In the **Controlling Data** tab, you assign cost element type **42** (assessment) to the G/L account (see Figure 4.80) to be able to use it in the overhead allocation.

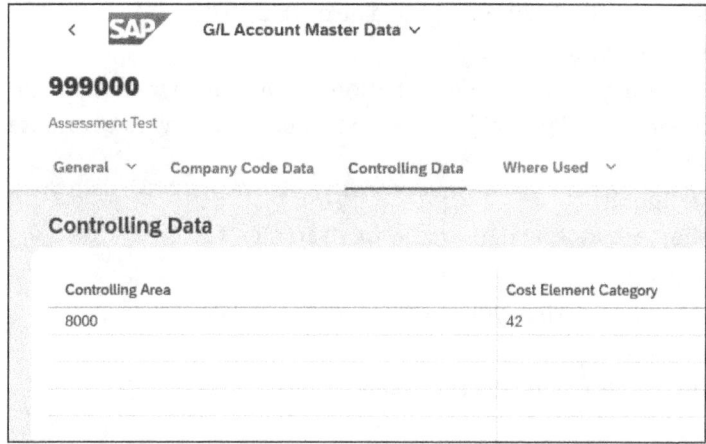

Figure 4.80 Review the G/L Account for Overhead Allocation to Margin Analysis

4.7.3 Create an Overhead Allocation Cycle

Now, let's create an overhead allocation cycle for margin analysis. To do so, go to the Manage Allocations app.

Click `Create` to create a new overhead allocation cycle. Figure 4.81 shows Cycle **TEST** being created. There's much more information to be maintained when creating an allocation than shown in SAP GUI. In addition to the cycle name, we're going to maintain the following fields:

Distribution key

Creating an overhead allocation cycle

- **Allocation Context:**
  Enter "Margin Analysis" to indicate that you're creating an allocation for margin analysis. The Manage Allocations app doesn't allow you to create allocation cycles for costing-based profitability analysis. Any costing-based profitability analysis allocation cycles have to be created with the SAP GUI transactions.

- **Allocation Type**
  Choose **Overhead Allocation**. In addition, you can also choose **Top-Down Distribution** and **Distribution**. **Overhead Allocation** uses a secondary cost element to post the allocation, whereas **Distribution** uses the original G/L account to post the allocation. **Top-Down Distribution** distributes values (revenues, costs, margins) from a higher-level record to lower-level dimensions using a reference data set or statistical key figure.

- **Ledger**
  You can create allocations for different ledger groups. We're creating our allocation in **Ledger 0L**.

- **Valid From/Valid To**
  This describes from when to when the allocation is valid. You can't change the start date after the allocation cycle is saved. We're maintaining a start date of **01.05.2025** and end date of **31.12.2025**.

- **Top-Down Template**
  If you're creating a top-down distribution (TDD), you have to maintain the **Top-Down Template** field for the system to know how to distribute the values.

- **Cycle Description**
  Enter a short description of the cycle, for example, "Test".

- **Controlling Area**
  This field populates when you enter the company code.

- **Company Code**
  Create the cycle for company code **8000**.

- **Actual/Plan**
  You can create allocations for **Actual** and **Plan**. Here, create the allocation for **Actual**.

Confirm the entries with [Enter] or by clicking on **Create**.

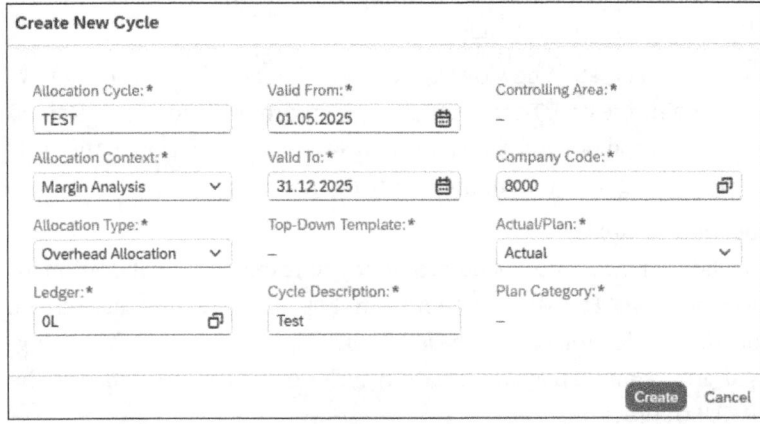

Figure 4.81 Create an Overhead Cycle for Margin Analysis

4.7 Overhead Allocation

Figure 4.82 shows the header of the cost center assessment cycle. On the left side of the screen, you can still see the selection bar of the Manage Allocations app, which allows you to search for allocation cycles. Maintain the following details in the **General Information** tab:

Creating the assessment cycle header

- **Cycle Run Group**
  If you have many cycles, you can determine a sequence in which the cycles should run. In this example, we're not maintaining any values in the cycle run group.

- **Tags**
  You can enter a tag for cycles here, which allows you to group cycles and makes it easier for you to search for them. In this example, we're not maintaining any tags.

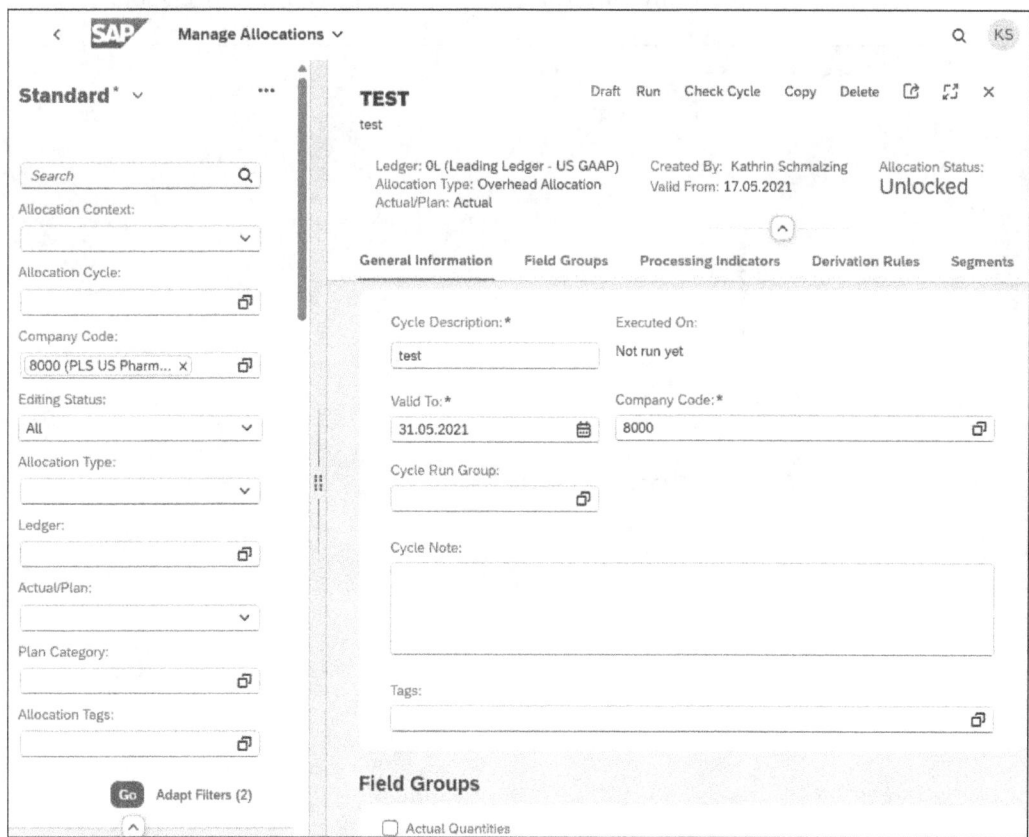

Figure 4.82 Maintain the Cost Center Assessment Cycle Header

Now, let's look to the right and review the additional tabs that you can enter in the **Cost Center Assessment Cycle** header in **General Information** (Figure 4.83):

4   Value Flows of Margin Analysis

- **Field Groups**
  - **Actual Quantities**: If checked, actual quantities will be included in the allocation.
  - **Balance Transaction Currency**: Ensures transaction currency is balanced during allocation (rarely used).
- **Processing Indicators**
  - **Iterative**: Cycles through the allocation until all amounts are allocated.
  - **Flexible Ledger**: Includes the flexible ledger in the allocation (advanced use).
- **Derivation Rules**
  - **Functional Area**: Enables derivation of the functional area during allocation, often used in reporting or controlling reconciliation.

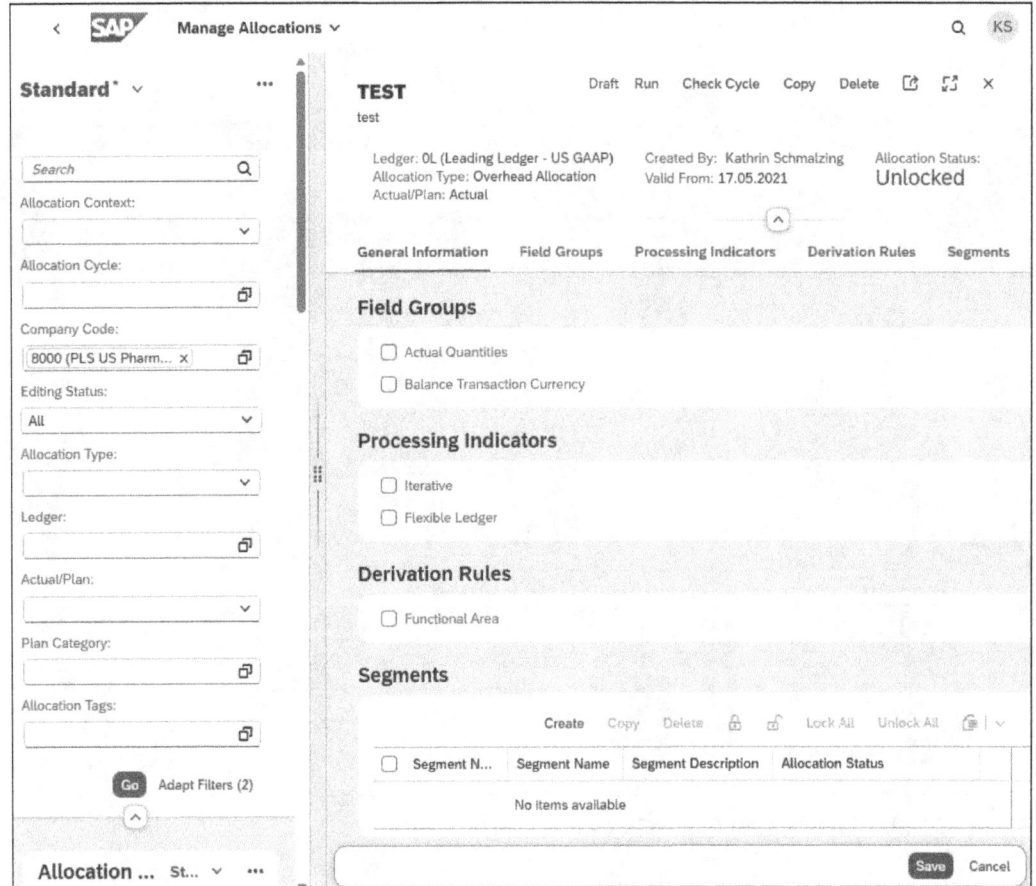

Figure 4.83 Maintain Header information in the Cost Center Assessment Cycle

Next, you need to create a segment with **Create**, as shown in Figure 4.84. Enter a **Segment Name** and **Segment Description** before you confirm your entries with [Enter].

You can assign any number of segments to cycles, but the orders play a critical role. For example, if you want to allocate depreciation costs to a separate value field and all other costs of the cost center to another value field, you have to assign the segment of the depreciation costs assessment before the segment of the remaining costs assessment; otherwise, the system can't find depreciation costs for the assessment because they have already been allocated with the remaining costs.

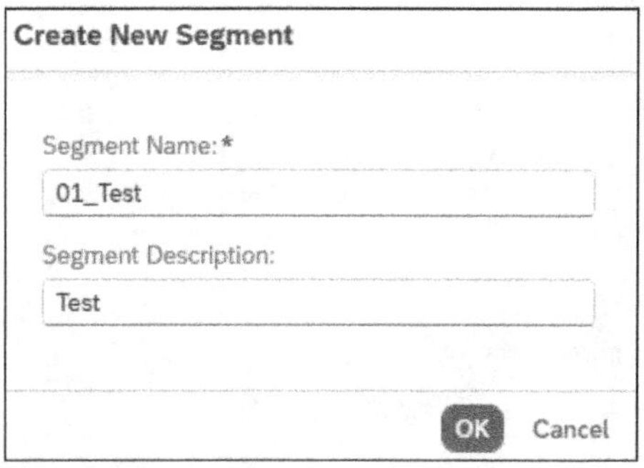

Figure 4.84 Maintain Rules in Assessment Cycle

Segments consist of various tabs. In Figure 4.85, the tab **Rules** contains the **Segments** header by defining the name ("01_Test") and the **Segment Description** ("Test") of the segment, as well as the following data:

- **Overhead Allocation Account**
  This is the G/L account with the G/L account type for **Secondary Costs** and cost element type **42** (**Allocation**). This is the G/L account with which the costs are allocated to margin analysis.
- **Sender Rule**
  In the **Sender Rule** section, you specify whether one of the following objects is allocated by the sender object and to which share in percentages the sender will be credited:
  - Posted amounts
  - Fixed amounts
  - Fixed prices

  In this example, **Posted amounts** is selected, and 100% of the posted amounts will be credited to the sender in the allocation in the **Send Share in %** field.

4   Value Flows of Margin Analysis

- **Receiver Rule**
  In the **Receiver Rule** dropdown, you determine according to which distribution key the sender values are distributed to the receiver. In this example, **Fixed percentages** is selected.

After the **Segment Rule** tab entries are made, move on to the **Senders** tab.

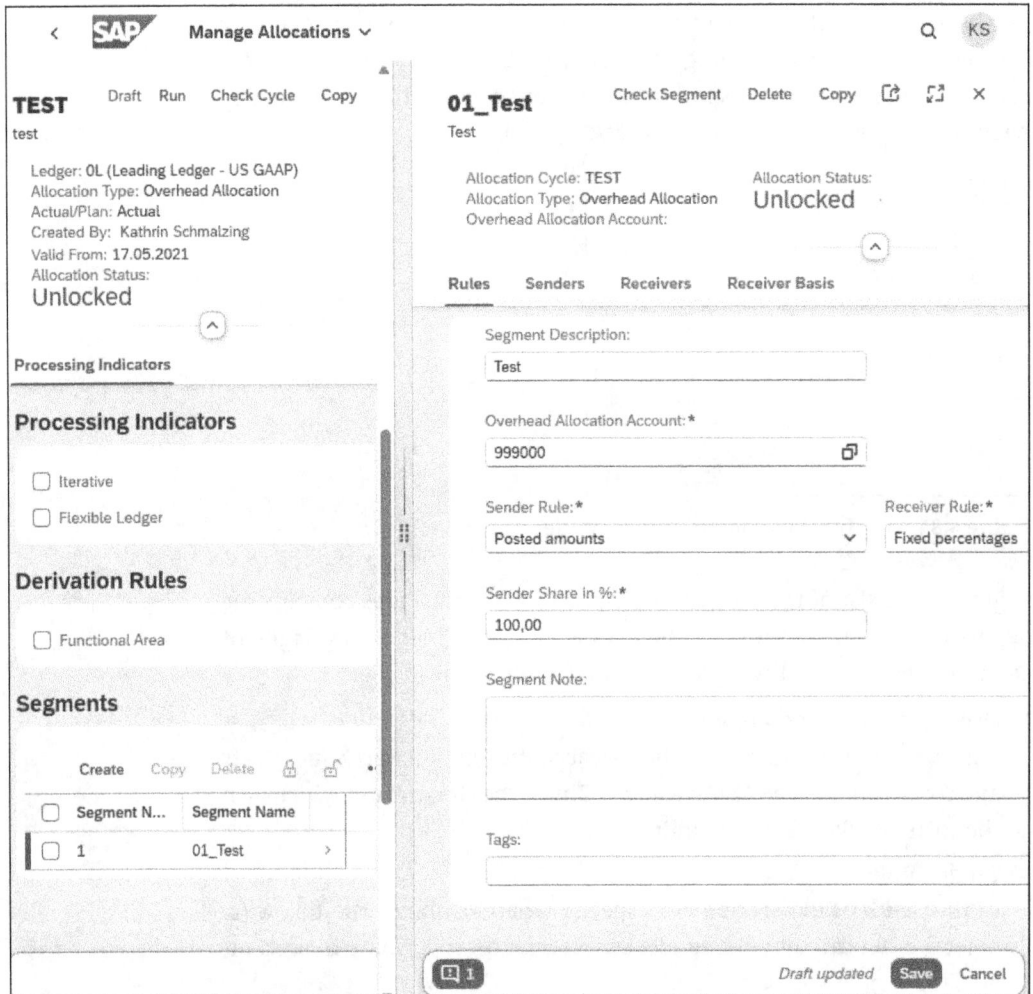

Figure 4.85 Create a Segment Header

Defining senders   In the **Senders** tab shown in Figure 4.86, you define the sender G/L accounts, and the sender profit centers will be allocated with this overhead allocation cycle. In this example, we maintain both for G/L accounts, profit centers, and an interval. Instead of an interval, we could also enter single values or a group. After you've made your entries to the **Senders** tab, move on to the **Receivers** tab.

4.7 Overhead Allocation

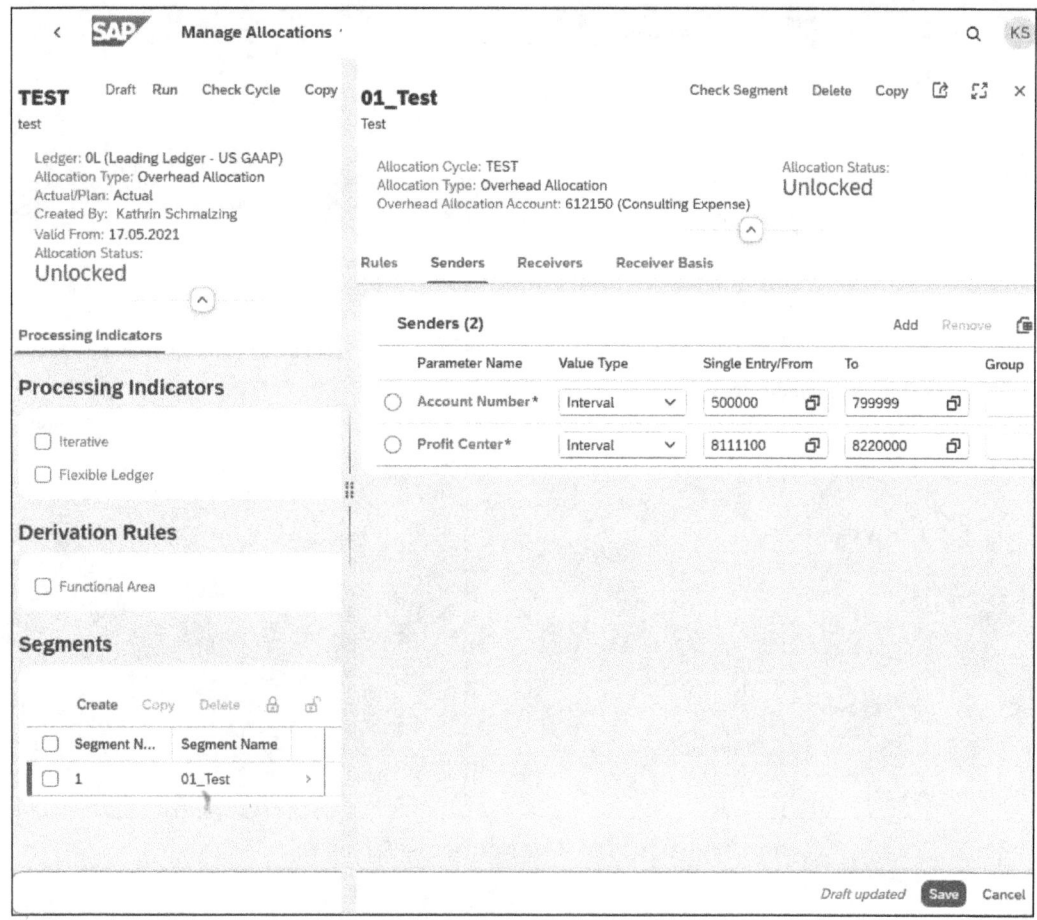

Figure 4.86 Maintain Senders/Receivers in the Cost Center Assessment Cycle

In the **Receiver** section, you specify the profit centers to which you want to allocate the cost. In the example, we enter profit center 840000. After maintaining the senders and receivers, move on to the **Receiver Basis** tab.

The **Receiver Basis** tab shown in Figure 4.87 displays all possible combinations of characteristics that you selected on the **Senders/Receivers** tabs as the receiver. Because **Fixed percentages** was selected in the screen shown earlier in Figure 4.85, you need to assign the percentage of costs it should receive to each combination of characteristics in the execution of the assessment cycle.

Defining the receiver basis

After assigning 100%, you can click on Check Segment to review whether the segment is consistent. In the popup shown in Figure 4.88, you can see whether the segment is free of any errors.

After the check has been successfully executed, save the segment with Save, and you're ready to execute the overhead allocation.

4 Value Flows of Margin Analysis

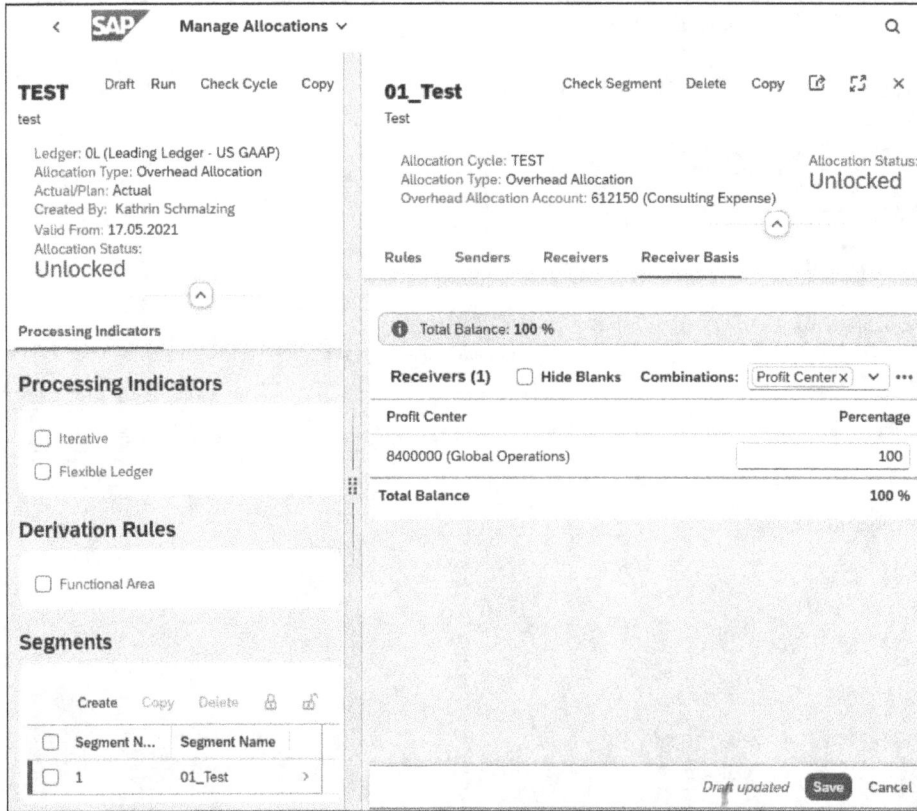

Figure 4.87 Maintain the Receiver Tracing Factor

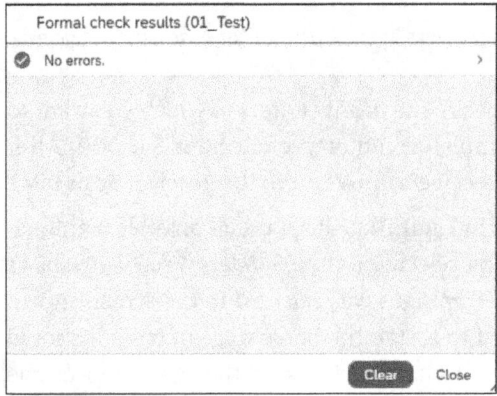

Figure 4.88 Review Segments for Consistency

4.7.4 Execute the Overhead Allocation Cycle

Executing the actual overhead allocation cycle

After saving the overhead cycle, you can see at the top of the screen that you have the possibility to run the cycle. If you don't want to run the cycle imme-

4.7 Overhead Allocation

diately, you can come back later and run it from the Manage Allocations app. As shown in Figure 4.89, click on Run to execute the allocation cycle.

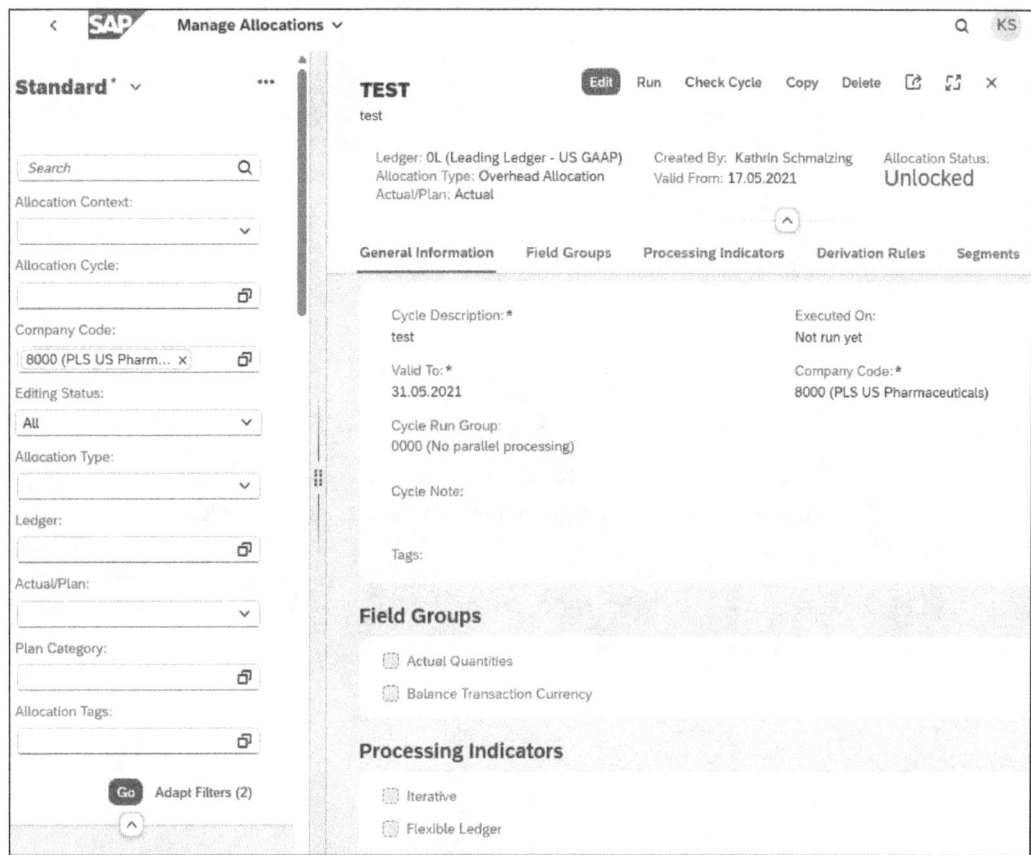

Figure 4.89 Manage Allocations App

In Figure 4.90, the Run Allocations app opens where you can execute a test run or execute the run via **Run**. In the popup, select the **Fiscal Period From** (**005.2025**) and **Fiscal Period To** (**012.2025**) to execute the run. As **Run Name**, enter "Test", and click on **Validate**.

The system validates the entries, and if the validation is successful, you can start executing the overhead allocation by clicking on **Run**.

After executing the run, click on **Allocation Result** to search for the run we just executed. You can see in Figure 4.91 that the allocation run was successfully executed.

Actual assessment basic list

The run is the last run in the list, which you can tell because it was executed as a live run. The Allocation Result gives you more detailed information as well. By clicking on ▶, you can see the number of senders and receivers as well as the journal entries that have been created during the execution.

4 Value Flows of Margin Analysis

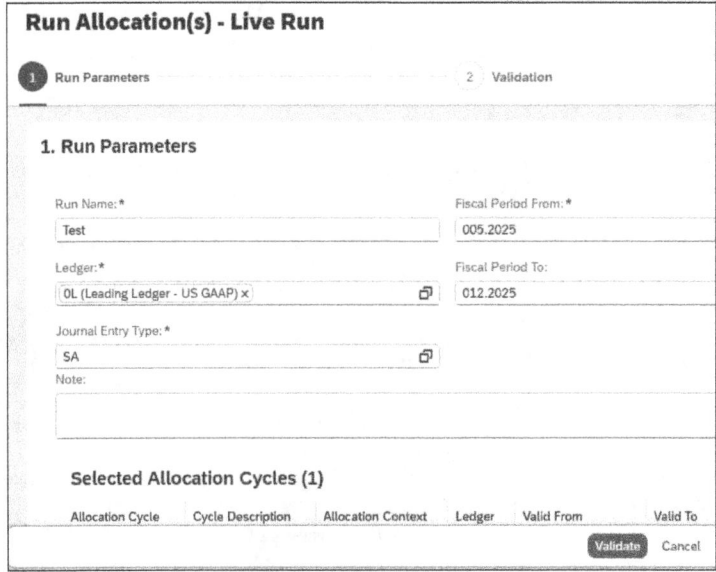

Figure 4.90 Validate Entries in the Cost Center Assessment Run

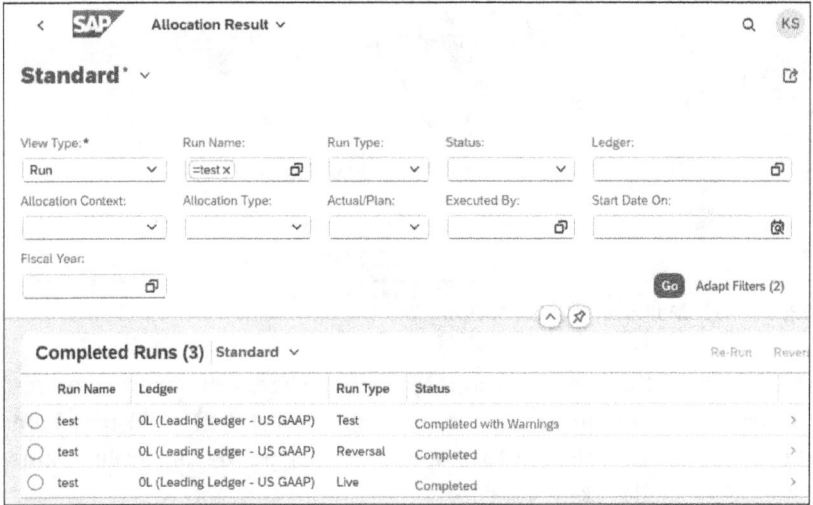

Figure 4.91 Review Allocation Results

4.8 Direct Account Assignment

Direct account assignment serves to assign values directly to the profitability segment. The account assignment won't be in cost centers or internal orders but the margin analysis directly. It's often used to post scrapping costs, for example. If you assign scrapping costs to cost centers, the information on the material is lost when the cost center is allocated to margin analysis at month end. If you assign scrapping costs directly to a profitability segment and hence margin analysis, you can also transfer the information on the material and quantity to margin analysis.

Before you configure the direct account assignment, you need to make sure that the field status group of the G/L account you want to assign to the direct account assignments has the profitability segment at least as an optional entry. The field status group is assigned to the G/L account in the **Create/Bank/Interest** tab. Double-click **Field Status Group**, and you can see the overview of the different sections of the field status group. Double-click on **Additional account assignments** to see an overview of the allowed account assignment objects. In Figure 4.92, you can see that **Profitability Segment** is an **Opt. entry** (optional entry).

Changing field status groups

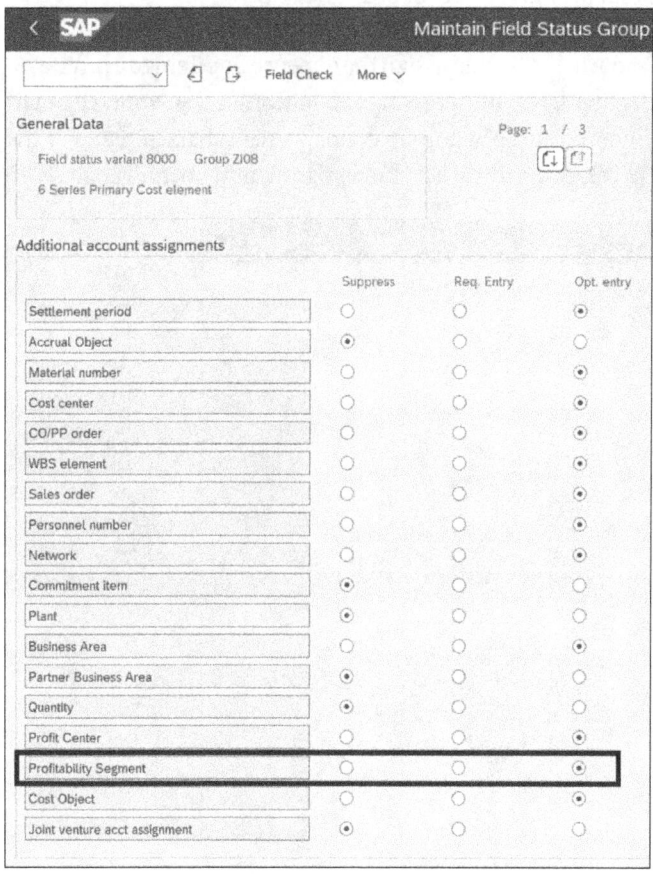

Figure 4.92 Review the Field Status Group

Because the field status group allows posting to the profitability segment, you can now configure direct account assignment with Transaction OKB9 or by following the configuration path **Controlling • Profitability Analysis •**

Account assignment to profitability segments

4 Value Flows of Margin Analysis

Flows of Actual Values · Direct Postings from FI/MM · Automatic Account Assignment. With **New Entries** or F5, you can maintain an entry for cost element **631000** in company code **8000**, as shown in Figure 4.93. Instead of assigning a cost center or an internal order, select the checkbox in the **PrfS** (profitability segment) column. Save the entries with **Save** or Ctrl+S.

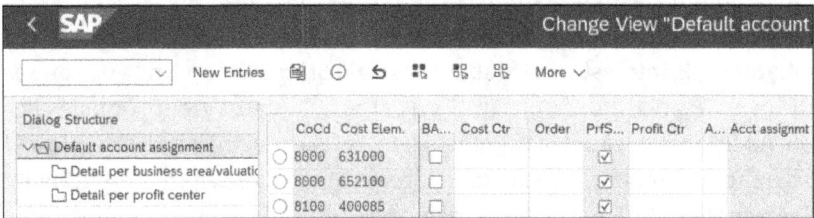

Figure 4.93 Maintain the Default Account Assignment

Create a journal entry

Now, let's create a journal entry and assign the expense line item to a profitability segment. In Figure 4.94, after pressing the button in the **Profit. segment** (profitability segment) column, a pop-up opens with all characteristics of the margin analysis to create the direct account assignment to margin analysis.

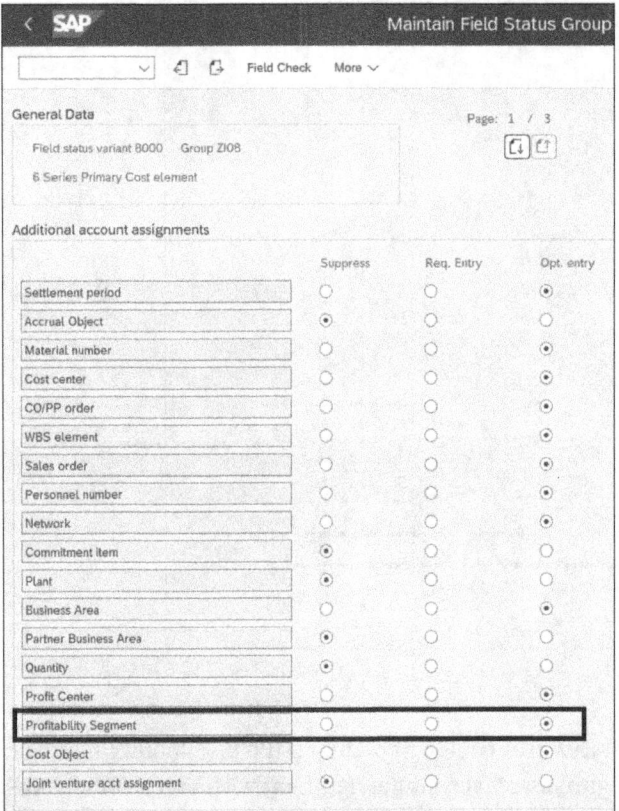

Figure 4.94 Create a Journal Entry with Direct Account Assignment

200

The system will open the **Assignment to a Profitability Segment** popup so you can select any characteristics assigned to the operating concern for the profitability segment. In Figure 4.95, maintain a characteristic for **Customer**, **Product**, and **Functional Area**, and then press `Enter`.

Assign characteristics

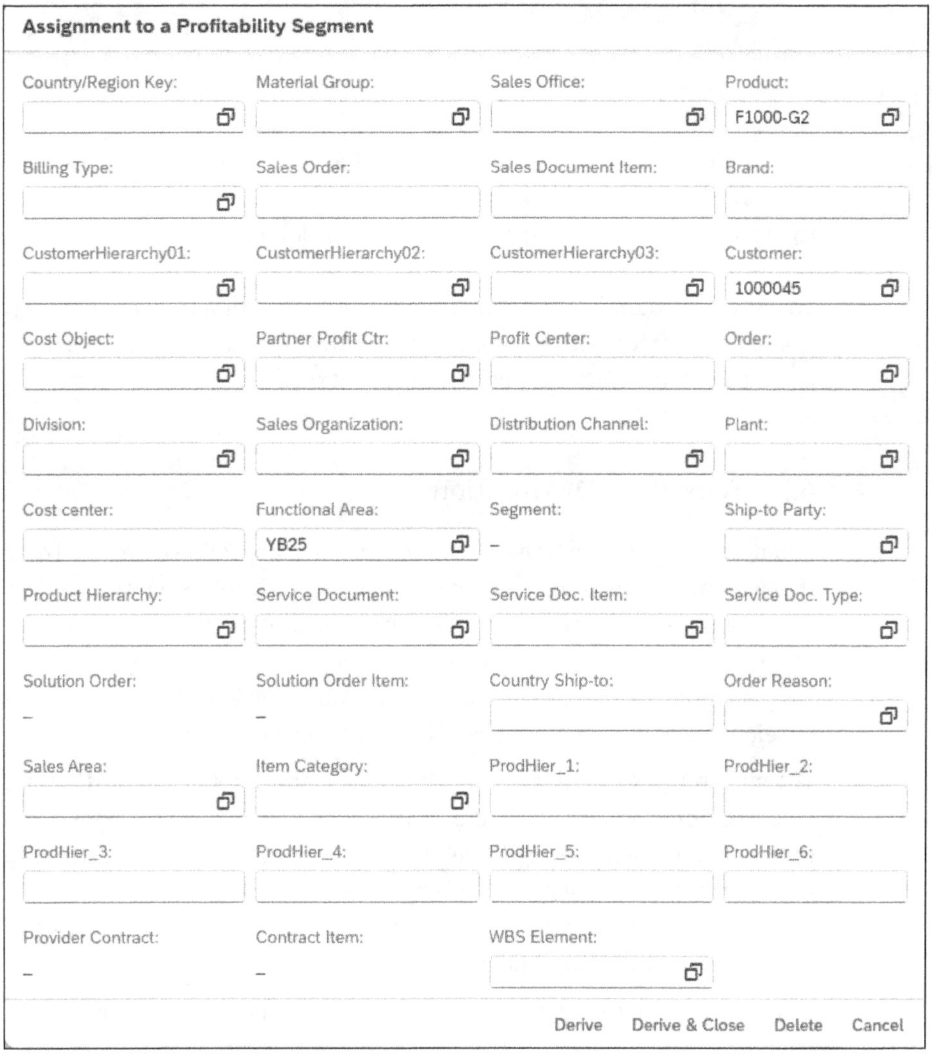

Assignment to a Profitability Segment			
Country/Region Key:	Material Group:	Sales Office:	Product: F1000-G2
Billing Type:	Sales Order:	Sales Document Item:	Brand:
CustomerHierarchy01:	CustomerHierarchy02:	CustomerHierarchy03:	Customer: 1000045
Cost Object:	Partner Profit Ctr:	Profit Center:	Order:
Division:	Sales Organization:	Distribution Channel:	Plant:
Cost center:	Functional Area: YB25	Segment: –	Ship-to Party:
Product Hierarchy:	Service Document:	Service Doc. Item:	Service Doc. Type:
Solution Order: –	Solution Order Item: –	Country Ship-to:	Order Reason:
Sales Area:	Item Category:	ProdHier_1:	ProdHier_2:
ProdHier_3:	ProdHier_4:	ProdHier_5:	ProdHier_6:
Provider Contract: –	Contract Item: –	WBS Element:	
		Derive Derive & Close Delete Cancel	

Figure 4.95 Enter Profitability Characteristics

After the creation of the journal entry, you can review it in the Universal Journal. You can see that profitability segment **77398** has been created for the expense line item and that the characteristics selected in the screen shown in Figure 4.96 have been assigned to the profitability segment.

Review the journal entry in Universal Journal

4 Value Flows of Margin Analysis

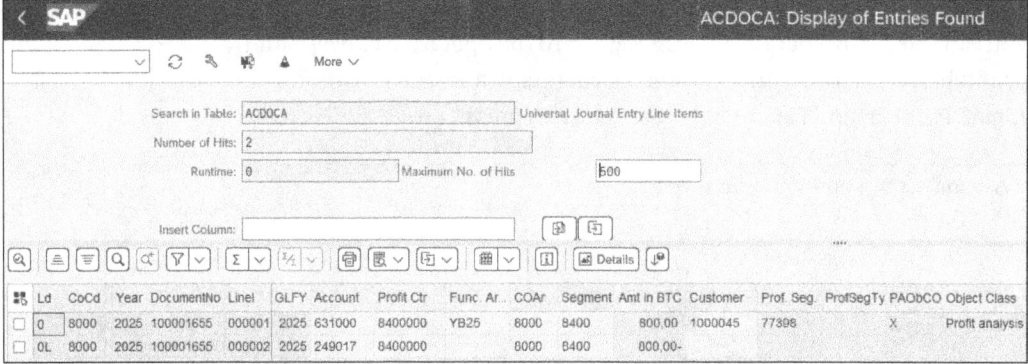

Figure 4.96 Review the Journal Entry in Universal Journal

The direct account assignment is especially useful for automatic postings as a lot of characteristics can be derived from the posting that would get lost if the posting was assigned to a cost center.

4.9 Top-Down Distribution

Top-down distribution (TDD) is a method used to break down financial data that was initially recorded at a broader, aggregated level—such as a product group or sales organization—into more detailed segments such as individual products or customers. This process allows companies to gain deeper insights into profitability by reallocating costs or revenues using historical or reference data, such as last year's sales figures or cost information.

The main purpose of this approach is to enable more detailed reporting and analysis. For instance, if certain costs—such as freight expenses—were originally recorded at a high level, TDD helps allocate these costs down to the product or customer level. This ensures that profitability is analyzed in a way that reflects the true performance of individual segments, leading to more informed decision-making.

TDD is guided by reference data and follows a specific distribution key, which could be based on quantities, percentages, or other relevant business metrics. It's typically executed on a regular basis, such as at the end of each month, to ensure that reports reflect a more accurate and granular financial picture.

To illustrate, consider a company that posts freight costs at the sales organization level. With TDD, these costs can be proportionally allocated to individual products or customers, using factors such as shipment volume or distance. This results in a clearer view of which products or customers are more or less profitable.

4.9 Top-Down Distribution

Throughout this process, the sender data represents the original aggregated values, while the reference data and distribution key determine how those values are spread across different segments. In some cases, certain characteristics from the original data are preserved in the breakdown, while others may be grouped or summarized to match the new structure.

To create a TDD cycle, go to the Manage Allocations app, and click on **Create**. In Figure 4.97, you create the header of the TDD cycle by choosing **Allocation Context** as **Margin Analysis** and **Allocation Type** as **Top Down Distribution**. The **Top-Down Template** field shows all templates that exist within the system for the TDD. They are controlling area–specific, so be mindful to choose the right template for the controlling area you're operating in. The TDD template defines how each characteristic is applied and processed within the universal allocation process. It plays a key role in setting up the TDD cycle, specifying which margin analysis fields appear on the various tabs of the segment definition. SAP predelivers several TDDs, but you can also create your own. Click **Create** to create the header of the TDD cycle.

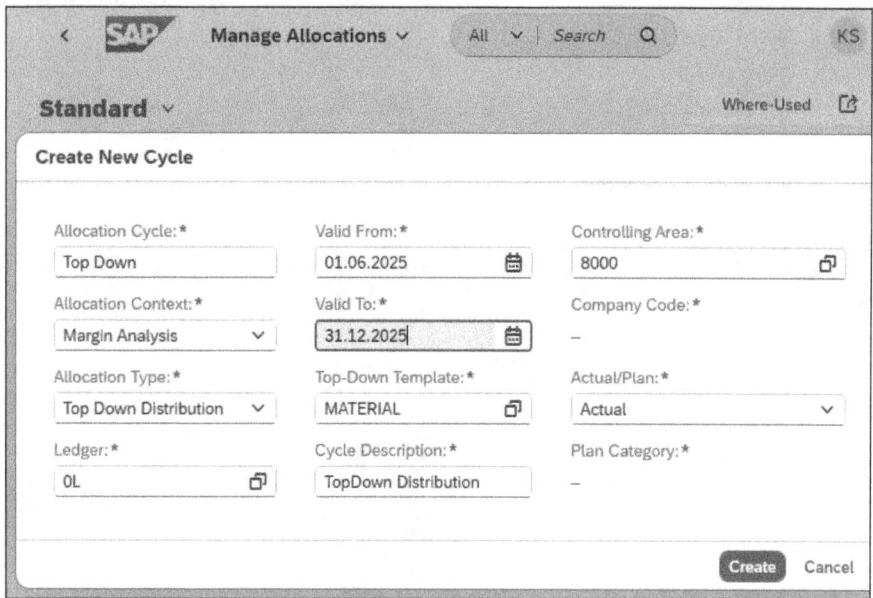

Figure 4.97 Create a Top-Down Distribution Cycle

The screen shown in Figure 4.98 appears where you can review your settings and then click on **Save**. Next, you create a segment for the TDD cycle.

In Figure 4.99, you can create a new segment. Enter the **Segment Name** and the **Segment Description**, and then confirm the entries with **OK** or Enter.

4   Value Flows of Margin Analysis

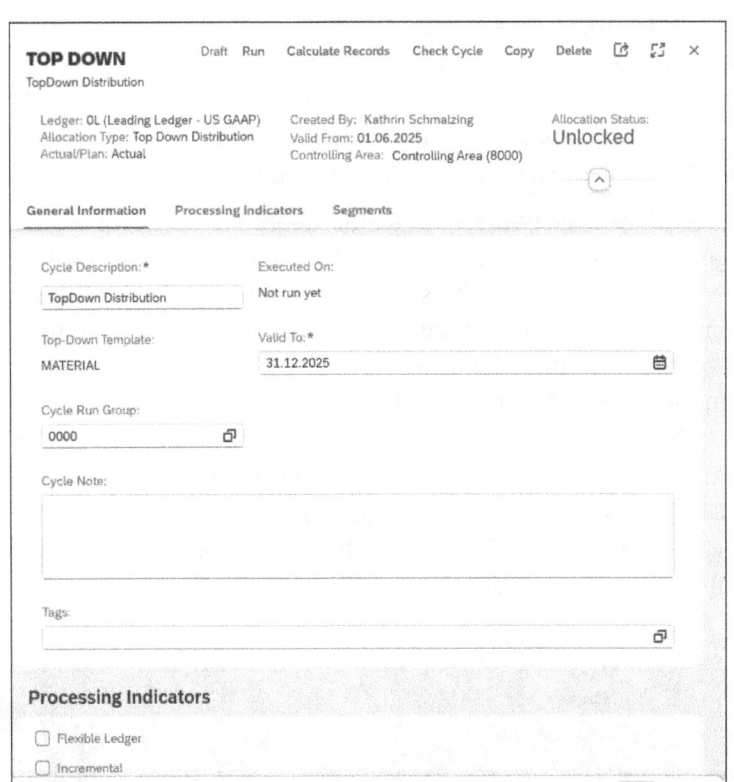

Figure 4.98 Review the Top-Down Distribution Header

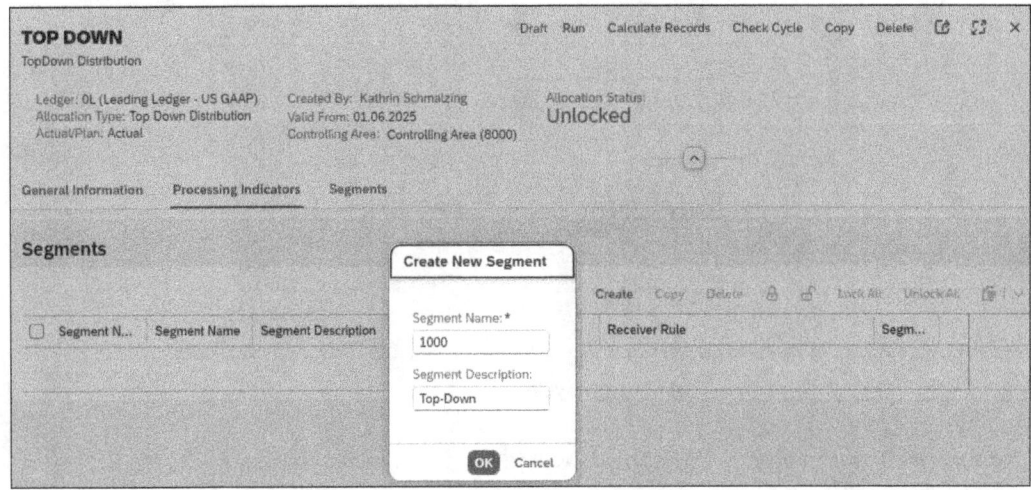

Figure 4.99 Create a Segment in the Top-Down Distribution Cycle

The next screens don't require any entries. In Figure 4.100, you can change the **Segment Description** or add additional notes in the **Rules** tab.

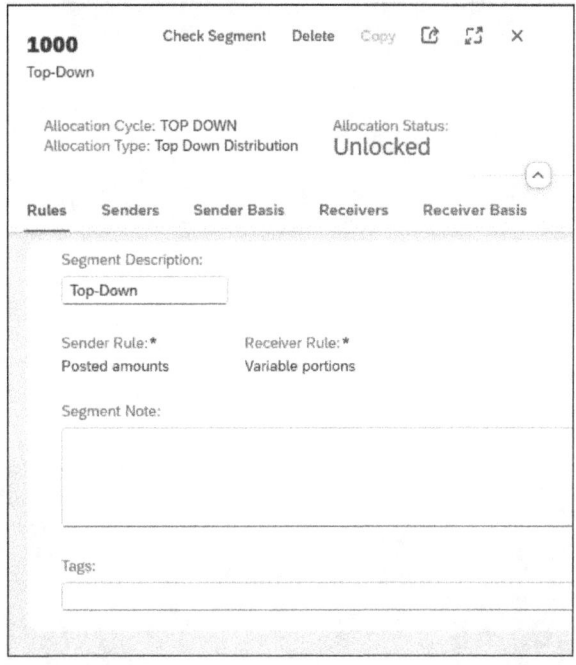

Figure 4.100 Create Rules in the Top-Down Distribution Segment

In the **Senders** tab shown in Figure 4.101, you see the characteristics per TDD. The values posted by **Customer** and **Product Sold** will be distributed based on the receivers maintained in the TDD.

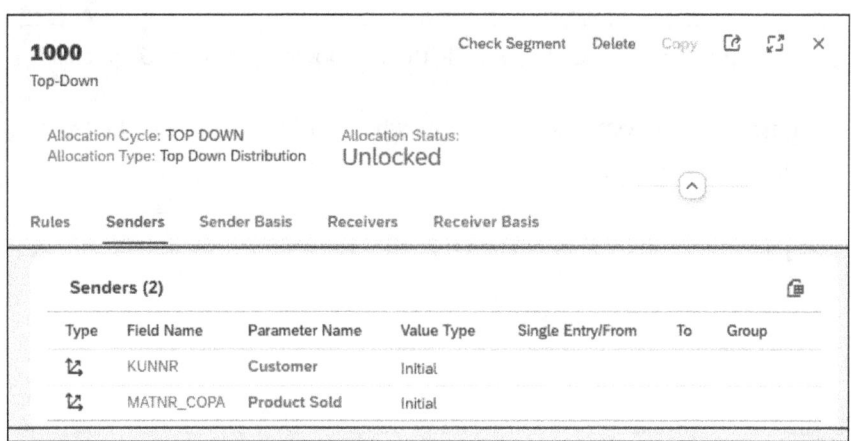

Figure 4.101 Create Senders in the Top-Down Distribution Segment

In the **Sender Basis** tab shown in Figure 4.102, you can further narrow down the senders by additional characteristics in margin analysis. No further entries need to be made in this tab because we want to select all postings with the combination of **Customer** and **Product Sold**.

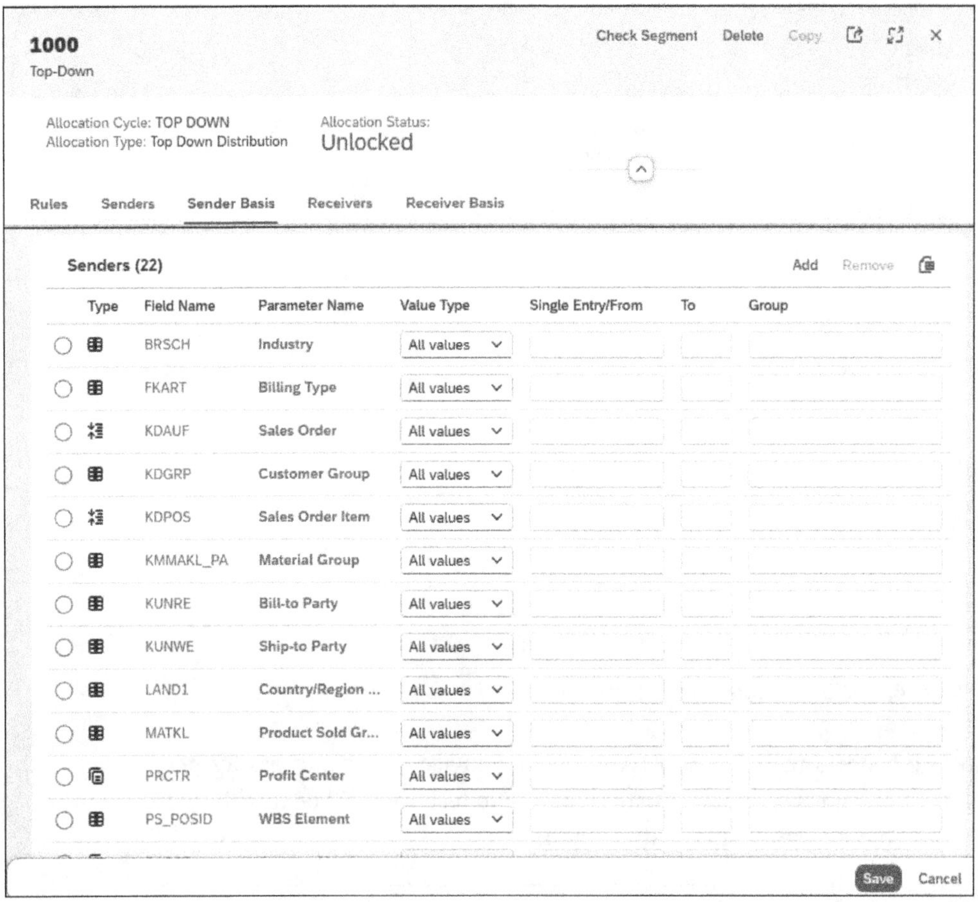

Figure 4.102 Create a Senders Basis in the Top-Down Distribution Segment

Figure 4.103 shows the **Receivers** tab, which looks the same as the **Senders** tab.

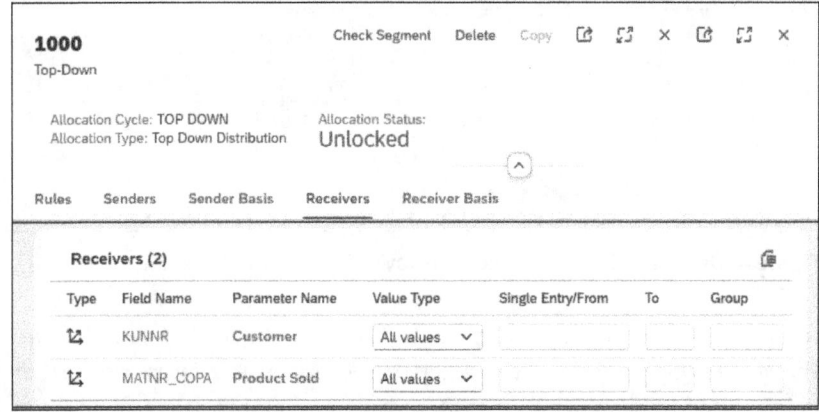

Figure 4.103 Create Receivers in the Top-Down Distribution Segment

4.9 Top-Down Distribution

Go to the **Receiver Basis** tab shown in Figure 4.104 to narrow down the basis on which to break down the freight cost posted by **Customer** and **Product Sold**.

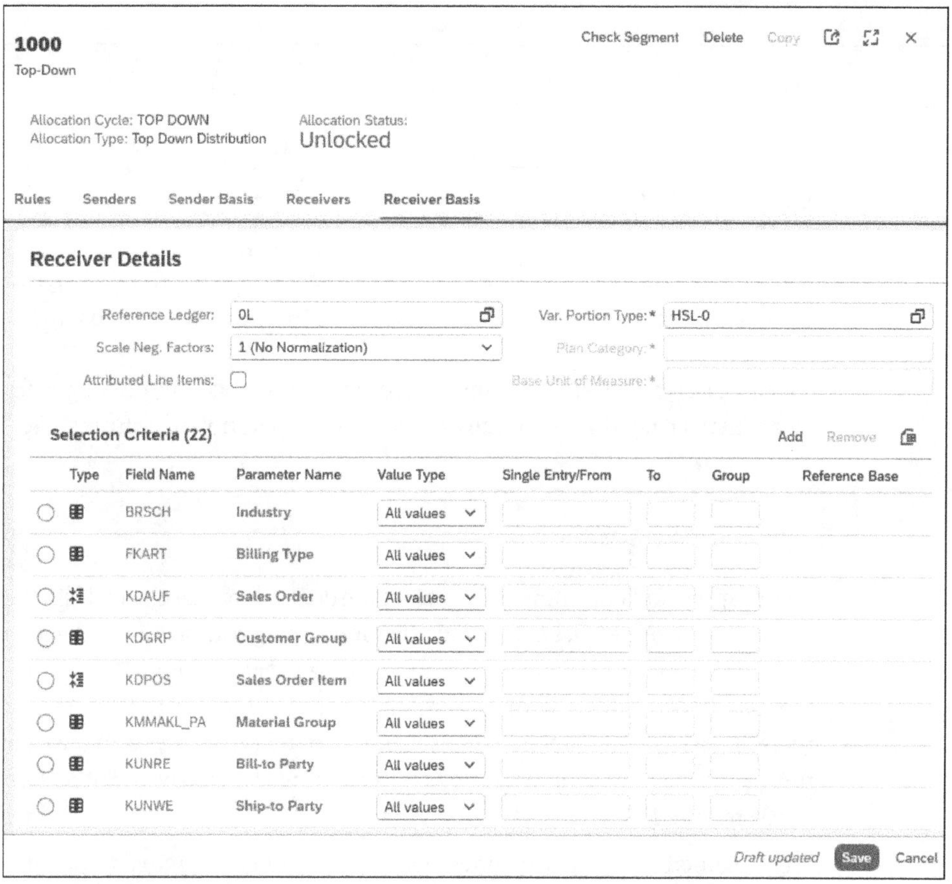

Figure 4.104 Create a Receiver Basis in the Top-Down Distribution Segment

After maintaining the entries, click on **Save**. Then, after you validate that all the settings are correct with **Check Segment**, click on **Run Cycle**. The screen in Figure 4.105 opens where you can click on **Run** to execute the cycle.

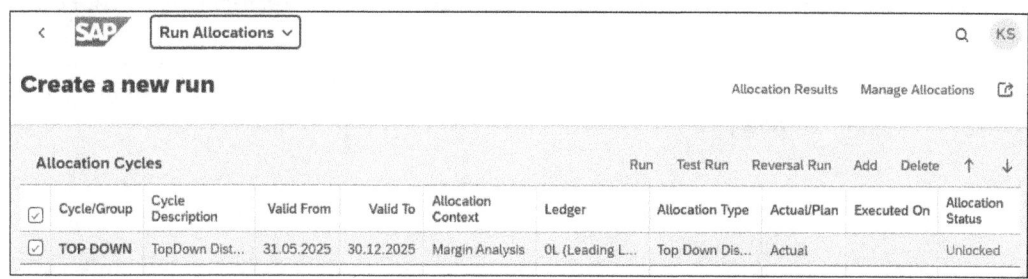

Figure 4.105 Run the Top-Down Distribution Cycle

4   Value Flows of Margin Analysis

In Figure 4.106, you can see the **Allocation Results**. The TDD read **3** senders and distributed the values to **4** different receivers. You can click into the **Senders** and **Receivers** tabs to review the details of how the values were distributed.

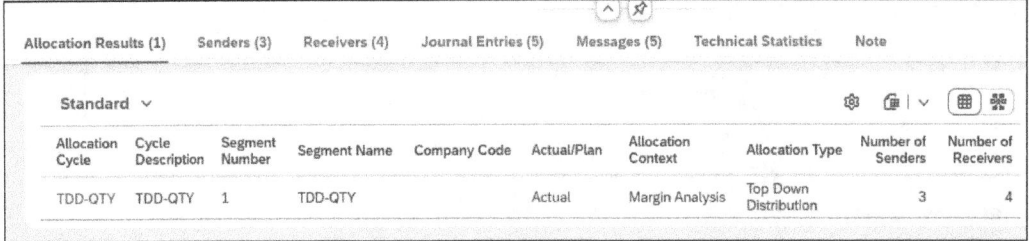

Figure 4.106 Review the Allocation Results of the Top-Down Distribution Cycle

Overall, TDD enhances the quality of profitability analysis by aligning high-level financial postings with the finer details of operational performance.

4.10 Summary

This chapter explained how actual values are derived for margin analysis. We explained the prerequisites for predictive accounting and gave an outlook on functionalities in predictive accounting that are yet to come. SAP recommends activating margin analysis when you transition to SAP S/4HANA as there are many more functionalities planned for margin analysis. In the near future, the creation of an operating concern for margin analysis will no longer be required.

We explained in this chapter how data is transferred to margin analysis and which configuration settings are required to integrate margin analysis in the Universal Journal and take full advantage of all existing functionalities. We also gave an outlook on functionalities yet to come in the future.

With this chapter, you should be able to activate margin analysis and configure the value flows to get a full P&L statement in margin analysis with all the characteristics you require to create meaningful reporting in your company.

In the next chapter, you'll learn how to configure costing-based profitability analysis.

Chapter 5
Costing-Based Profitability Analysis

Value fields and quantity fields are only required in costing-based profitability analysis. They are considered master data and define the structure of costing-based profitability analysis. The actual value flow determines how data from predecessor processes is transferred and displayed in costing-based profitability analysis. This chapter shows you how to set up the actual value flow correctly to get the most out of costing-based profitability analysis and be able to reconcile it with financial accounting.

The basic settings of costing-based profitability analysis not only set the foundation for the activation of costing-based profitability analysis but also define its structure. The characteristics, their derivation rules, and value fields define the master data of costing-based profitability analysis. This is why it's so important to put some time and effort in defining the foundation of costing-based profitability analysis. The removal of characteristics and value fields from the operating concern is only possible when there are no profitability documents with the characteristic and value field that you intend to delete. The characteristic and value fields have to be deleted from any configuration as well to be able to remove them from the operating concern.

Basic settings of costing-based profitability analysis

This chapter explains how to set up the actual value flow for costing-based profitability analysis. The actual value flow defines how costing-based profitability analysis receives data. This chapter describes how to transfer billing documents, cost of goods manufactured (COGM), and variances from production to costing-based profitability analysis. It also provides information on how to settle or allocate overhead costs, for example, internal orders and cost centers, in costing-based profitability analysis. We'll discuss the required configuration settings and confirm our configuration is working as intended with numerous examples.

Actual value flow

Costing-based profitability analysis won't be developed further and can only be activated on SAP S/4HANA—not SAP S/4HANA Cloud. Therefore, there are no new functionalities with SAP S/4HANA. It's highly recommended to activate margin analysis in parallel with costing-based profitability analysis instead of only activating costing-based profitability analysis in SAP S/4HANA.

Future of costing-based profitability analysis

5 Costing-Based Profitability Analysis

Reconciling with financial accounting In costing-based profitability analysis, the definition of actual value flows is a very important process to allow for reconciliation with financial accounting. All financial information comes together in costing-based profitability analysis based on your configuration. There are hardly ever any postings created directly in costing-based profitability analysis. Therefore, the prerequisite for a reconciliation with the general ledger (G/L) is that the value flows are correct and transparent.

Configuring costing-based profitability analysis When configuring costing-based profitability analysis, you define the interfaces to other components, for example, sales and distribution in SAP S/4HANA, but the prerequisites for an accurate configuration are properly setup processes in the components that trigger the document flow. It's critical that you understand and include the setup of these processes in the conceptual design of your costing-based profitability analysis, or you'll probably have to master various challenges when reconciling profitability analysis with financial accounting due to improperly setup processes in preliminary SAP components.

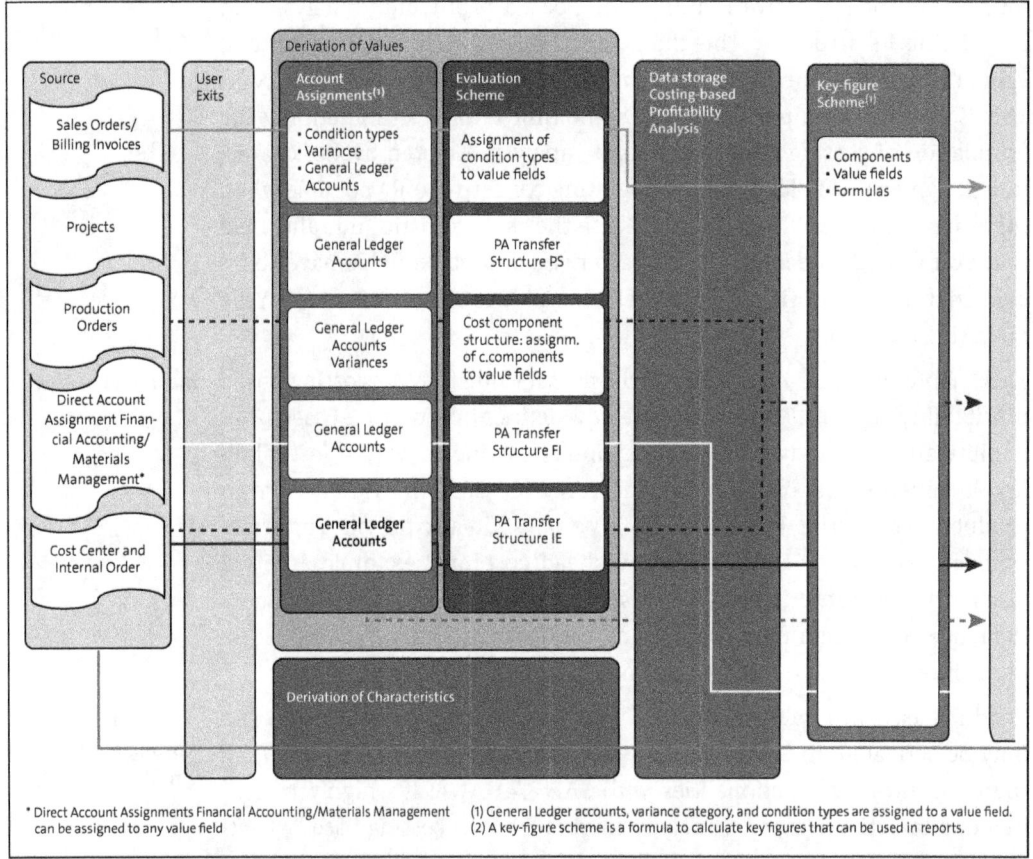

Figure 5.1 Upstream Processes in Costing-Based Profitability Analysis

Costing-Based Profitability Analysis

For costing-based profitability analysis, value fields are considered essential master data. The value fields represent the sales and costing blocks. Costing-based profitability analysis doesn't use G/L accounts to display amounts; instead, amounts are structured by function in value fields. Therefore, creating a 1:1 relationship between G/L accounts and value fields isn't recommended. Amounts on value fields can be broken down by characteristics. The value fields structure the profit contribution in costing-based profitability analysis.

Figure 5.1 shows an example of upstream processes and how they will result in postings on profitability segments in a sample report in costing-based profitability analysis, as shown in Figure 5.2.

Upstream processes

Figure 5.2 Sample Report in Costing-Based Profitability Analysis

In this chapter, we'll explain all the **Source** objects shown in Figure 5.1 on the left side, such as sales order, billing invoices, cost center, and so on, and how they are transferred to costing-based profitability analysis. The transfer

211

5 Costing-Based Profitability Analysis

happens in the **Derivation of Values**, which is also shown in this figure; you can see different **Account Assignments** and **Evaluation Schemes** that are the basis for the transfer of the values.

In parallel with the transfers of values, the **Derivation of Characteristics** occurs. We explained how characteristics are created and derived in Chapter 3. The **Data Storage Costing-Based Profitability Analysis** happens in separate data tables that are created when the operating concern is activated for costing-based profitability analysis. The value fields and formulas you can create are available in the reporting of costing-based profitability analysis. We show you how to do this in Chapter 7. A sample report is displayed in Figure 5.2. Here, BU stands for *business unit*.

Transferring data to costing-based profitability analysis

In this chapter, you'll learn how to configure costing-based profitability analysis to transfer data from its upstream processes. At the beginning of each section, we'll describe the upstream processes before we explain the configuration. We think it's critical for a correct configuration that you understand the processes to create a meaningful and correct costing-based profitability analysis.

> **Number of Value Fields**
>
> Technically you can assign up to 200 value fields to the operating concern.

5.1 Value Fields

Value fields: essential master data

Value fields are created in configuration, and they define the reporting structure in costing-based profitability analysis. There are two types of value fields:

- Value fields for amounts
- Value fields for quantities (also known as quantity fields)

In this chapter, you'll learn how to create value fields for amounts. In Chapter 5, you'll learn how you can assign amounts to those value fields.

Create value fields: Transaction KEA5

To create value fields, call Transaction KEA5, or follow the configuration path **Controlling** • **Profitability Analysis** • **Structures** • **Define Operating Concern** • **Maintain Value Fields**. Transaction KEA5 is used for the creation, change, and display of both characteristics and value fields. After calling the transaction, click on **Value fields** next to the box where you can enter a transaction, or press [Shift]+[F2] to switch the screen to the value field maintenance.

In Figure 5.3, you see the selection screen to create, change, or display a value field. There are two areas in the screen.

Value field maintenance

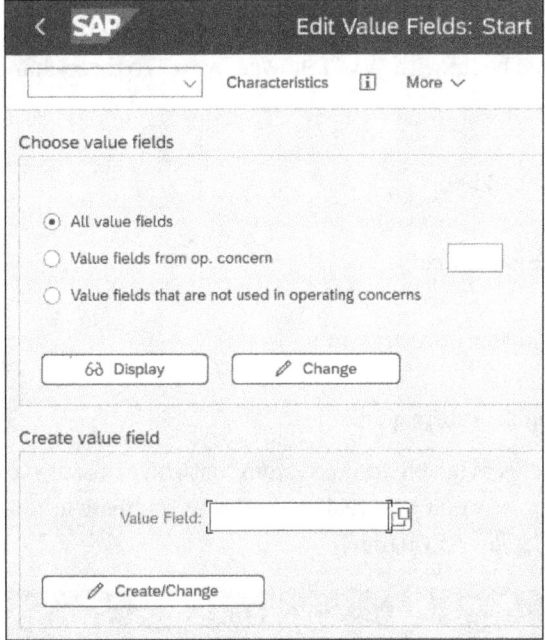

Figure 5.3 Selection Screen to Create/Change and Display Value Fields

You can jump directly to the lower **Create value field** area if you know that you want to create a new value field. All custom-created value fields have to start with "VV". The value field must be between four and five characters long. If you first want to take a look at what value fields are available in the system, go to the upper **Choose value fields** area where you can choose to display the following:

- **All value fields**
  This option will display all value fields that are available in the client, whether they are used or not.
- **Value fields from op. concern**
  This option will display value fields that are used in an operating concern that you specify.
- **Value fields that are not used in operating concerns**
  This option displays all value fields that aren't assigned to any operating concern available in the system.

In the example in Figure 5.3, choose to display **All value fields** that are available in the client, and click on **Display**.

5 Costing-Based Profitability Analysis

Value field details
In Figure 5.4, you can see a list of all the value fields available in the system. The list has the following columns:

- **Value Field**
  Four- to five-character-long technical name of the value field. All value fields that you've created begin with "VV" followed by two to three characters.

- **Description**
  Description of the value field.

- **Short text**
  Short description of the value field.

- **Amount**
  Selected if the value field will store amounts.

- **Qty**
  Selected if the value field will store quantities.

Change value fields
Take a look at the value fields available in the system that you can reuse to build up your operating concern. If you realize you want to create a new value field, click on 🖉 (Display <-> Change).

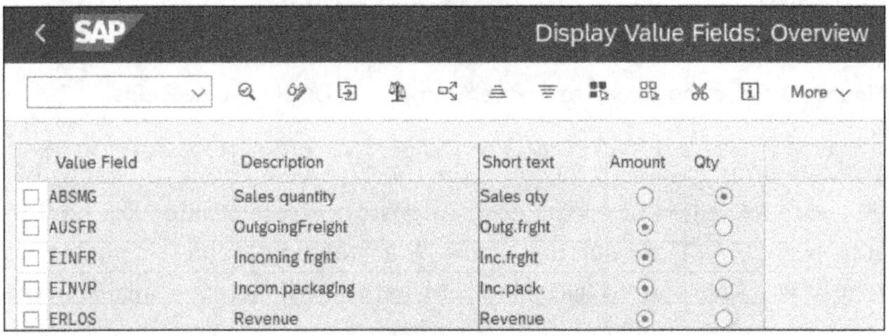

Figure 5.4 List of All Value Fields Available in the Client

Impact of changing cross-client data structures
You'll see the popup shown in Figure 5.5 warning that you're processing cross-client data structures. That means if you're creating a new value field in a development client of a productive system, all processes that involve costing-based profitability analysis, such as the creation of an invoice, the settlement of an order, and so on, will be locked. This is why the maintenance of the operating concern and its data structure must be treated with caution. In addition, configuration transports lock any processes that involve costing-based profitability analysis, which is why you should transport when no one is executing processes to avoid stopping processes or, even worse, data inconsistencies.

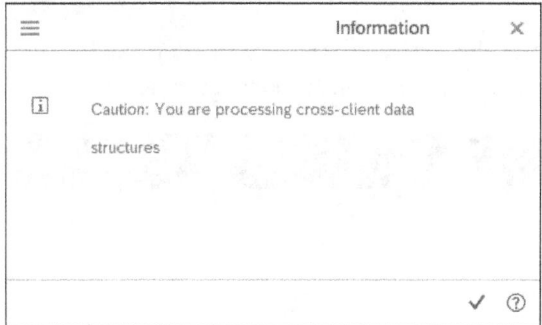

Figure 5.5 Warning Message for Cross-Client Data Structures

You can confirm the warning message with [Enter]. The menu bar will look slightly different, but you can click on 🗋 (Create) or press [F7] to create a new value field. In Figure 5.6, enter the technical name of the value field (here, "VV010") and the description "Disposal". In addition, select the **Amount** radio button to create a value field for amounts.

Create value field

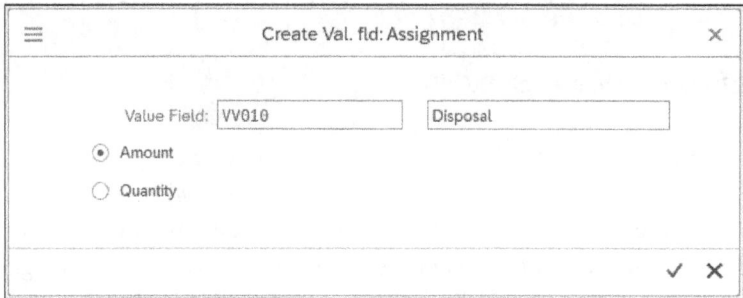

Figure 5.6 Create a Value Field

Confirm your entries with [Enter], and you'll be able to maintain further details for the value field VV010. In Figure 5.7, enter a **Short text** ("Disposal") and the **Aggregation** ("SUM") of the value field.

Maintain value field details

The aggregation defines how the amounts are saved in the value field. SAP offers the following types of aggregation for amount value fields:

- **SUM – Summation**
  The value field adds all values that are stored. This is the default setting for value fields and should be used for all amount value fields.

Maintain the aggregation

- **AVG – Average (All Values)**
  The value field determines the average value of all values. This is useful for certain key figures, such as the average number of employees.

5 Costing-Based Profitability Analysis

- **LAS – Last Value**
  The value field only stores the last value. This aggregation can be used, for example, if interest rates or other monthly key figures are stored.

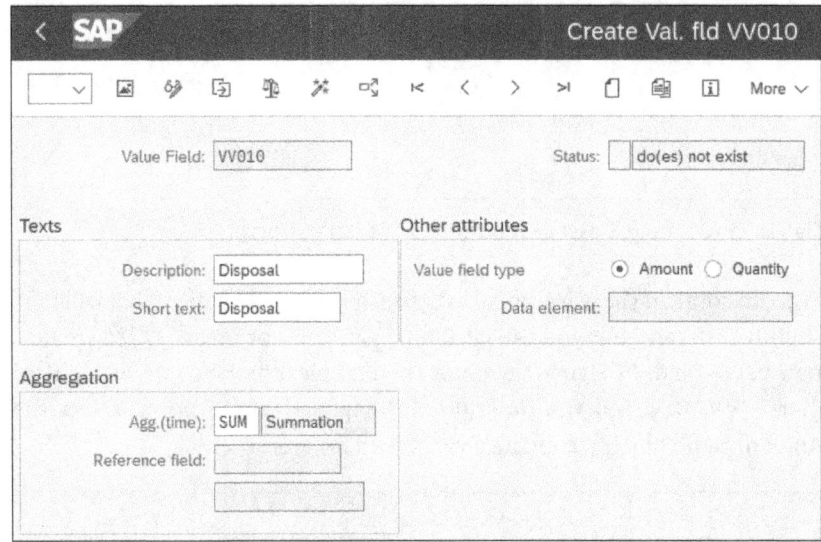

Figure 5.7 Maintain Value Field Details

Now, you can save your entries with **Save** or Ctrl+S. Before you leave the screen, don't forget that you also have to activate your newly created value field with (Activate) or Ctrl+F3. Before you can use the value field in the configuration of costing-based profitability analysis, you need to assign it to the operating concern, which you'll learn how to do in Section 5.3.

5.2 Quantity Fields

Value field for quantities
In costing-based profitability analysis, you have to create a value field for quantities to be able to derive quantities. You create the quantity value field in the same way as the amount value field.

Create quantity field: Transaction KEA5
To create quantity value fields, call Transaction KEA5, or follow the configuration path **Controlling • Profitability Analysis • Structures • Define Operating Concern • Maintain Value Fields**.

In Figure 5.8, go to the **Create value fields** area, and enter the technical name of the value field to be created in **Value Field**. All custom value fields must begin with "VV". This example is creating a quantity field for the unit carton. The technical name of the quantity field is **VVKAR**. After entering the technical name, click on **Create/Change** to create the quantity field.

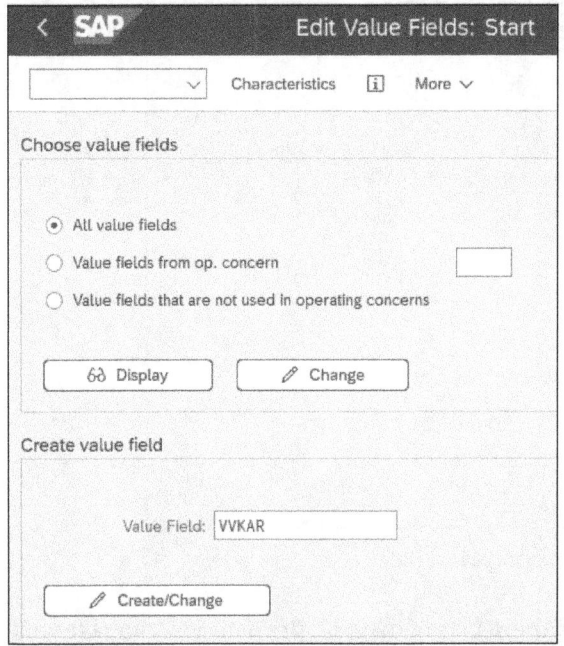

Figure 5.8 Create the Quantity Field

> **Quantity Fields**
> In costing-based profitability analysis, you can define up to nine quantity fields and assign them to the operating concern. For each quantity field, you create a value field with value field type **Quantity**.

In Figure 5.9, you can maintain the details of the quantity field as follows:

- **Description**
  Enter a description for the quantity field (in this example, "Carton").
- **Short text**
  Enter a short text with a maximum of 10 characters (in this example, "Carton").
- **Agg.(time)**
  Define how the quantities will be saved. In this example, choose **SUM – Summation** so that the quantities on this quantity field will be added up by period/fiscal year. In other words, the quantities in **VVKAR** will be added up with **SUM** as the aggregation.
- **Value field type**
  Define whether the value field receives amounts or quantities. The value field **VVKAR** is defined as a **Quantity** field.

Maintain quantity field details

5 Costing-Based Profitability Analysis

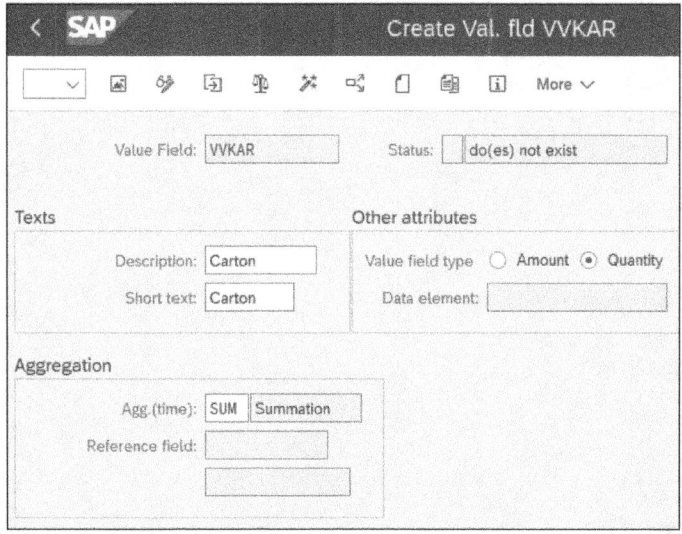

Figure 5.9 Maintain the Quantity Field Details

Activate the quantity field

After you've maintained the details of the quantity field, you can save your changes. Don't forget to activate the quantity field with ![icon] (Activate) or [Ctrl]+[F3]; otherwise, you won't be able to assign the quantity field to the operating concern. The assignment of the quantity field to the operating concern is essential to use the quantity field in costing-based profitability analysis.

5.3 Assigning Value Fields and Quantity Fields to the Operating Concern

Change the operating concern data structure: Transaction KEAO

To use value fields and quantity fields in your costing-based profitability analysis, you have to assign them to the operating concern by calling Transaction KEAO or following the configuration path **Controlling • Profitability Analysis • Structures • Define Operating Concern • Maintain Operating Concern**. To assign a value field to an operating concern, you follow the same steps as for assigning characteristics to the operating concern, which we explained in Chapter 3.

5.4 Transferring Incoming Sales Orders

Definition of incoming sales orders

When the customer orders goods, the sales team creates a sales order in SAP. The sales order is the purchase order for the customer. It has all the information about the customer, who is receiving the goods, who is paying for the invoice, and so on. The sales order also has information about which

218

5.4 Transferring Incoming Sales Orders

goods the customer ordered and in which quantity. When the sales order is saved, no financial document is created because there has been no delivery or invoice created yet.

In costing-based profitability analysis, you can transfer the incoming sales orders to profitability analysis. The incoming sales order gives you some predictive data in costing-based profitability analysis, and you can see how sales are developing and can react in a timely manner if you see the numbers don't align with your plan. The incoming sales orders are transferred with a separate record type to not mix up actual sales with incoming sales orders.

Before we start looking at the configuration, let's look at a sales order. In Figure 5.10, you see the sales data of a sales order. In the header of the sales order, you can see that a **Sold-to Party (1000043)** and **Ship-To Party (1000043)** have been assigned. The **All Items** area shows the **Material (364)** and quantity that the customer ordered. Entering the material and ordered quantity triggers the pricing process in the sales order. Select the item **10** in the sales order, and click on 🖼 (Item Conditions) to see an overview of the conditions used for pricing.

Sales order structure

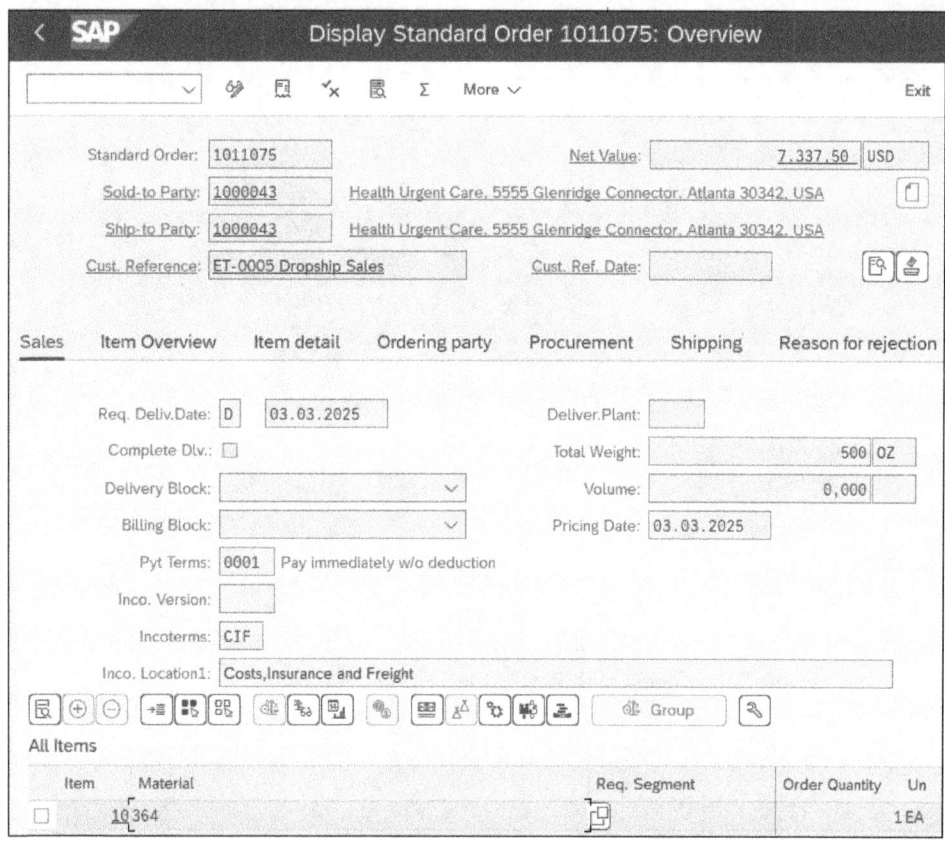

Figure 5.10 Sales Data of the Sales Order

5 Costing-Based Profitability Analysis

Pricing conditions in the sales order In Figure 5.11, you see the pricing conditions that make up the price of the material. Each condition is a pricing element, and the corresponding costing sheet that makes up the pricing structure is defined in the sales order type. In sales and distribution, the condition values are defined as master data in condition maintenance. Some costing conditions calculate the values on the basis of a formula (e.g., percentage surcharge), and there are also conditions to which you can assign values manually.

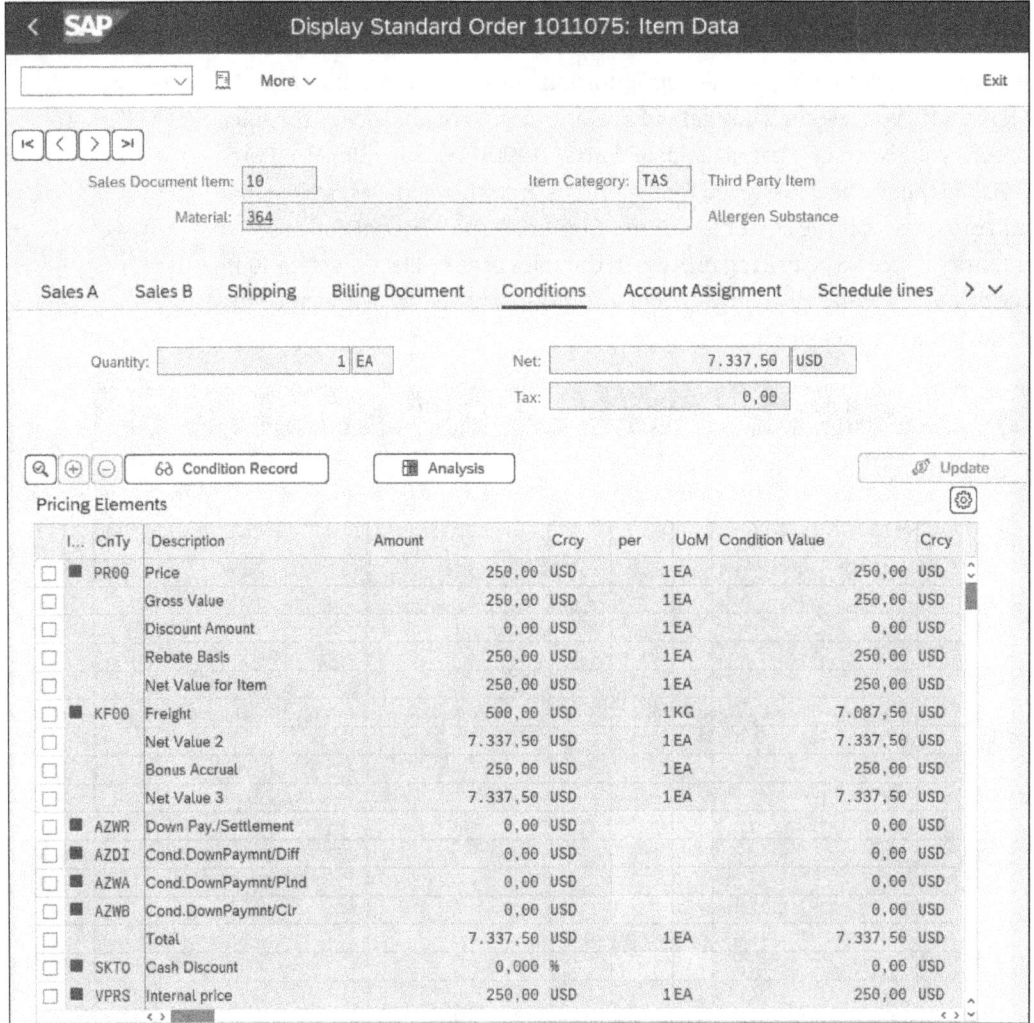

Figure 5.11 Pricing Conditions in the Sales Order

In Figure 5.11, the following conditions are used to calculate the price:

- **PR00 – Price**
  This is the sales price to the customer.

- **KF00 – Freight**
  This is a condition that calculates freight based on the weight of the goods to be shipped. The weight is maintained in the material master.
- **SKTO – Cash Discount**
  The cash discount is based on the payment term that is maintained in the sales view of the customer. In this example, no cash discount is calculated.
- **VPRS – Internal price**
  The cost of the material is the copied price of the material master used also for inventory valuation.

For the sales order to transfer incoming sales orders to costing-based profitability analysis, all relevant conditions have to be assigned to a value field in configuration. We'll show you in this section how to assign value fields to sales conditions.

In Figure 5.12, the **Account Assignment** tab shows whether the correct account assignment has been derived. For this example, a profitability segment has been created as you can see in the **Profit. Segment** field.

Account assignment in the sales order

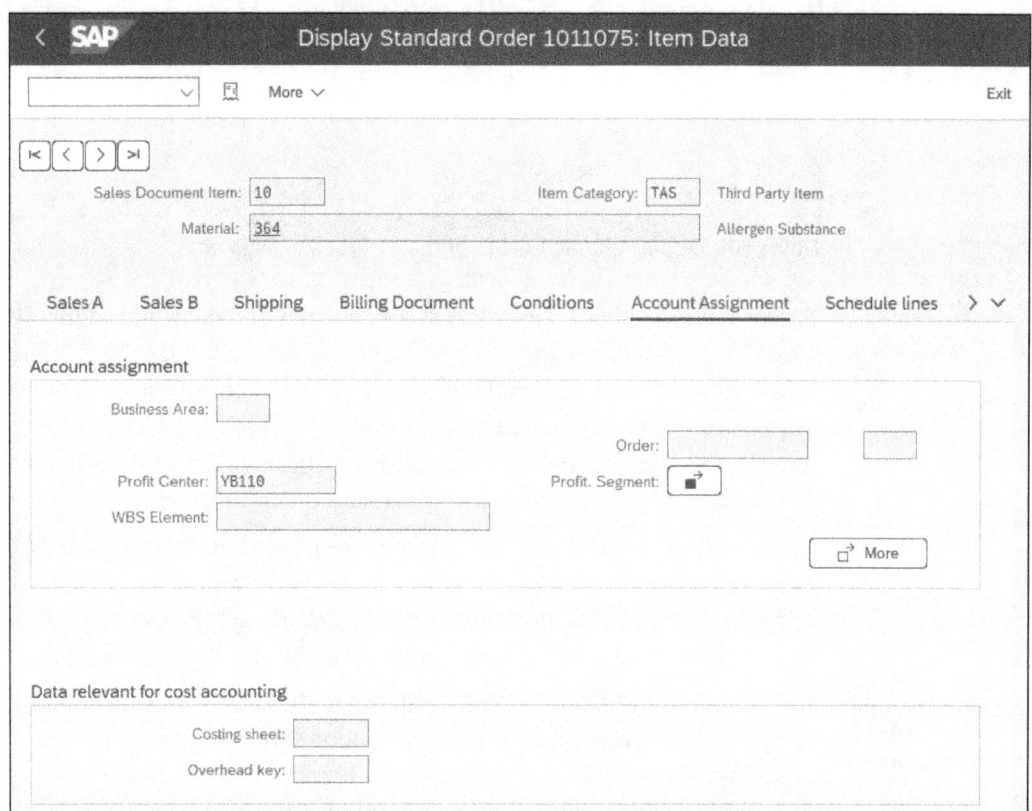

Figure 5.12 Profitability Segment in the Sales Order

5   Costing-Based Profitability Analysis

Assignment to the profitability segment

With ▣ (Display Account Assignment), you can display all characteristics that were derived from the sales order in Figure 5.13. If any data in the sales order changes for a field that is also a characteristic for costing-based profitability analysis, the profitability segment will be updated.

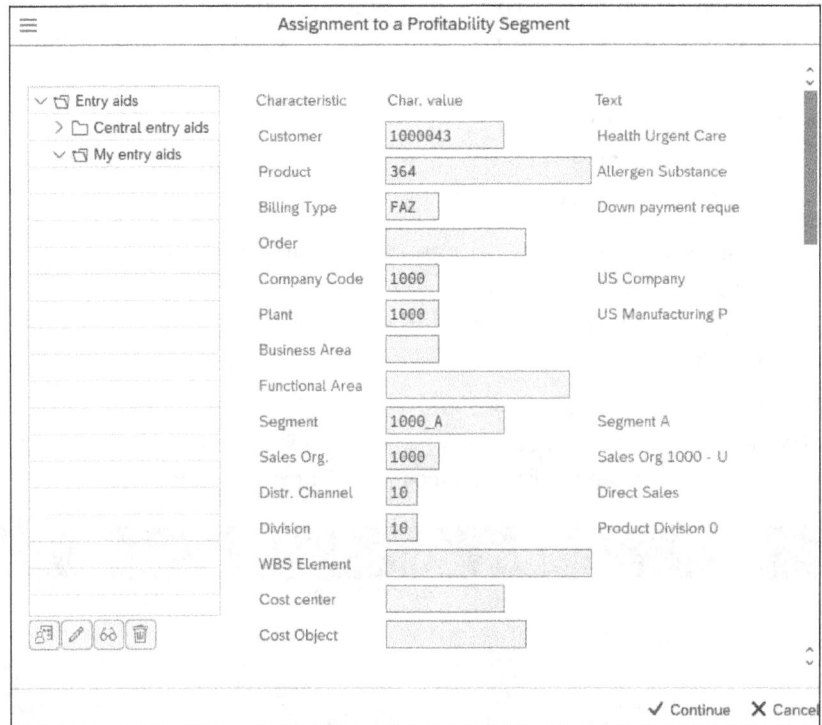

Figure 5.13 Display Characteristics of the Profitability Segment

Assign pricing conditions to value fields: Transaction KE4I

To transfer incoming sales orders as well as sales and distribution in SAP S/4HANA billing documents to costing-based profitability analysis, value fields need to be assigned to all sales conditions. Incoming sales orders are mapped to the same value fields as revenues but are stored in a separate record type when transferred to costing-based profitability analysis so that you can differentiate between incoming sales orders and real sales.

To transfer the incoming sales orders to costing-based profitability analysis, go to Transaction KE4I, or use the configuration path **Controlling • Profitability Analysis • Flows of Actual Values • Transfer Incoming Sales Orders • Assign Value Fields • Maintain Assignment of SD Conditions to CO-PA Value Fields**.

Assign material price to value fields: Transaction KE4I

In Figure 5.14, value fields are being assigned to the conditions shown earlier in Figure 5.11 with **New Entries** or F5. Condition **PR00 – Price** is also assigned to transfer the material cost with the sales to calculate a margin in costing-based profitability analysis. Save your entries with **Save** or Ctrl+S.

5.5 Transferring Quantity Fields

Figure 5.14 Assigning Pricing Conditions to Value Fields

5.5 Transferring Quantity Fields

In addition to pricing conditions, you can also transfer the quantity of the incoming sales orders to costing-based profitability analysis. You can transfer various quantity fields from sales and distribution in SAP S/4HANA.

Assign quantities to quantity fields: Transaction KE4M

To transfer quantity fields, go to Transaction KE4M or follow the configuration path **Controlling · Profitability Analysis · Flows of Actual Values · Transfer Incoming Sales Orders · Assign Quantity Fields**.

Types of quantity fields

You can transfer the following quantity fields to profitability analysis:

- **BRGEW (Gross Weight)**
  The gross weight of the materials is transferred from the sales order. The system calculates the gross weight on the basis of the material master basic data.

- **FKIMG (Billed Quantity)**
  The system uses the billed quantity from the customer billing document after the sales order has been billed. You should transfer this quantity as the sales quantity to costing-based profitability analysis.

223

- **FKLMG (Billing Quantity in SKU)**
  The billing quantity in stock keeping unit (SKU) is transferred from the billing document. The unit of measure in sales and distribution may deviate from the unit of measure in the warehouse. The SKU corresponds to the base unit of measure in the material master.

- **KBMENG (Cumul. Confirmed Quantity)**
  The cumulative confirmed quantity in the sales unit of measure is the quantity from the sales order that has been confirmed via the availability check.

- **KLMENG (Cumul. Confirmed Quantity)**
  The cumulative confirmed quantity in the standard unit of measure is the quantity from the sales order that has been confirmed via the availability check.

- **KWMENG (Order Quantity)**
  The cumulative order quantity in the sales unit of measure is the order quantity from the sales order.

- **LSMENG (Required Delivery Quantity)**
  The required delivery quantity is the sales order quantity that is relevant to deliveries and confirmed for delivery.

- **NTGEW (Net Weight)**
  The net weight is the gross weight minus the weight of the packaging. This default value is transferred from the basic data of the material master and multiplied by the order quantity.

- **VOLUM (Volume)**
  The volume is calculated by multiplying the order quantity of the item by the volume from the basic data of the material master.

In Figure 5.15, the **SD qty field KWMENG – Order Quantity** and **FKIMG – Invoiced Quantity** are assigned to **CO-PA qty field ABSMG** in costing-based profitability analysis. Save your entries with **Save** or Ctrl + S.

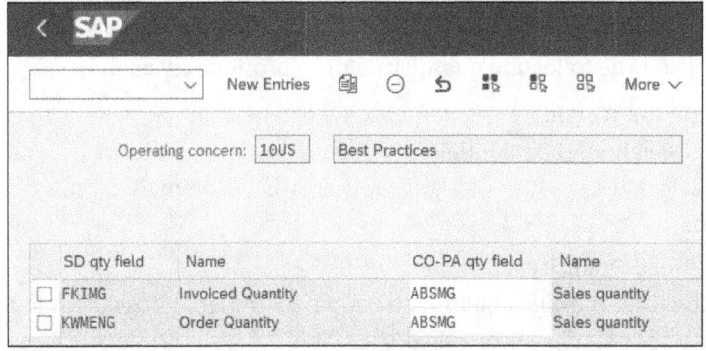

Figure 5.15 Assign SD Quantity Fields to Quantity Fields in Profitability Analysis

5.5 Transferring Quantity Fields

After assigning the sales and distribution in SAP S/4HANA conditions and sales and distribution in SAP S/4HANA quantity fields to costing-based profitability analysis value and quantity fields, you have to activate the transfer of incoming sales orders in costing-based profitability analysis. Go to Transaction KEKF or follow the configuration path **Controlling • Profitability Analysis • Flows of Actual Values • Transfer Incoming Sales Orders • Activate Transfer of Incoming Sales Orders**.

Activating incoming sales orders: Transaction KEKF

In Figure 5.16, you see an overview of all controlling areas (**COAr**), including from which year forward they are valid (**From FY**) as well as their operating concerns (**Op.co**). To activate the transfer of incoming sales orders, you have to select a status in the **Inc.SO** (incoming sales orders) column.

Options for activating incoming sales orders

The following options are available:

- **(blank/no entry): Inactive**
  If the **Inc.SO** column is empty, the transfer of incoming sales orders is inactive.
- **1 (Active with Date of Entry)**
  Incoming orders during the entry period are transferred to profitability analysis when the sales order is entered and saved.
- **2 (Active with Deliv.Date/Billing Plan Deadline [Using KWMENG])**
  The cumulative order quantity is transferred to profitability analysis during the delivery date period.
- **3 (Active with Deliv.Date/Billing Plan Deadline [Using KBMENG])**
  The cumulative order quantity is transferred to profitability analysis during the delivery date period confirmed by the availability check.

In Figure 5.16, you can also see the **Final Billing Update** column. If you set the checkmark in this column, the sales order quantity is updated when the quantity in the sales order is changed even if the sales order item has already been fully billed.

Final billing update

You can enable the **Hist. Rate f. Orders** (historical rate for order changes) column if you always want to use the exchange rate with the date of the order creation to convert the incoming sales orders. Otherwise, SAP standard uses the exchange rate that was valid on the date of creation of the sales order if no rate has been defined in the sales order. If you don't want to create exchange rate differences when adjusting the order quantity, you should set this checkmark.

Exchange rate for order changes

In Figure 5.16, the transfer of incoming sales orders in costing-based profitability analysis is activated for controlling area **1000**, which is assigned to operating concern **10US**. You activate the transfer for incoming sales orders with the date of entry = **1**. Save your entries with **Save** or [Ctrl]+[S].

Incoming sales orders with date of entry

5   Costing-Based Profitability Analysis

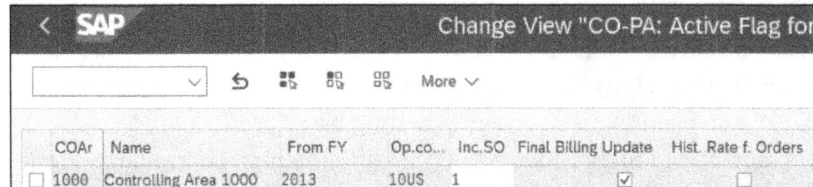

Figure 5.16   Activate Transfer of Incoming Sales Order in Costing-Based Profitability Analysis

Document flow of sales order

The incoming sales order should now have been created for costing-based profitability analysis. Let's take a quick look at the document flow of the sales order in Figure 5.17 to ensure no finance document has been created, which isn't the case because there are no additional documents assigned to the sales order.

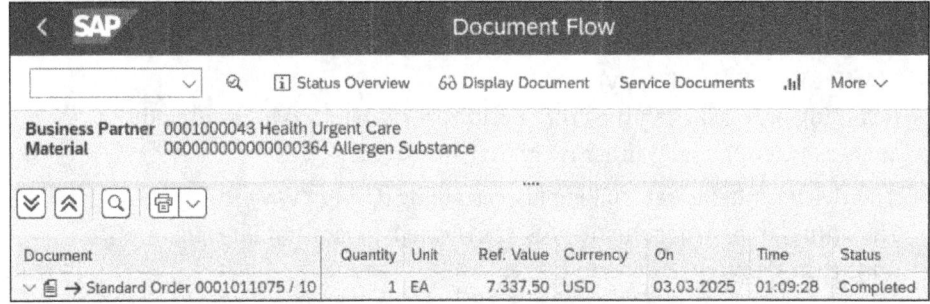

Figure 5.17   Document Flow of the Sales Order

Record type A (incoming sales order)

Let's check the document for the incoming sales order in costing-based profitability analysis. Incoming sales orders are transferred with record type A (incoming sales order) to be able to differentiate it from real sales, which are transferred with record type F (billing data).

Figure 5.18 shows the costing-based profitability analysis document that was generated when the sales order from Figure 5.10 was saved.

Display actual line items: Transaction KE24

You can display costing-based profitability analysis documents with Transaction KE24 or by following the menu path **Accounting • Controlling • Profitability Analysis • Information System • Display Line Item List • KE24 – Actual**. You can restrict the selection screen on **Record Type A**. In Figure 5.18, you see the **Characteristics** tab of the profitability segment. All characteristics that have been derived from the sales order and characteristics derivations of the operating concern are assigned to the profitability segment.

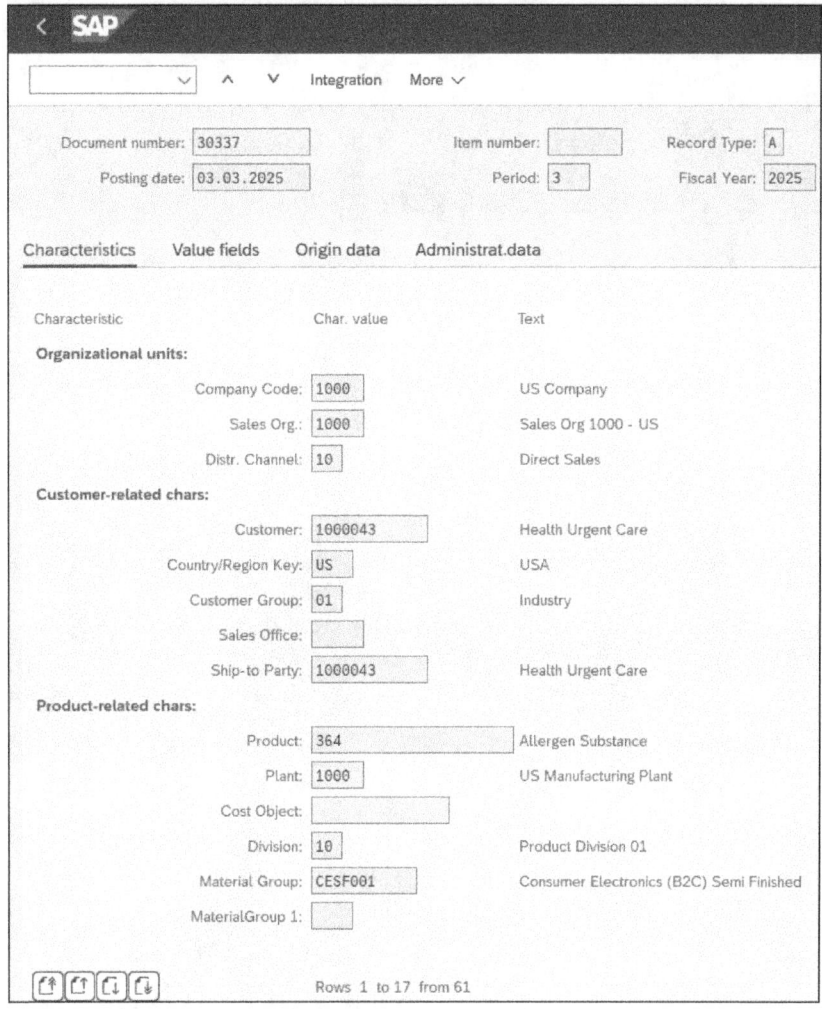

Figure 5.18 Profitability Segment Characteristics

In the **Value fields** tab in Figure 5.19, you see all value and quantity fields of the operating concern. We can see that the revenue has been derived into value field **Revenue (250.00)**. All other values and quantities have been derived in the respective fields as well, but they aren't visible in this screenshot.

Displaying profitability segment value fields

> **Note**
> Both cost and revenue in costing-based profitability analysis are displayed as positive values.

5   Costing-Based Profitability Analysis

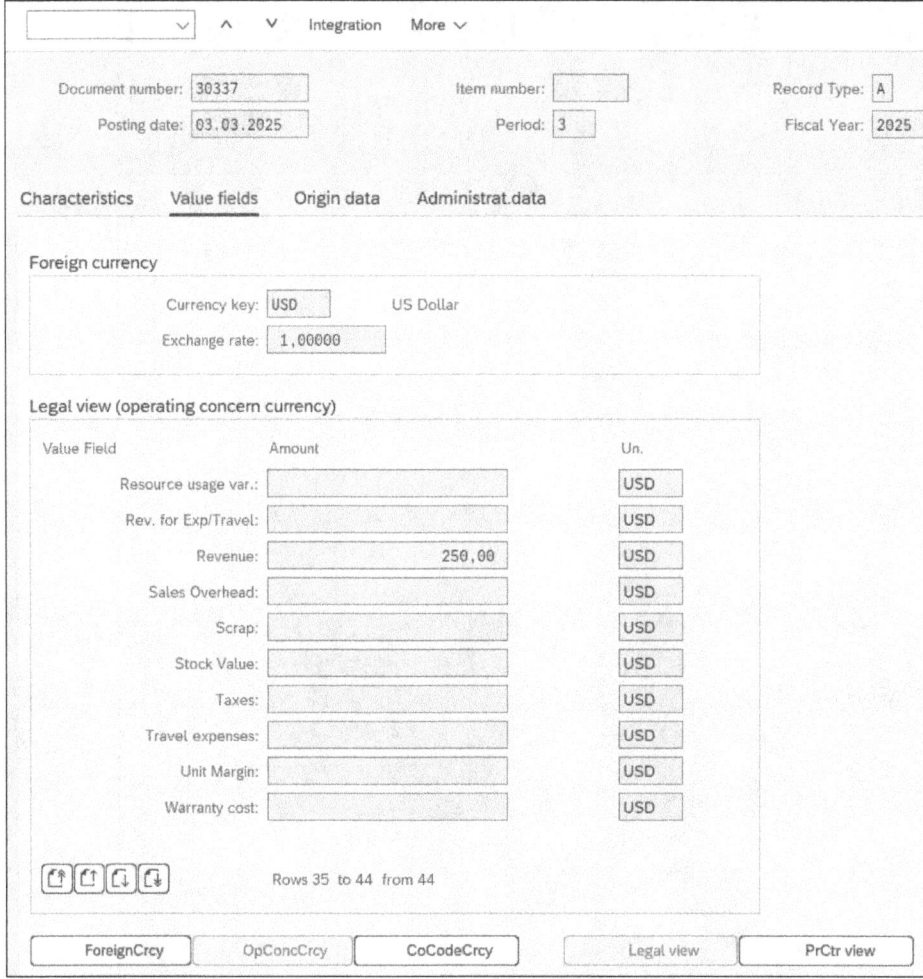

Figure 5.19 Profitability Segment Value Fields

Displaying profitability segment origin data

In Figure 5.20, you see the **Origin data** tab of the profitability segment, which includes the original documents from sales and distribution or controlling that generated the creation of the profitability segment.

Displaying profitability segment administrative data

The next tab, **Adminstrat.data**, of the profitability segment shows the reference document that, in this case, is also the sales order (see Figure 5.21). For a billing invoice, the system would show here the billing invoice number, and for a settlement of an internal order, the system would show the internal order number. You also see information on when the profitability segment was created and by whom.

5.5 Transferring Quantity Fields

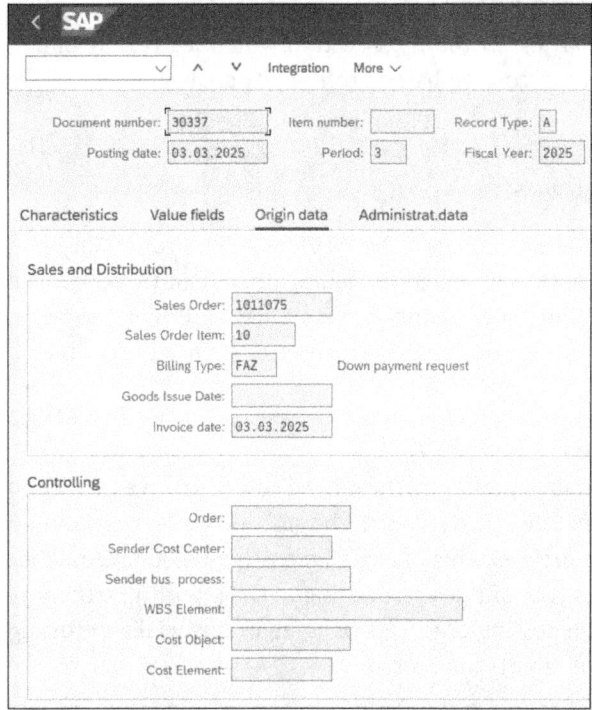

Figure 5.20 Profitability Segment Origin Data

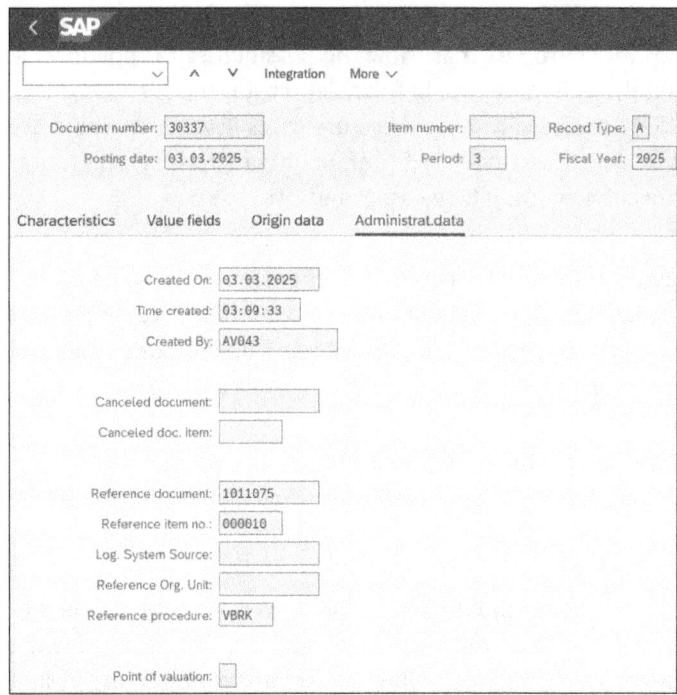

Figure 5.21 Profitability Segment Administrative Data

5 Costing-Based Profitability Analysis

Now you've learned how incoming sales orders get transferred to costing-based profitability analysis. In the next section, you'll learn how billing invoices are transferred to costing-based profitability analysis.

5.6 Billing Data Transfer

Billing document in costing-based profitability analysis

In costing-based profitability analysis, the system creates profitability segments for the billing document when the billing document is posted. The configuration for the transfer of billing data is similar to the configuration for the transfer of incoming sales orders, which we explained in Section 5.4.

5.6.1 Transferring Cost of Goods Sold

Cost of sales accounting

In costing-based profitability analysis, the related cost of goods sold (COGS) is transferred to profitability analysis with the sales revenue of the billing documents (at the same time). This means that costing-based profitability analysis only supports cost of sales accounting in the reporting structure, and there's a lag in the posting of the goods issue in financial accounting and costing-based profitability analysis.

Cost of goods sold (COGS)

The transfer of COGS together with sales revenue is a challenge in the reconciliation with the G/L, especially when the goods issue is already posted in financial accounting but isn't transferred to costing-based profitability analysis because the billing invoice hasn't yet been created. There are different scenarios in Figure 5.22 that show the possibilities for posting the goods issue and the billing invoice in financial accounting versus costing-based profitability analysis. Among these use cases listed here, there are two scenarios in which costing-based profitability analysis can be reconciled with financial accounting (Use Case A and Use Case D):

- **Use Case A**
  The goods issue in financial accounting gets posted in the same month as the billing invoice. The billing invoice is transferred to costing-based profitability analysis with the COGS in the same month: financial accounting and costing-based profitability analysis can be reconciled.

- **Use Case B**
  COGS and sales are posted in different months in financial accounting. Financial accounting and costing-based profitability analysis can't be reconciled.

- **Use Case C**
  The goods issue is posted in a different month from the billing invoice in financial accounting. The goods issue gets accrued and reversed when the billing invoice is posted. The billing invoice and the COGS are posted in costing-based profitability analysis, but the COGS is evaluated with

the price that is valid at the date of the billing invoice. This can lead to differences between financial accounting and costing-based profitability analysis as the price for the material being sold can change on a monthly basis. A reconciliation in Use Case C between financial accounting and costing-based profitability analysis isn't guaranteed.

- **Use Case D**
  This is similar to Use Case C with the difference that the valuation of the COGS in costing-based profitability analysis occurs with the date of the goods issue. So, if your goods issue valuation changes on a monthly basis, Use Case D will guarantee you the reconciliation between financial accounting and costing-based profitability analysis.

To summarize, Use Case A and Use Case D will deliver the basis for a reconciliation between financial accounting and costing-based profitability analysis. You will learn in Section 5.8 how to ensure that the COGS will be valuated with the date the goods issue was posted.

Use Case	Date	Financial Accounting (VA: Goods Issue)	Profitability Analysis (VA: Billing Invoice)	Recon. Financial Accounting/ Profitability Analysis*
Use Case A Goods issue and billing invoice in the same month	August = WA + F	Sales and COGS Posting	Sales and COGS Posting	✓
Use Case B SAP ERP FI does not post an accrual for Goods in Transit	August = WA	COGS	–	✗
	September = F	Sales	Sales and COGS Posting	✗
Use Case C SAP ERP FI does post an accrual for Goods in Transit	August = WA	Accrual of Goods in Transit	–	✓
	September = F	Sales Posting + Reversal of Accrual (= COGS at the time of the goods issue)	Sales and COGS Posting (=COGS at the time of billing invoice)	✗

Goods in Transit has to be accrued; due to time of assessment there are still differences (Use Case C).

Use Case	Date	Financial Accounting (VA: Goods Issue)	Profitability Analysis (VA: Billing Invoice)	Recon. Financial Accounting/ Profitability Analysis*
Use Case D Change of time of assessment in profitability analysis is to date of goods issue and accrual in SAP ERP FI	August = WA	Accrual of Goods in Transit	–	✓
	September = F	Sales + Reversal of Accrual (= COGS at the time of goods issue)	Sales and COGS Posting (= COGS at the time of billing invoice)	✓

✗ difference between financial accounting and profitability analysis   F = Billing invoice   VA = Record type   WA = Goods issue
✓ Financial accounting and profitability analysis reconcile

Figure 5.22 Scenarios for COGS and Sales Posting in Costing-Based Profitability Analysis

5.6.2 Assigning Value Fields to Pricing Conditions

Map pricing conditions to value fields

Because costing-based profitability analysis doesn't use G/L accounts, you have to map all revenues and costs to value fields. The values from sales and distribution in SAP S/4HANA billing documents are transferred to profitability analysis the same way as values from incoming sales orders. The pricing conditions in a sales order have to get assigned to value fields in configuration. You can transfer both statistical conditions and "real" conditions to costing-based profitability analysis. To check if a condition is statistical or not, you can double-click on a condition, as shown earlier in Figure 5.11. In Figure 5.23, look at the pricing condition **SKTO** for the **Cash Discount**. In the **Control** area, you can see that the **Statistical** checkbox is checked, indicating that this is a statistical condition that doesn't trigger a posting in financial accounting.

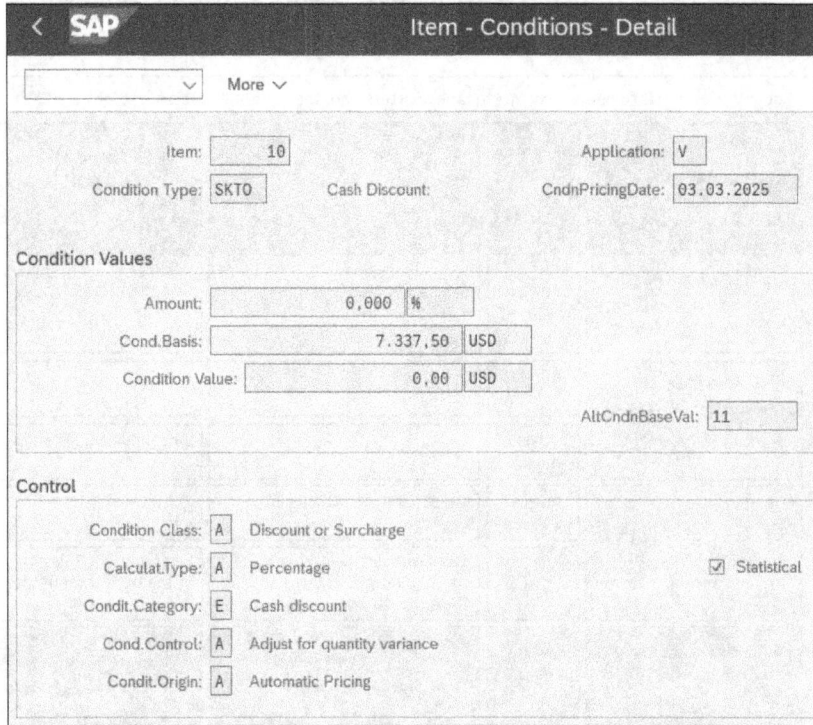

Figure 5.23 Statistical Condition in Sales and Distribution in SAP S/4HANA

Statistical conditions

The payment term of the customer master is derived in the sales order and displayed in the sales data, as shown earlier in Figure 5.10. In this example, the payment term is **0001 – Pay immediately w/o deduction**, so we don't see the statistical condition for cash discount in Figure 5.11. Based on the payment term, the cash discount is calculated in the sales order and deducted from the profit margin in Figure 5.11. No financial document is triggered because this is a statistical condition. You can transfer the statistical condition to costing-based profitability analysis to analyze the cash discount at the customer level. Don't mix the statistical values with "real" values though because you'll never be able to reconcile financial accounting with costing-based profitability analysis if you combine statistical and real values in one value field. In financial accounting, the cash discount is posted when the client's payment is received. As clients often combine the payment of multiple billing invoices in one payment, it's not possible to divide the cash discount by sales order. Usually, the cash discount is posted with direct account assignment or via a cost center assessment to costing-based profitability analysis. You have to make sure that you don't consider any statistical values in the reconciliation of financial accounting with costing-based profitability analysis.

Consider the condition **PR00** for the **Price**, as shown in Figure 5.24. There is no **Statistical** checkbox, but there is a new **Account Determination** area with an **Account Key** field. The account keys are used in the sales and distribution in SAP S/4HANA account determination to trigger the correct G/L account for the financial posting.

Details of the non-statistical condition

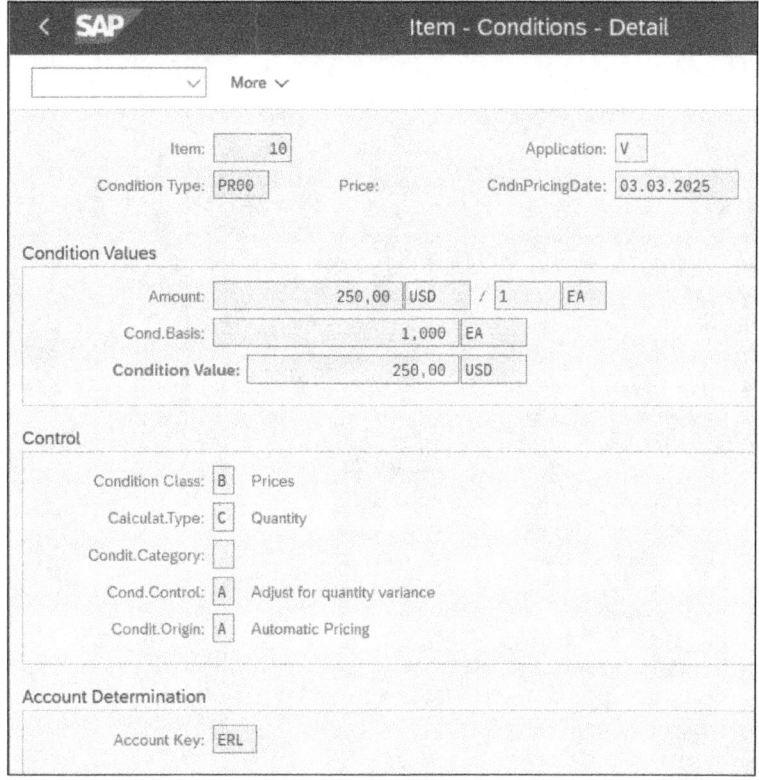

Figure 5.24 Nonstatistical Condition in Sales and Distribution in SAP S/4HANA

5   Costing-Based Profitability Analysis

Collaboration with the sales team

It's important that you collaborate with your colleagues from sales and distribution to create the conditions in a way that allows you to display the required level of detail both in financial accounting and costing-based profitability analysis. All conditions that you don't assign to a value field won't be transferred with the billing invoice to costing-based profitability analysis, which can lead to difficulties in the reconciliation of costing-based profitability analysis with financial accounting as the sales values will be either too high or too low.

Assigning conditions to value fields: Transaction KE4I

To assign sales and distribution in SAP S/4HANA conditions to value fields, go to Transaction KE4I or follow the configuration path **Controlling • Profitability Analysis • Flows of Actual Values • Transfer of Billing Documents • Assign Value Fields • Maintain Assignment of SD Conditions to CO-PA Value Fields**.

You can assign new conditions to value fields in Figure 5.25 with **New Entries** or F4. In the **CnTy** (condition type) column, you enter the condition type and assign the profitability analysis value field in the **Val. fld** (value field) column.

Figure 5.25 Assign Pricing Conditions to Value Fields

Normally, the system lists all values with a positive sign in the value fields. However, in exceptional cases, it may be necessary to enable the transfer of plus/minus signs for condition values to profitability analysis. In this case, you need to select the checkbox in the **Transfer +/-** column. As already mentioned, you should only use this indicator in exceptional cases, such as if multiple condition types in one value field are supposed to be balanced with different signs (e.g., transfer price surcharge or loss) or conditions with differing signs are added to a value field (e.g., in the case of bonus agreements). There are conditions that can have negative and positive values, for example, bonus conditions.

Transfer plus/minus sign

5.6.3 Assigning Quantity Fields to Quantities

In addition to values, you can also transfer quantities to costing-based profitability analysis. To transfer quantities, go to Transaction KE4M or follow the configuration path **Controlling • Profitability Analysis • Flows of Actual Values • Transfer of Billing Documents • Assign Quantity Fields**.

Transferring quantity fields: Transaction KE4M

You can add quantities to quantity fields in costing-based profitability analysis in Figure 5.26 with **New Entries** or F5. The following sales and distribution in SAP S/4HANA quantity fields are available for selection:

Add quantities for transfer

- BRGEW (Gross Weight)
- FKIMG (Billed Quantity)
- FKLMG (Billing Quantity in SKU [in Stock Keeping Unit])
- KBMENG (Cumul. Confirmed Quantity)
- KLMENG (Cumul. Confirmed Quantity [in Stock Keeping Unit])
- KWMENG (Order Quantity)
- LSMENG (Required Delivery Quantity)
- NTGEW (Net Weight)
- VOLUM (Volume)

All quantities are derived from the sales order. Figure 5.26 shows sales and distribution in SAP S/4HANA quantity field **FKIMG (Invoiced Quantity)** assigned to quantity field **Sales quantity**. This quantity is filled with amounts when billing documents are transferred to costing-based profitability analysis. Sales and distribution in SAP S/4HANA quantity field **KWMENG (Order Quantity)** is also assigned to quantity field **ABSMG (Sales quantity)**. This quantity field is filled with amounts when incoming sales orders are transferred to profitability analysis. Incoming sales orders are posted with record type A and billing documents with record type F. This ensures that you can still differentiate between **Order Quantity** and **Invoiced Quantity** in costing-based profitability analysis.

5 Costing-Based Profitability Analysis

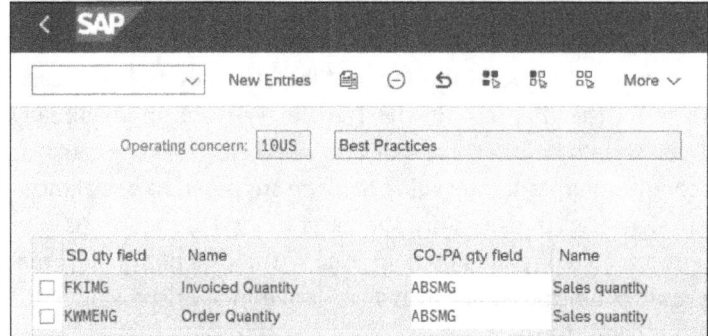

Figure 5.26 Assign Sales and Distribution Quantity Fields to Profitability Analysis Quantity Fields

Transfer alternative unit of measures: Transaction KE4MS

In addition to the sales and distribution quantities, you can also transfer alternative units of measure. To do so, both configuration and master data maintenance are required. First, let's look at the configuration. We assume you have at least two quantity fields created and assigned to your operating concern. With Transaction KE4MS or by following the configuration path **Controlling • Profitability Analysis • Flows of Actual Values • Initial Steps • Store Quantities in CO-PA Standard Unit of Measure**, you can maintain a different unit of measure.

In Figure 5.27, the **Source quantity** field **ABSMG** for the **Sales quantity** is assigned to the **Quantity** field **VVKAR** for **Carton**, which is the quantity field for the alternative unit of measure. For this example, all of our sales quantities will be displayed in the quantity of the material master as well as in the quantity unit **Carton**, which is maintained in the **Unit (CAR)** column. Save your entries with **Save** or [Ctrl]+[S].

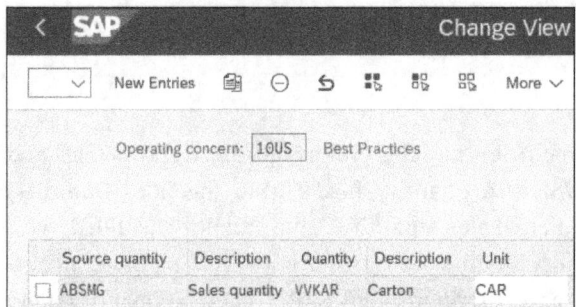

Figure 5.27 Maintain an Alternative Unit of Measure in Costing-Based Profitability Analysis

Display the material master: Transaction MM03

For the system to calculate the alternative quantity unit, you have to maintain an alternative unit of measure in the material master. Go to Transaction MM03 or follow the menu path **Logistics • Material Management •**

5.6 Billing Data Transfer

Inventory Management • Environment • Information • MM03 – Material to figure out if an alternative quantity unit is maintained.

Alternative quantity units are maintained in **Basic Data 1** of the material master. Follow →**Additional Data** (Go to Additional Data), and go to the **Units of Measure** tab. In Figure 5.28, you can see that an alternative unit of measure is maintained: 1 Carton contains 10 kg.

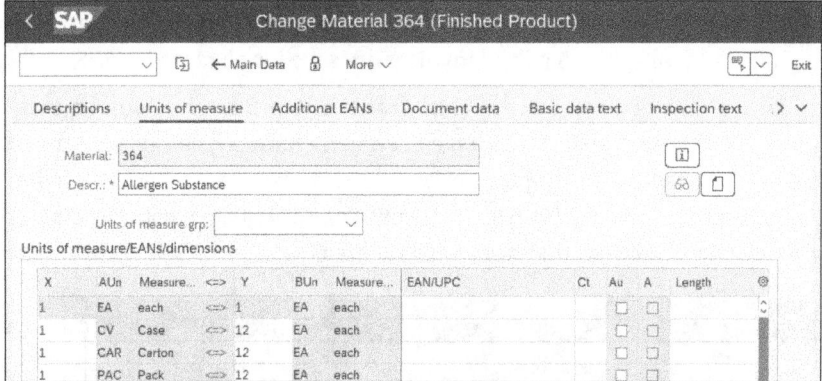

Figure 5.28 Display Alternative Units of Measure in the Material Master

Let's look at a sales order to be invoiced. In Figure 5.29, you can see that customer **1000043** has ordered 1 EA of **Material 364** that will be shipped to customer **1000043**.

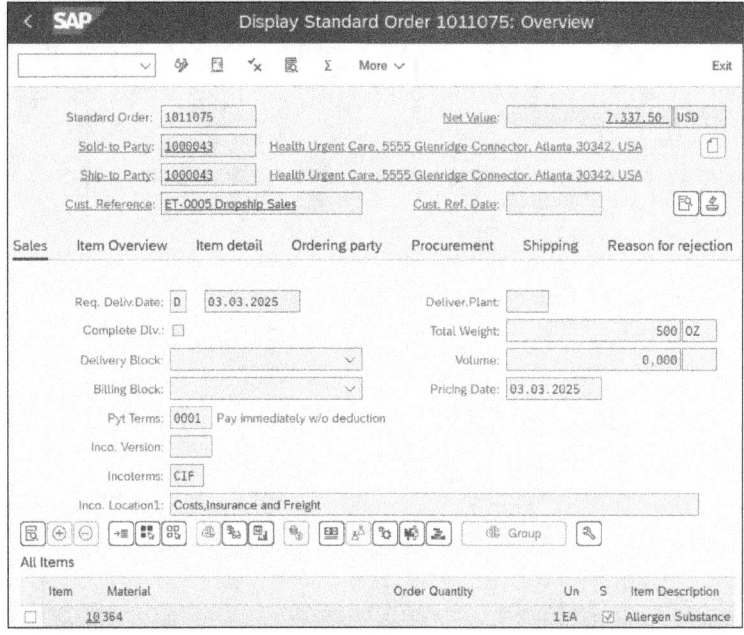

Figure 5.29 Sales Order to Be Invoiced

237

5.6.4 Reviewing Profitability Document

Display profitability document: Transaction KE24

After the billing document has been created and successfully transferred to both financial accounting and costing-based profitability analysis, you display the profitability document in costing-based profitability analysis with Transaction KE24 or by following the menu path **Accounting • Controlling • Profitability Analysis • Information System • Display Line Item List • KE24 – Actual**. You can enter the billing invoice number in the selection screen in the **Reference document number** field and execute the line item report with ⓖ (Execute) or F8 . All billing invoices are saved with record type **F**.

In Figure 5.30, you can see the **Characteristics** tab that gives you an overview of all characteristics that have been derived from the sales order and billing document. The system derives characteristics that are assigned to the operating concern. With ⓖⓖ (Next Page/Last Page), you can display further characteristics that have been derived.

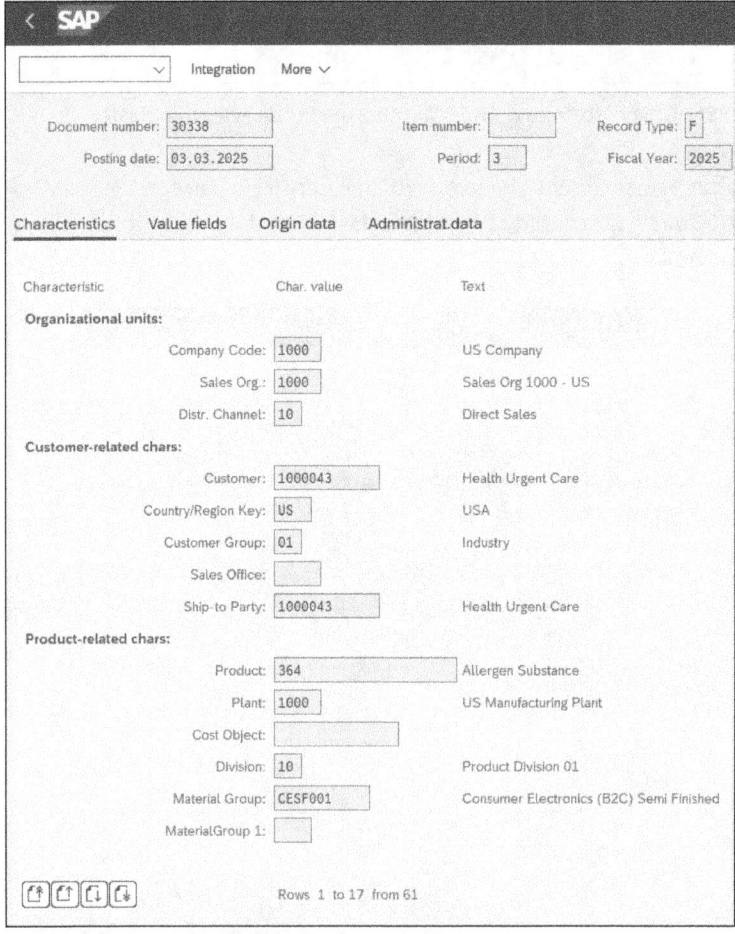

Figure 5.30 Display Characteristics of the Profitability Segment

5.6 Billing Data Transfer

The **Value fields** tab shown in Figure 5.31 provides an overview of the value fields and quantity fields that were populated when the billing document was transferred.

Display value fields in the profitability document

In Figure 5.27, shown previously, you maintained an alternative unit of measure. In Figure 5.31, you can see that the sales quantity of 1 EA got transferred in quantity field **Sales quantity** and the carton quantity is blank, as the quantity the customer ordered wasn't enough to translate it in a carton. If the customer orders 12 EA, the quantity in the quantity field **Carton** will increase to **1**. The system will apply the conversion for the quantity that is maintained in the material master in Figure 5.28. In addition to the quantity fields, the value fields got transferred as configured earlier in Figure 5.25 (not visible in the screenshot).

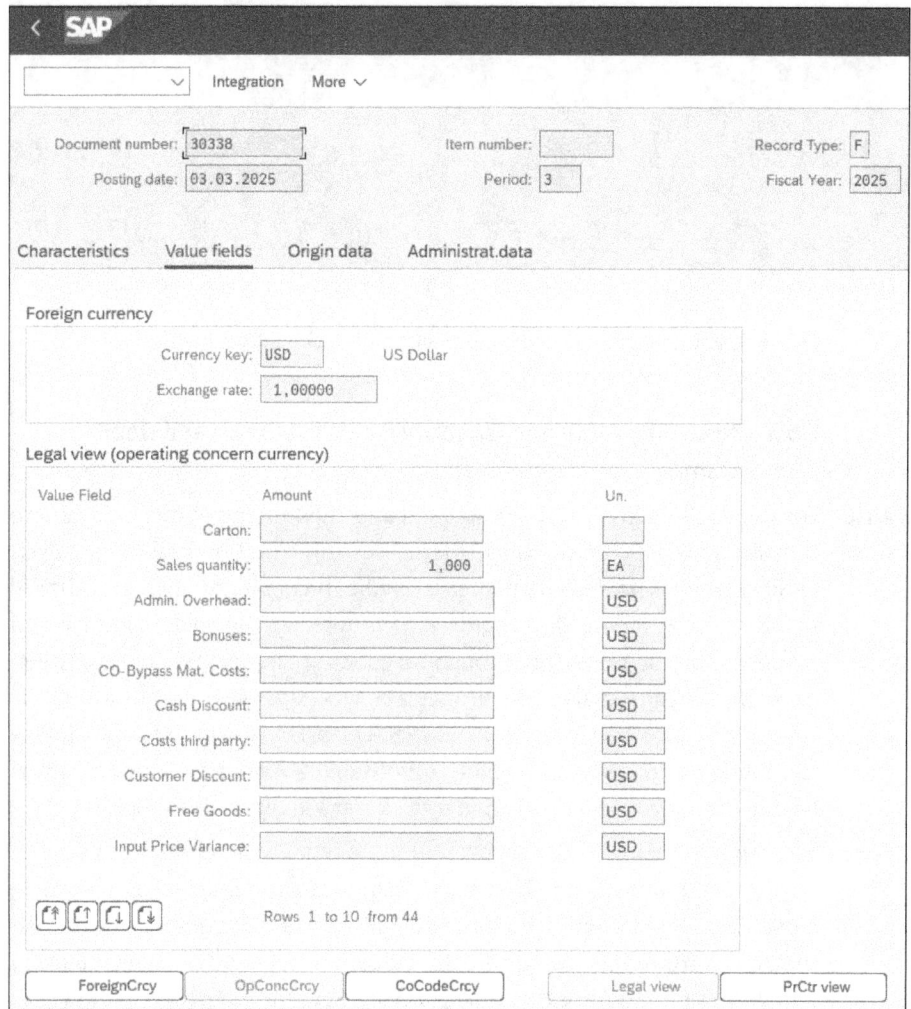

Figure 5.31 Display Value Fields of the Profitability Segment

239

5 Costing-Based Profitability Analysis

Display additional data with integration

With **Integration**, you can display master data (e.g., material master) and customer or transactional data (e.g., sales order and billing invoice). In addition, you can also display the accounting document. In Figure 5.32, you see the accounting document of the billing invoice. You can check whether the postings were made to the correct accounts according to the revenue account determination. When the financial accounting document was created, no COGS were updated. The goods issue was posted when the delivery was created.

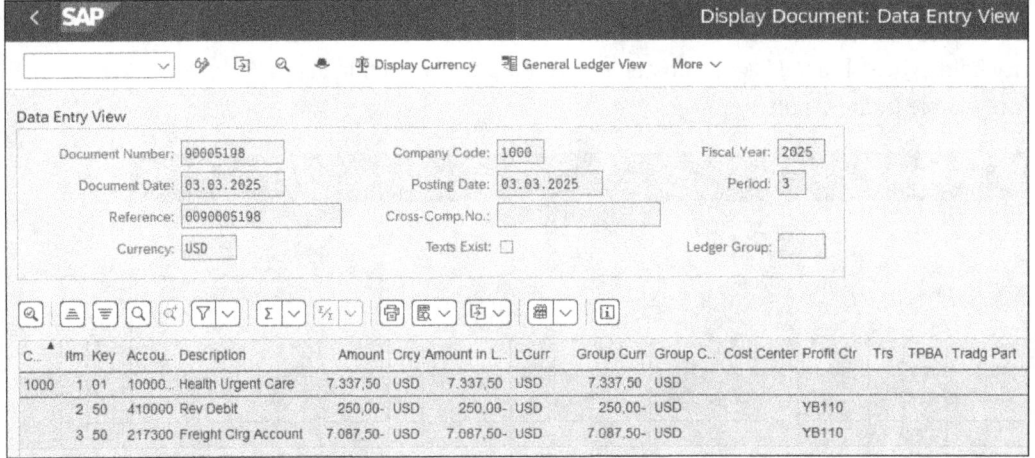

Figure 5.32 Display the Accounting Document of the Billing Invoice

5.6.5 Resetting Value and Quantity Fields for Credit and Debit Memos

Credit and debit memos

For some billing invoices, such as credit and debit memos, you don't want to transfer material cost and quantities to costing-based profitability analysis because there is no delivery related to credit and debit memos. For technical reasons, those billing documents always include the sales quantity and planned material price even though no goods are issued. When the costing-based profitability analysis document is created, the sales quantity and COGS are derived automatically. To avoid that, go to Transaction KE4W or follow the configuration path **Controlling • Profitability Analysis • Flows of Actual Values • Transfer of Billing Documents • Reset Value/Quantity Fields**.

Reset value fields and quantity fields: Transaction KE4W

You can reset value fields for specific billing types with **New Entries** or [F5]. Entries for Billing Type **G2 (Credit Memo)** are maintained as shown in Figure 5.33. When a billing invoice with billing type **G2** is transferred to costing-based profitability analysis, the value fields **ABSMG – Sales quantity** and **VVMEP – Material external Pr** are reset. Save your changes with **Save** or [Ctrl]+[S].

5.6 Billing Data Transfer

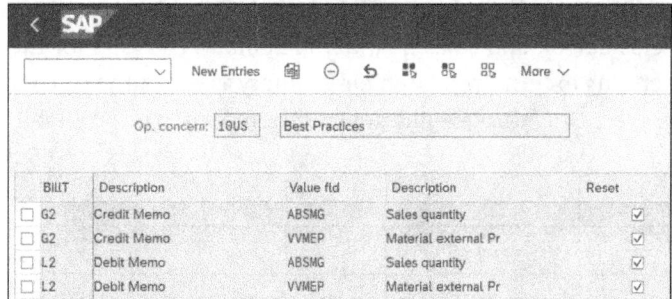

Figure 5.33 Resetting Value Fields

Let's create a credit memo and transfer it to costing-based profitability analysis to check the profitability segment. In Figure 5.34, you can see the pricing elements of a credit memo we just created. You can see that there is a **Quantity** and an **Internal price** assigned to the credit memo. The purpose of the credit memo is to credit the customer 50% of the sales price of 1 kg of material 369 received because the merchandise was damaged during transport.

Pricing elements of credit memos

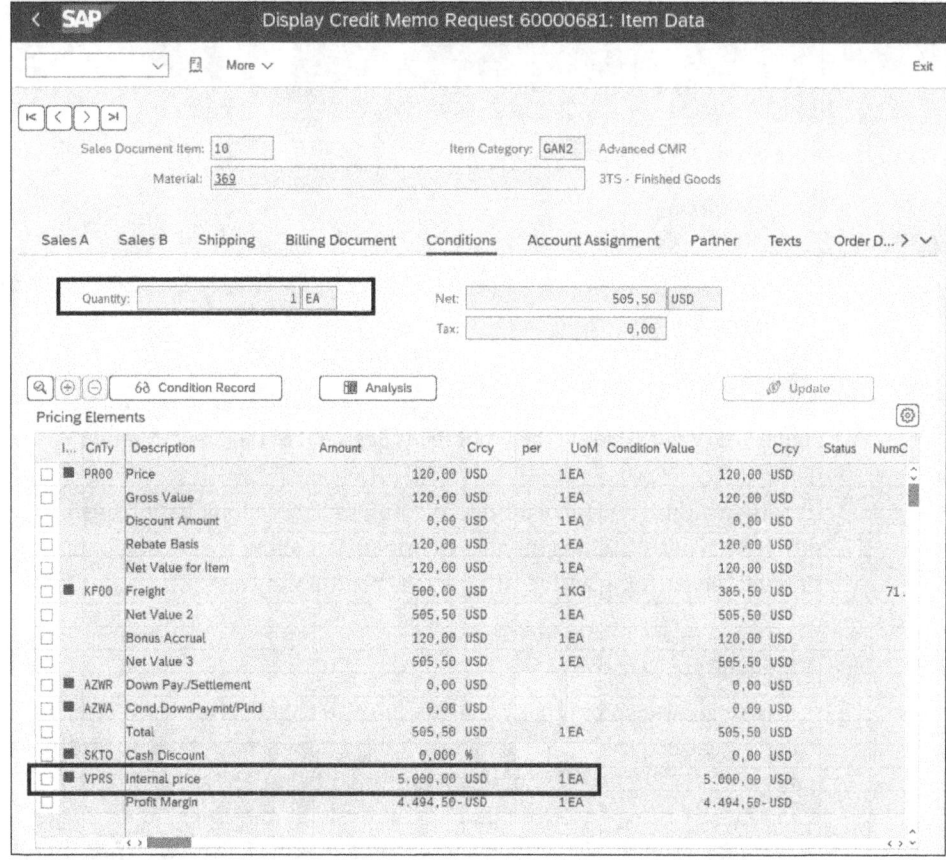

Figure 5.34 Display the Credit Memo

241

Next, let's check the profitability segment. In Figure 5.35, you can see that the quantity fields **Carton** and **Sales quantity** are empty. The field for the **Price – Manufacturing** external procurement is empty as well, although not visible on the screen. Only the **Revenue** got transferred with the profitability segment.

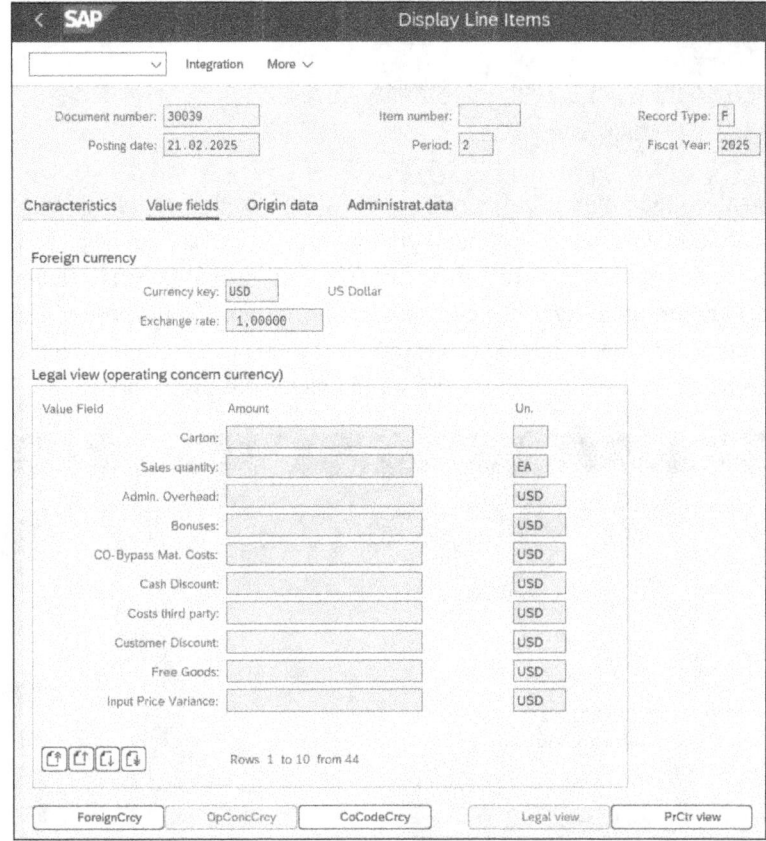

Figure 5.35 Value Fields in the Profitability Segment of the Credit Memo

In this section, you learned how to configure the billing data transfer for different types of billing invoices correctly to allow for reconciliation of financial accounting with costing-based profitability analysis.

> **Billing Data Transfer**
>
> In costing-based profitability analysis, billing documents are transferred to costing-based profitability analysis via the assignment of sales and distribution in SAP S/4HANA conditions to value fields. The COGS are transferred to costing-based profitability analysis at the same time as the revenues, which leads to challenges regarding reconciliation with financial accounting.

5.6 Billing Data Transfer

To help you reconcile your operating profit in financial accounting with costing-based profitability analysis, you can go to Transaction KEAT or follow the menu path **Accounting** • **Controlling** • **Profitability Analysis** • **Tools** • **Analyze Value Flows** • **Check Value Flow in Billing Document Transfer**.

Profitability analysis value flow: Transaction KEAT

The report gives you an overview of all billing invoices transferred to costing-based profitability analysis and compares the values in sales and distribution in SAP S/4HANA with the values in financial accounting and costing-based profitability analysis.

Checking the sales and distribution in SAP S/4HANA

In the selection screen in Figure 5.36, you specify which data you want to analyze, which, for this example, is in **Company Code** 1000. The selected **Currency type** is 10, which corresponds to the company code currency. You can display the data in two currencies: document currency and company code currency. In the **Selection of FI values or PCA values** area, choose **Display FI values** because the profit center valuation isn't activated. Execute the value flow analysis report with **Execute** or F8.

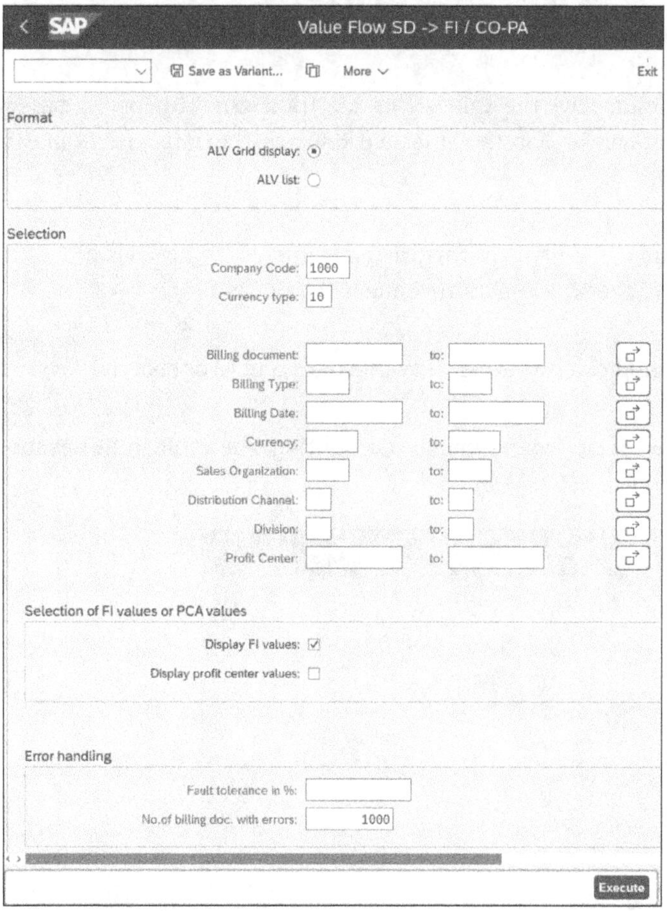

Figure 5.36 Selection Screen of the Value Flow Analysis Report

5   Costing-Based Profitability Analysis

Analyzing the results

The value flow analysis report in Figure 5.37 shows the sum of all the billing invoices selected and is organized in various columns. Let's look at the report and analyze the results:

- **Value field/cond. type/account (value field/condition type/G/L account)**
  In this column, you see the value field name, which is **Revenue**, on the first level. When you expand the view of the value field with [>], you can see the pricing elements/condition types from sales and distribution that you've seen in the sales orders created earlier, for example, in Figure 5.11. Figure 5.37 shows that the sales conditions **PR00** is listed. When you expand the view of the sales conditions with [>], you can see on which G/L accounts the values have been posted; in this case, this is the G/L account **410000**.

  At the bottom of the screen, you see the **Other condition types** folder in which you'll see condition types that aren't assigned to a value field but are price elements in a sales order.

- **Crcy (Currency)**
  You can see the currency of the values displayed. In this example, all values are displayed in the company code currency **USD**.

- **CO-PA value**
  This column displays the values that are transferred to costing-based profitability analysis. For the value field **Revenue**, the transferred value is **250.00 USD**.

- **SD value**
  The **SD value** is the value of the pricing elements in the sales order. In this example, the **SD value** for revenue is **250.00 USD**.

- **+/-**
  The +/- sign indicates whether the value is a credit (-) or debit (+).

- **FI value**
  The **FI value** got posted in financial accounting. The value in this example for revenue is **250.00** USD.

Value field/cond. type/account	Crcy	CO-PA value	SD value	+/-	FI value
∨ Revenue	USD	250,00	250,00	—	250,00-
∨ PR00			250,00	—	250,00-
0000410000			250,00	—	250,00-
∨ Material external Pr	USD	378,00	378,00		0,00
∨ VPRS (G)			378,00		0,00
3-party trans			378,00		
> Other condition types			0,00		

Figure 5.37 Analyze the Value Flow Analysis Report

The report helps you analyze any differences for the billing invoices transfer to costing-based profitability analysis. There are some additional helpful functionalities in the report. With (Hide Error List) or F7, you can display billing invoices with errors. They aren't transferred to financial accounting or costing-based profitability analysis yet.

Analyzing differences

With (Statistical Information), you can display statistical information. In Figure 5.38, you see that there has been one billing document with five relevant conditions that were posted. If there were billing documents created where a financial accounting document is missing, these billing invoices would need to be analyzed before month-end closing and either canceled or corrected and transferred to financial accounting and costing-based profitability analysis.

Display statistical information

> **Value Flow Analysis**
>
> Transaction KEAT helps you analyze the value flows from sales and distribution in SAP S/4HANA to costing-based profitability analysis. When setting up costing-based profitability analysis, it's critical to understand the value flows and configure them properly to minimize reconciliation issues with financial accounting.

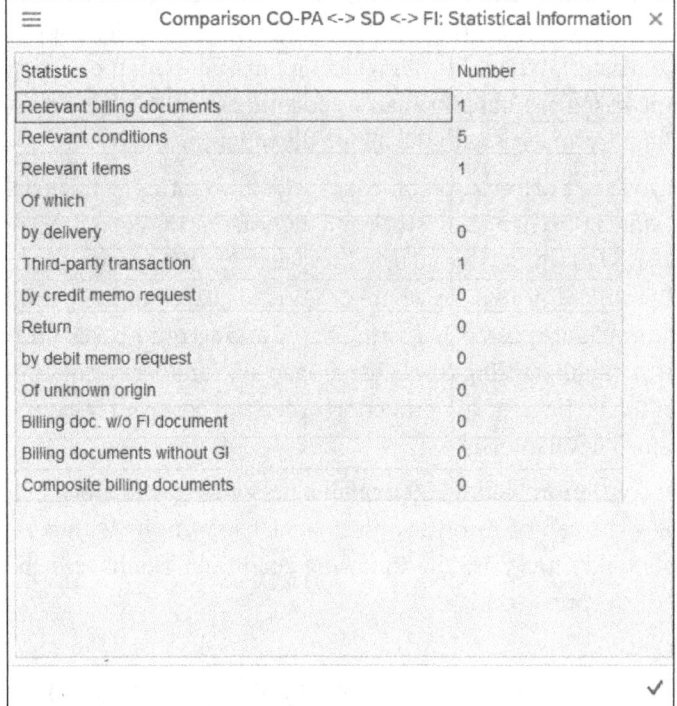

Figure 5.38 Comparison CO-PA <-> SD <-> FI: Statistical Information Screen

5.7 Costs of Goods Manufactured

Deriving costs of goods manufactured (COGM)

Costing-based profitability analysis and account-based profitability analysis differ considerably in the derivation of COGM. Account-based profitability analysis can be mapped with period accounting and cost of sales accounting. Costing-based profitability analysis, in contrast, only supports cost of sales accounting.

Goods issue posting

In costing-based profitability analysis, COGS information is transferred to profitability analysis with the billing document. As already mentioned in Section 5.6, this is a challenge when reconciling profitability analysis with financial accounting. The difference between COGM and COGS is transferred to costing-based profitability analysis with the variance calculation, which we'll explain in Section 5.9.

In the billing document, condition PCIP (Internal Price) corresponds to the COGM. It's derived from the price according to the price control in the material master when the sales order is created. Condition PCIP (Internal Price) is assigned to a value field for goods usage when the sales and distribution in SAP S/4HANA conditions are assigned to the value fields (Transaction KE4I). When the goods issue for delivery is posted, the system creates the financial accounting document and posts the goods issue with the value of the standard price that is valid at the time of the goods issue posting or with the moving average price (depending on the price control in the material master). When the billing document is posted, the system creates a costing-based profitability analysis document, and COGS is transferred to profitability analysis with the billing document.

Cost of sales accounting versus period accounting

In Figure 5.39, you see a comparison of cost of sales accounting in financial accounting in which the changes to stock and manufacturing costs are displayed in the period in which they incurred, and cost of sales accounting in costing-based profitability analysis where COGS is displayed in the period in which the revenues are posted. In costing-based and account-based profitability analysis, manufacturing costs are mapped via variance calculation on the basis of the settlement of production order and cost center assessment, as explained in Chapter 4.

Reconciliation of financial accounting

The **Transition** column in Figure 5.39 explains how changes to stock and manufacturing costs can be reconciled in financial accounting with costing-based profitability analysis. The following costing blocks have to be considered in the reconciliation:

- **COGS (VAX)**
  Basically, COGS transferred to costing-based profitability analysis with the billing document correspond to the goods issue postings linked to a

sales order. In financial accounting, they are posted to the account that is defined with record GBB-VAX in materials management in SAP S/4HANA account determination.

- **Balance production orders**
  Depending on the order status, the balances of the production orders are settled either in financial accounting as work in process (WIP) and therefore don't affect net income or in costing-based profitability analysis as variances on the input or output side. In financial accounting, these variances are posted as price differences that correct changes to stock as offset accounts and therefore have no effect on the operating income.

- **Variance from subcontracting**
  The SAP system posts subcontracting costs to a cost center, which is then settled via cost center assessment in profitability analysis.

- **Price/inventory difference**
  The SAP system posts price differences that result from stock revaluations due to new determinations of standard prices, for example, and inventory differences usually to different cost centers for each company code and/or profit center and settles them via cost center assessment in costing-based profitability analysis.

- **Balances on manufacturing cost centers**
  The system uses manufacturing cost centers to plan costs for employees, material overhead costs, and depreciations for the machines used in production and post them as actual costs. Cost centers are used to plan prices for each activity type that maps the price for a specific activity, for example, 1 personal hour = 20.00 USD. The time for manufacturing the individual products is specified in the routing. When the product is manufactured, the time is recorded on the production order, and the cost center is credited the number of recorded hours multiplied by the price for the activity. At the end of the month, cost centers usually show a balance. This balance (difference between debit and credit postings on the cost center) is settled via cost center assessment to costing-based profitability analysis.

- **Scrapping**
  Like inventory differences, the SAP system posts scrapping via default account assignments to cost centers or internal orders. At the end of the month, these costs are settled via cost center assessment or order settlement in costing-based profitability analysis.

Depending on your industry and your enterprise-specific requirements, the illustration in Figure 5.39 can look very different; we tried to generalize the value flow in this example.

5   Costing-Based Profitability Analysis

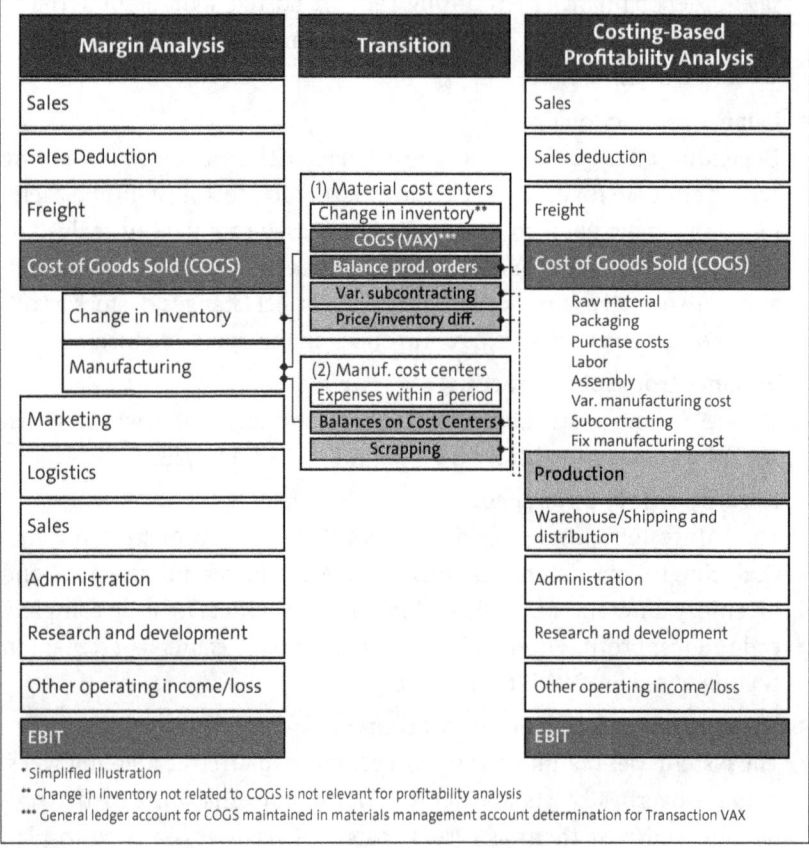

Figure 5.39 Comparison of Account-Based and Costing-Based Profitability Analysis

Sample posting

In Figure 5.40, you can see a sample posting that illustrates how COGM is posted in financial accounting compared to costing-based profitability analysis. It maps the value flow from purchasing to sales. The left part of the figure illustrates postings in financial accounting, mapped in T-accounts, while the right part of the figure shows how postings are displayed in the profit and loss (P&L) statement in financial accounting in period accounting and costing-based profitability analysis.

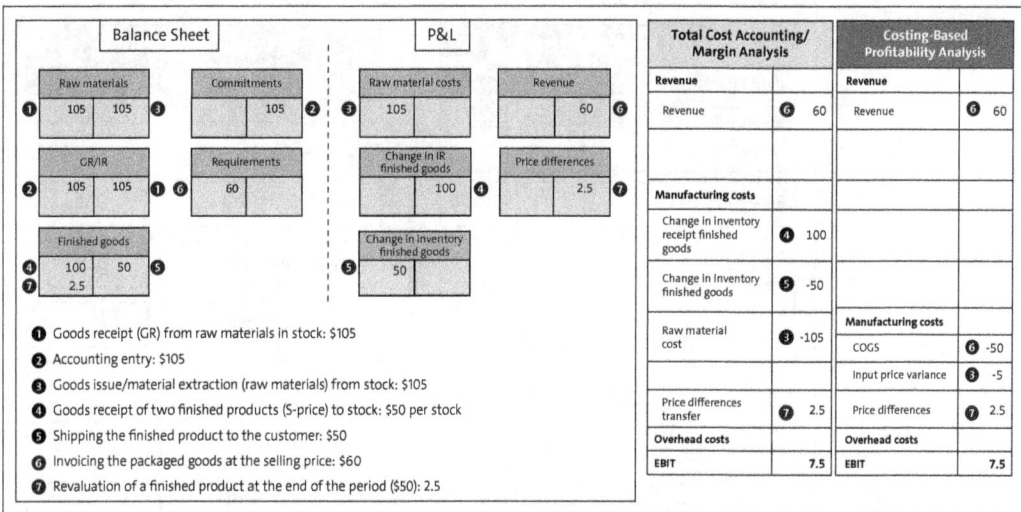

Figure 5.40 Posting of COGS and COGM in Costing-Based Profitability Analysis

Now that you've learned that costing-based profitability analysis only displays COGS and you know how to reconcile COGM in financial accounting with costing-based profitability analysis, we want to show you how to split the COGS according to the product costing.

5.8   Cost of Goods Sold Transfer and Split

In Section 5.4, you saw how to assign the pricing condition VPRS (internal cost) to a value field in costing-based profitability analysis. With that configuration, you made sure that COGS was getting transferred to costing-based profitability analysis. Because this example assigned the internal price = COGS to one value field, there aren't a lot of possibilities to analyze the COGS in costing-based profitability analysis. However, you can split the COGS in the components of product costing. In Figure 5.41, you can see how an internal price of a finished product that has a valid product costing in Product Cost Controlling can be split on value fields in costing-based profitability analysis.

COGS split

5   Costing-Based Profitability Analysis

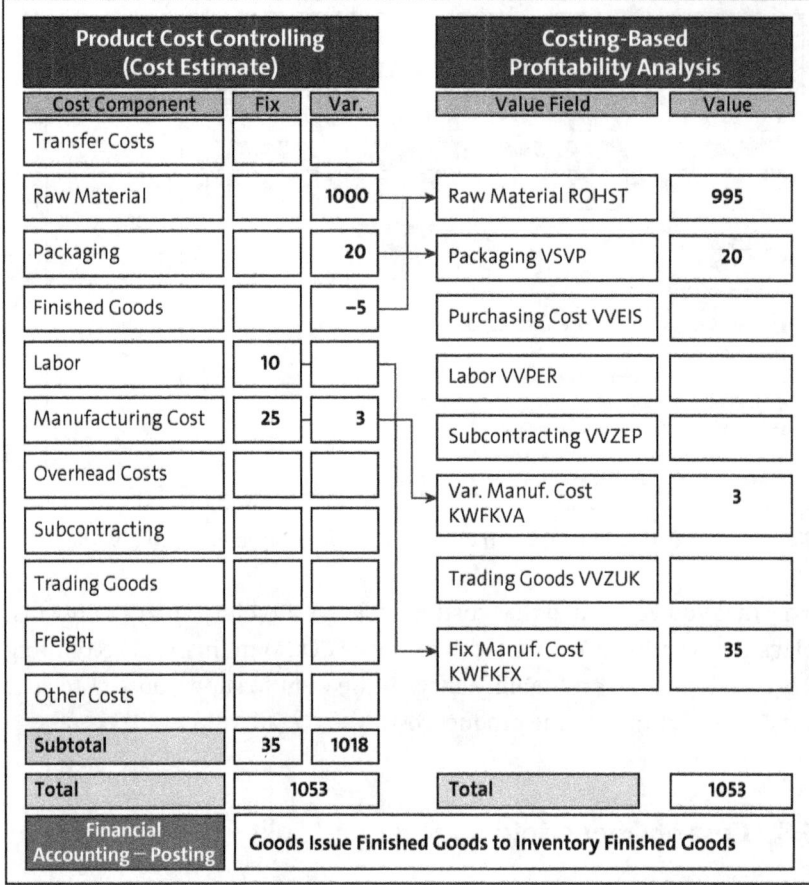

Figure 5.41 Transfer of Cost Estimate to Costing-Based Profitability Analysis

Before we go into the configuration details, let's look at Figure 5.42, which provides an overview of all influencing factors or components of COGS in costing-based profitability analysis. These are useful when introducing value flow analysis to allow for reconciliation of costing-based profitability analysis with financial accounting:

❶ **Reconciliation process**

We recommend creating a checklist for the reconciliation of financial accounting with costing-based profitability analysis to map the origin of costs to the individual value fields in detail. You can use this checklist, for example, for month-end closings to facilitate and accelerate the reconciliation process.

❷ **Master data maintenance**

To determine COGS correctly, you have to maintain your master data properly. The costing for finished products is based on master data, for example, bills of material (BOMs) and routings. The more detail with which they are maintained, the fewer variances occur in production and the better you can reconcile the COGS in costing-based profitability analysis with financial accounting.

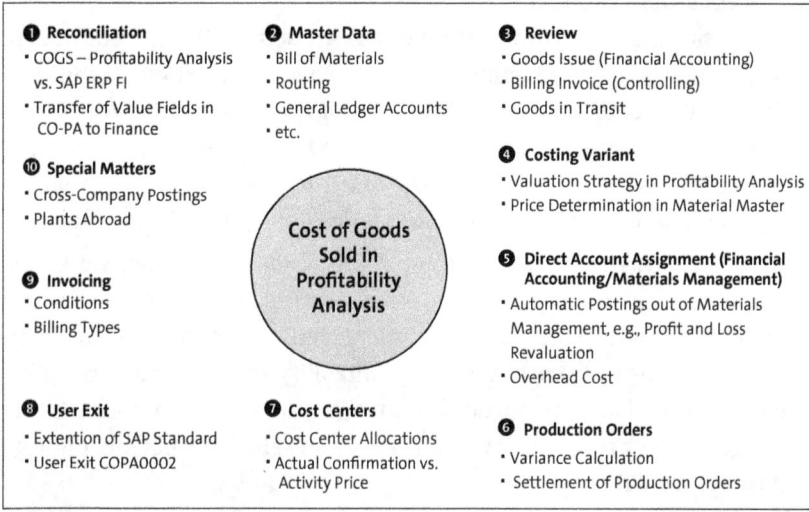

Figure 5.42 Influencing Factors for COGS Reconciliation

❸ **Time of valuation**

By valuating the sales quantity using various costing variants in costing-based profitability analysis, you can also define the time of valuation. The time of valuation determines the validity date on which the cost estimate is transferred. Various times of valuations are available; to ensure reconciliation of COGS with financial accounting, select the goods issue date as time of valuation. Depending on the processes in logistics, you have to analyze and accrue the goods in transit. This is the case, for example, if a goods issue is posted in September, and the corresponding revenue is posted in October. In costing-based profitability analysis, COGS is posted with the revenue in October, which leads to discrepancies with financial accounting.

❹ **Costing variant**

In costing-based profitability analysis, you can valuate COGS using different costing variants. To ensure reconciliation with financial accounting, you have to use the costing variant that is used to determine the standard price to determine COGS.

❺ **Direct account assignment of financial accounting in SAP S/4HANA/ materials management in SAP S/4HANA**

You can post individual costs directly to a value field in costing-based profitability analysis using the direct account assignment. This is often done for costs such as expenses/income from scrapping and price differences. For reconciliation or value flow analysis purposes, you should create documentation that defines which costs are directly assigned to which value field.

❻ **Process orders/production orders**
Before process orders/production orders are settled, WIP is determined. All process orders/production orders that have neither status **Technically Completed** (**TECO**) nor status **Delivered** (**DLV**) are marked as WIP. When process orders/production orders are settled, the order balance is capitalized as WIP in the balance sheet. SAP S/4HANA Finance postings that are generated when orders with WIP are settled don't affect the net income.

If an order has the **TECO** or **DLV** status, the system determines variances. Variance calculations analyze the order balance and divide it into different variance categories. When the order is settled, the balance is updated to the value fields in costing-based profitability analysis, and price differences are updated in financial accounting. The offset account is the change to stock account, so the posting doesn't affect the net income (e.g., postings for WIP).

❼ **Cost centers**
Usually, cost centers in production show a balance at the end of the month. The balances on cost centers are the sums of debits and credits that will be settled via cost center assessment in costing-based profitability analysis. Similar to direct account assignments, you should list senders and receivers of cost center assessments to accelerate the reconciliation process.

You may have to adapt the activity prices in the cost centers if the cost center balance is rather high every month. You should analyze whether labor hours are actually confirmed in production. In real-world scenarios, you frequently come across orders with missing confirmations.

Other overhead cost centers, for example, administration cost centers, are settled 100% to the operating profit.

❽ **User exit**
The COPA00002 user exit for the valuation of the sales quantity during the transfer of billing documents to costing-based profitability analysis allows you to manipulate the valuation. Many companies don't know that this user exit is active in the system. You should therefore check whether this user exit is activated and consider whether it's actually required.

❾ **Billing**
You should also check whether all conditions that are used in sales and distribution in SAP S/4HANA are assigned to a value field. Statistical conditions should be assigned to a separate value field so that actual values aren't mixed with statistical values.

⑩ **Special matters**

The SAP system provides various special processes that require particular attention in costing-based profitability analysis. These include, for example, the function of plants abroad or cross-company postings. Usually, finished products are moved to nonmanufacturing plants with these processes.

Besides analyzing the sales conditions, you should also analyze the billing types. For billing types without goods movement, you have to reset the value fields for the sales quantity and transfer of COGS.

Now, we're going to explain how to transfer COGS to costing-based profitability analysis and split it into the components of the material cost estimate. You can evaluate your COGS in costing-based profitability analysis with up to six different costing variants as well as the actual costing from the material ledger. Let's look at the different configuration components that are necessary to transfer material cost estimates and split them according to their component structure.

Valuation with costing variants

5.8.1 Defining Valuation Strategies

First, you need to define a valuation strategy. The valuation strategy plays a major role for the transfer of costing variants to costing-based profitability analysis. It defines which record types get valuated with a material cost estimate and defines the time of valuation.

To define a valuation strategy, go to Transaction KE4U or follow the configuration path **Controlling** • **Profitability Analysis** • **Master Data** • **Valuation** • **Valuation Strategies** • **Define and Assign Valuation Strategy**.

Defining valuation strategies: Transaction KE4U

Create valuation strategy **YB1** – **Valuation Strategy**, as shown in Figure 5.43, with **New Entries** or F5 . Now, mark the valuation strategy by clicking the checkbox at the beginning of the line, and navigate in the **Dialog Structure** to the **Details** folder.

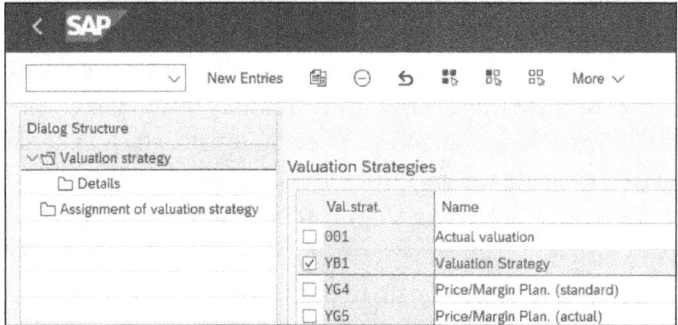

Figure 5.43 Create the Valuation Strategy

5   Costing-Based Profitability Analysis

Valuation types You can define details for the valuation strategy in Figure 5.44 with **New Entries** or [F5]. You can define different valuation strategy details such as the following:

- **Valuation using profitability analysis conditions**
  The valuation is performed using conditions of a sales and distribution in SAP S/4HANA costing sheet. Each condition is assigned to a value field.

- **Valuation using material cost estimates**
  One or several material cost estimates are used to evaluate the sales quantity. This scenario is used most often to valuate sales quantities in costing-based profitability analysis and will be described in detail in this section.

- **Valuation using user-defined valuation routines**
  A user exit is available that can be used to program the valuation according to user-defined/customer-specific requirements.

- **Valuation using transfer prices**
  In planning, you can valuate sales quantities using transfer prices; however, this scenario can't be used for actual valuations.

You can also apply a mix of these valuations. In the example in Figure 5.44, one entry is created for the valuation with material cost estimates.

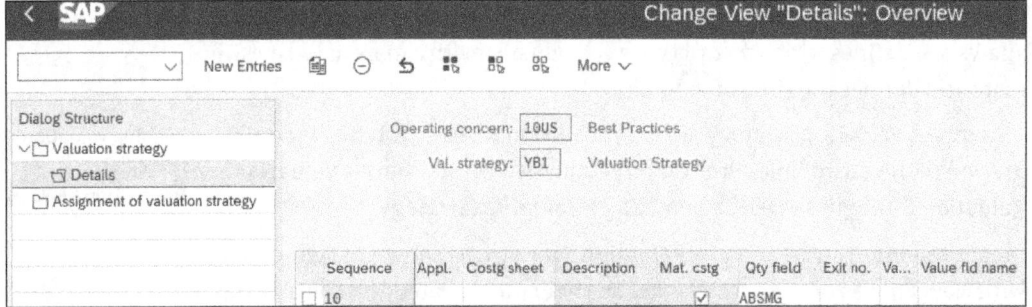

Figure 5.44 Define Details of the Valuation Strategy

Check the checkbox for **Mat. cstg.** (material cost estimate) and maintain the quantity field **ABSMG** (Sales Quantity) in the **Qty field** column. The system now will take the material that is used in the profitability segment and look for a material cost estimate whose price it will multiply by the sales quantity in the **ABSMG** field. The system doesn't yet know with which material cost estimate it will evaluate the quantities; this configuration step will be explained next. Save the valuation strategy with **Save** or [Ctrl]+[S].

Assign the valuation strategy to record types Now, you have to assign the valuation strategy to record types and a point of valuation. To do that, mark the details of the valuation strategy by clicking the checkbox at the beginning of the line, and navigate in the **Dialog Structure** to the **Assignment of valuation strategy** folder.

5.8 Cost of Goods Sold Transfer and Split

With **New Entries** or F5, you can assign the valuation strategy to record types and a point of valuation. In Figure 5.45, assign valuation strategy **YB1** (created earlier in Figure 5.43) to record types **A** (for incoming sales orders) and **F** (for billing invoices) because we want the COGS split in cost components for both incoming sales orders and billing invoices. For the point of valuation (**PV**), choose **01** (real-time valuation of actual data) because we want the split of the COGS to happen after the profitability segments are created.

You can choose from different point of valuations, however:

- **01** (Real-time valuation of actual data)
  The material cost estimate data is derived in real time when the actual data is created, for example, when billing documents are transferred.

- **02** (Periodic valuation of actual data)
  The material cost estimate data is derived during the revaluation process, for example, at the end of the month when billing documents are revaluated using Transaction KE27.

- **03** (Manual planning)
  The material cost estimate data is derived when the planning process is executed manually, for example, if you use Transaction KEPM for planning.

- **04** (Automatic planning)
  The material cost estimate data is derived when an automatic planning process is created, for example, if planning data is transferred from a predecessor module.

Points of valuation

Now, save the assignment of the valuation strategy with **Save** or Ctrl+S.

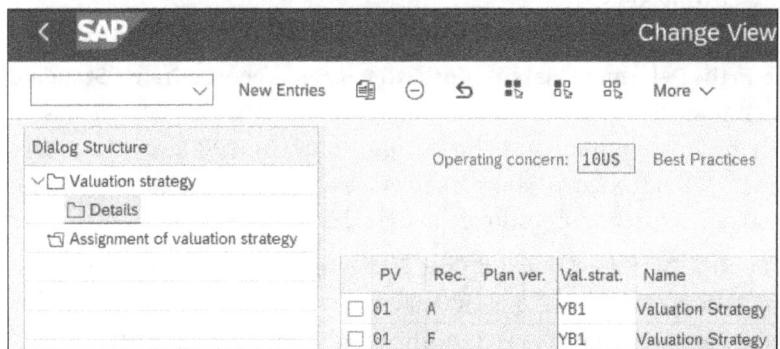

Figure 5.45 Assign the Valuation Strategy to the Point of Valuation and Record Type

The next step after defining and assigning the valuation strategy is to create a costing key that determines which costing variant is assigned to the valuation strategy, among other things.

Create a costing key

5.8.2 Setting Up Valuations Using Material Cost Estimates

Usually, several costing variants are used within a company. Every costing variant uses a different valuation approach. In costing-based profitability analysis, you can evaluate your sales quantities with up to six costing variants.

To do this, the following prerequisites must be fulfilled:

- Material cost estimates were created with this costing variant.
- A valuation strategy for the valuation of sales quantities using material cost estimates has been defined in the configuration of costing-based profitability analysis.
- The value fields for the cost component split have been created and assigned to the operating concern.

Define Access to Standard Cost Estimates

Accessing standard cost estimates

If all these prerequisites are met, the next step is to define the access to the material cost estimates by following configuration path **Controlling • Profitability Analysis • Master Data • Valuation • Set Up Valuation Using Material Cost Estimate • Define Access to Standard Cost Estimates**.

A number of costing keys already exist in the SAP standard. A new costing key is created in Figure 5.46 with **New Entries** or F5. To reconcile costing-based profitability analysis with financial accounting, you need to assign the costing variant that is used for the release of planned costs in the material master to the costing key. Therefore, it's important that you know which costing variant is used for the calculation of the standard prices.

Create Costing Keys

Creating costing keys

The example in Figure 5.46 is creating **Costing Key YB1 – Standard Cost Estimate**. In the **Determine Material Cost Estimate** area, choose **Transf. Standard Cost Estimate** to transfer the cost estimate according to the material master. This is the most common scenario for make to stock (MTS). If you have make to order (MTO) scenarios, you probably need an additional costing variant that accounts for the cost estimate in the sales or production order.

The **Control Data for Standard Cost Estimate and Production Order** Cost Estimate area contains several subareas that continue in Figure 5.47. In the **Costing Data** area shown in Figure 5.46, make the following entries:

- **Costing Variant**
  Enter "YPC1", which is used to calculate the standard price that is released to the material master on a monthly basis.

- **Costing Version**
  Every material cost estimate can have different costing versions if you want to execute the material cost estimate various times and save the

result in the database. Mixed cost estimates also work with costing versions. Usually costing version 1 is used to create cost estimates that are released to the material master, so enter "1" here.

- **Period Indicator**
  The period indicator is very important if you want to reconcile your costing-based profitability analysis with financial accounting. The only way to reconcile is if you choose the period indicator **Released standard cost estimate matching goods issue date**. If you evaluate your sales quantity with different cost estimates, you can choose different entries here as the reconciliation won't be your main focus.

- **Additive costs**
  Additive costs are used in material cost estimates to add additional costs manually to a cost estimate because they can't be determined automatically, such as freight costs. If you select the **Additive costs** checkbox in **Costing Data**, the system will only transfer additive costs. Therefore, for this example, don't select this checkbox because the system will be looking for material cost estimates, and if additive costs exist, they will be transferred together with the material cost estimate.

In Figure 5.47, you can see more options for costing key maintenance.

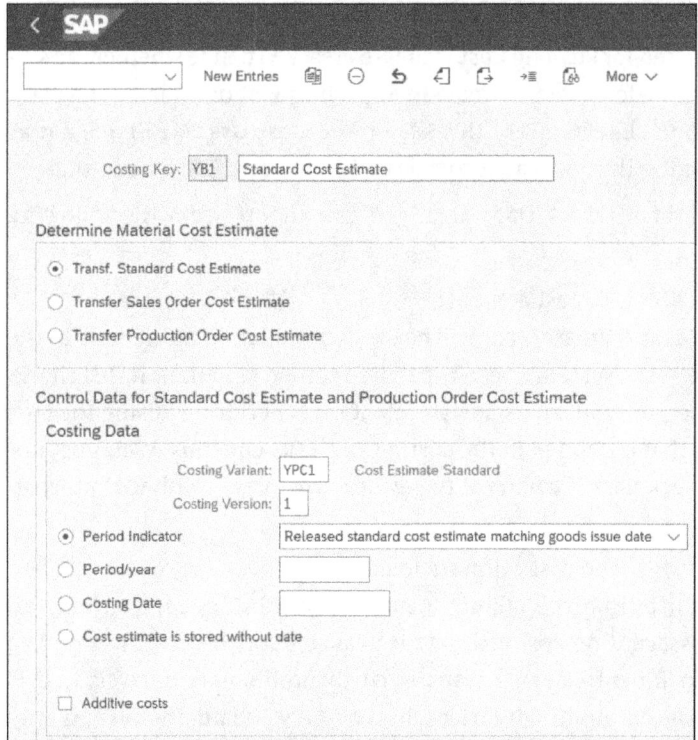

Figure 5.46 Maintain the Costing Key

5   Costing-Based Profitability Analysis

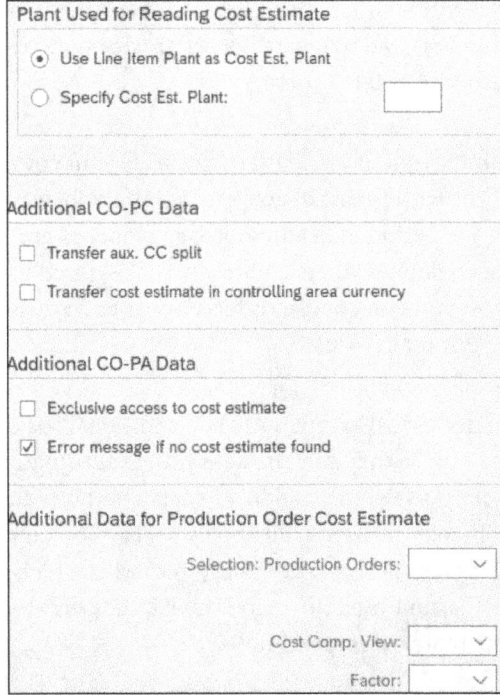

Figure 5.47 Maintain the Costing Key (continued)

In the **Plant Used for Reading Cost Estimate** area, you define whether the system is searching for a cost estimate in a specific plant or is taking the plant according to the line item. For this example, choose **Use Line Item Plant as Cost Est. Plant** so that you can reconcile COGS with financial accounting.

Maintaining additional settings

In the **Additional CO-PA Data** area, you can choose from the following options:

- **Exclusive access to cost estimate**
  If several cost estimates that will be used for valuation are defined in the costing key, the system doesn't search for other valuations if this checkmark is set; instead, an exclusive valuation is performed using the cost estimate that is defined in the costing key. Consequently, you must not set the checkmark if you want to perform a valuation with various costing variants.

- **Error message if no cost estimate found**
  When transferring the billing document, the SAP system outputs an error message if no cost estimate is available for this costing key. The billing document can't be transferred to profitability analysis and is stuck in the accounting interface. In the SAP standard, the system only throws an error message if no cost estimate can be found for any costing

key at all. Because you want to ensure data consistency and use all costing variants for derivation, you set the checkmark for the error message.

In the **Additional Data for Production Order Cost Estimate** area, you can make additional entries for the search of production order cost estimates. As the costing key in this example isn't relevant for production order cost estimates, don't maintain any entries here.

Now that you've maintained all the required fields, save the costing key with **Save** or [Ctrl]+[S].

Assign Costing Keys

After defining the costing key, you have to assign it so that it can be used for transferring the billing documents to costing-based profitability analysis. The assignment of costing keys is similar to the creation of a characteristic derivation. To maintain the assignment of costing keys, go to Transaction KEPC or follow the configuration path **Controlling · Profitability Analysis · Master Data · Valuation · Set Up Valuation Using Material Cost Estimate · Assignment of Costing Keys to Any Characteristics**.

Assigning costing keys: Transaction KEPC

To create a new assignment, click on [icon] (Display <-> Change) in Figure 5.48, and then you'll be able to create a new entry with [icon] (Create Step) or [F5]. Enter the **Step Description** as "Costing Derivation YBC1". The screen is split into two areas. In the upper **Source Fields** area, you can see the following characteristics:

Defining source fields of costing keys

- **BWFKT – Point of valuation**
  This characteristic is assigned by default and determines for what point of valuation the derivation of the costing key will take place. You saw the point of valuation earlier when assigning the valuation strategy to a point of valuation in Figure 5.45.

- **VRGAR – Record Type**
  This characteristic is also assigned by default and determines for which record types the derivation of the costing key will take place.

- **VERSI – Plan version (CO-PA)**
  The plan version is also a characteristic that is assigned by default to the derivation of the costing key. The plan version determines from which version the plan data will be read in case data is derived for planning.

- **MTART – Material Type**
  This characteristic was added manually to restrict the entries by material type. You can add any characteristic that is assigned to your operating concern. The system has to be able to derive this characteristic when the profitability segment is created; otherwise, you'll never be able to derive the costing key.

5 Costing-Based Profitability Analysis

In the lower **Target Fields** area shown in Figure 5.48, maintain the following characteristics:

- **KALAW1 – First costing key (CO-PA)**
  You can assign up to six costing keys for the derivation. We only created one costing key, so we only assign one characteristic for the costing key to the derivation.

- **VALUE_FLD1 – CO-PA: value field that is reset**
  In Figure 5.14, shown earlier, the **WEINS – Good usage** value field was assigned to a sales condition to transfer COGS of the sales order to costing-based profitability analysis. This field needs to reset when the system finds a material cost estimate based on the costing key; otherwise, you would have COGS twice in the profitability segment: from the **WEINS** value field and from the costing key.

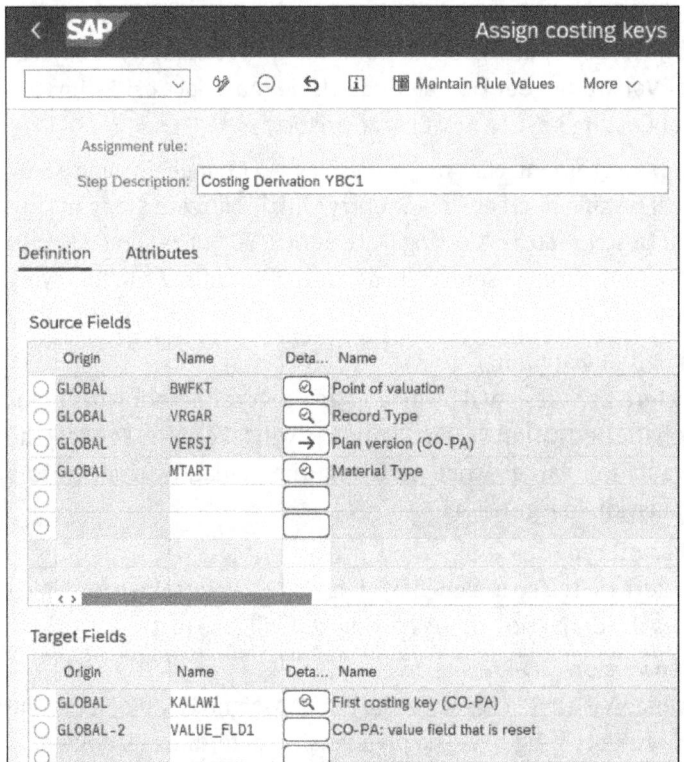

Figure 5.48 Maintain Derivation of the Costing Key

Maintaining rule entries

Let's move on to the next tab: **Attributes**. In Figure 5.49, you can set a couple of checkboxes as follows:

- **Issue error message if no value found**
  If you activate this checkbox, the system will throw an error message if

no cost estimate is found according to the costing key. The document processing can't be completed, and you can't transfer billing invoices to costing-based profitability analysis. This checkbox isn't set for this example because it's possible to derive the costing key at a later point if there have been issues during the document processing. You'll learn how to derive costing keys after a profitability segment is created at the end of this section.

- **End strategy processing when a value has been found**
  If this checkbox is activated, the system will stop processing the derivation after an entry has been found. Because this example is only working with one costing key, it doesn't make sense to set this checkbox. If you're working with several costing keys, however, you can set this checkbox, and the system will process the derivation until a costing key with a value for the material in the profitability segment has been found.

- **Maintain entries using validity date**
  If you activate this checkbox, you can maintain your entries with a validity date, which especially makes sense if you're deriving costing keys for plan data or if you're planning to switch your costing variant.

- **User-defined step ID (optional, see F1 help)**
  If you're working with user exit EXIT_SAPLKEAB_004 and SAP enhancement COPA0002, you can enter a step ID here. During processing of the derivation, the system will jump directly to the derivation maintained step in the user exit. User exits are a development, and it's hard for users to retrace how values have been determined. The derivation of costing keys is very flexible, and you should think twice if you really need to work with a development here.

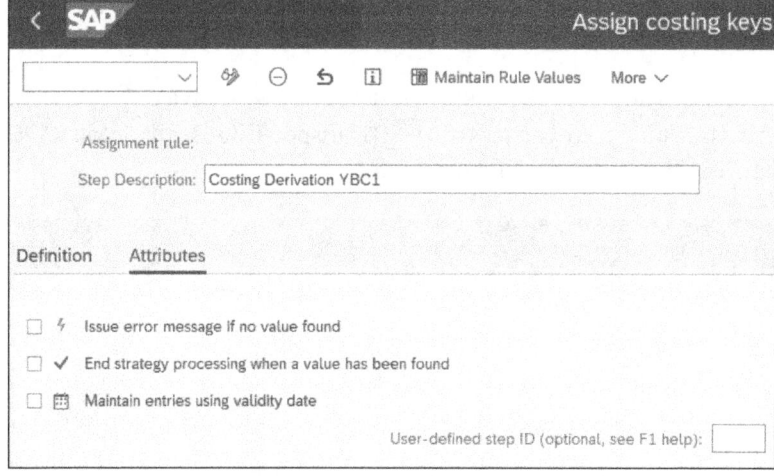

Figure 5.49 Maintain Attributes in Costing Key Derivation

5 Costing-Based Profitability Analysis

Maintaining rule values

Next, maintain the derivation rules with :≡ **Maintain Rule Values** or F9. In Figure 5.50, maintain two lines with the following characteristics:

- **Point of valuation**
  Both lines are for the point of valuation **01** (real-time valuation of actual data), which means that values are only derived in real time. If, for any reason, the system can't find a material cost estimate, you can't re-derive a material cost estimate at a later point as the derivation is only maintained for real-time valuation.

- **Record Type**
  For the record type, enter "A" for incoming sales orders and "F" for billing invoices to derive the material cost estimate for both incoming sales orders and billing invoices.

- **Plan version (CO-PA)**
  An entry in this column is only necessary if you're deriving planning data. This example is only deriving actual data, so don't maintain any entries here.

- **Material Type**
  For the material type, enter "FERT" for finished goods as this is the only material type for which material cost estimates have been created. For all other material types, you want the system to transfer the price according to the material master.

- **First costing key (CO-PA)**
  In this column, enter the costing key "YB1" that you created earlier in Figure 5.43.

- **CO-PA: value field that is reset**
  If a material cost estimate can be derived, you want the system to reset the value field **VVMEP** for goods usage in which you transfer the price according to the material master.

After you've maintained all the entries, save them with **Save** or Ctrl+S. You can transport your entries with 🚚 (Transport), but Transaction KEPC is also accessible in the production system.

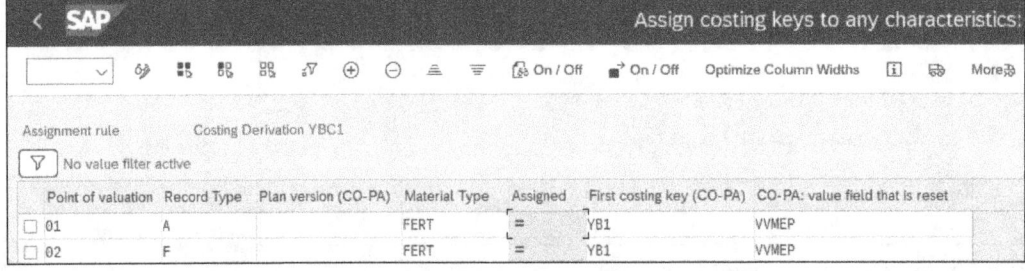

Figure 5.50 Assign Costing Keys to Characteristics

Assign Cost Components to Value Fields

Now, the missing piece in the configuration is the assignment of cost components to value fields. Go to Transaction KE4R or follow the configuration path **Controlling • Profitability Analysis • Master Data • Valuation • Set Up Valuation Using Material Cost Estimate • Assign Value Fields**. The assignment of the value fields to cost components depends on the operating concern and cost component structure. This example uses **Operating concern 10US** and **Cost component structure Y1**, as shown in Figure 5.51. Be sure to maintain the entries for the cost component structure that is assigned to the costing variant that you use for creating material cost estimates. Press Enter to confirm your choices.

Assign cost components to value fields: Transaction KE4R

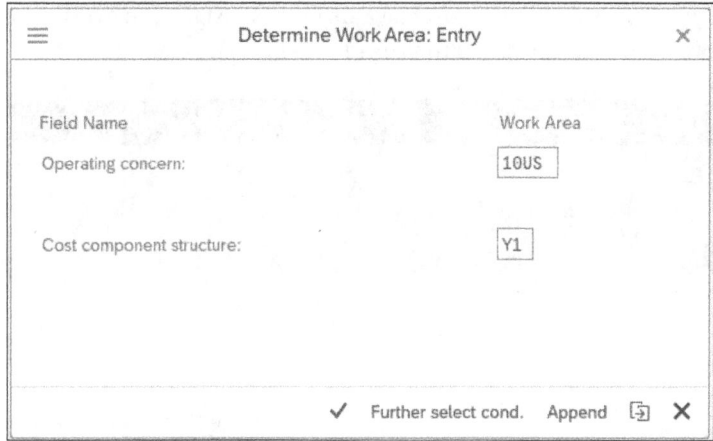

Figure 5.51 Operating Concern and Cost Component Structure for Assigning Cost Components to Value Fields

In Figure 5.52, maintain the following entries in the various columns:

- **PV**
  The point of valuation is the same as the point of valuation for costing keys. Point of valuation **01** refers to the real-time valuation of actual data.

- **CCo**
  In the cost component column, you enter all the cost components of your cost component structure. Make sure that the data you provide is complete; otherwise, the missing cost components won't be transferred to costing-based profitability analysis, and your COGS won't be correct.

- **Name of Cost Comp.**
  The name of the cost component gets derived automatically in this column.

5   Costing-Based Profitability Analysis

- **F/V**
  In the fixed/variable flag column, you can split variable and fixed costs on different value fields. The following entries are available:
  - **1** (fixed amounts)
  - **2** (variable amounts)
  - **3** (totals of fixed and variable amounts)

  For this example, choose **3** because you're not working with fixed and variable costs in controlling.

- **Fld name 1**
  In the field name column, you assign the cost component to a value field. With the [F4] Help, you can display both the available cost components and the available value fields. After a value field is assigned to every cost component, save the entries with **Save** or [Ctrl]+[S].

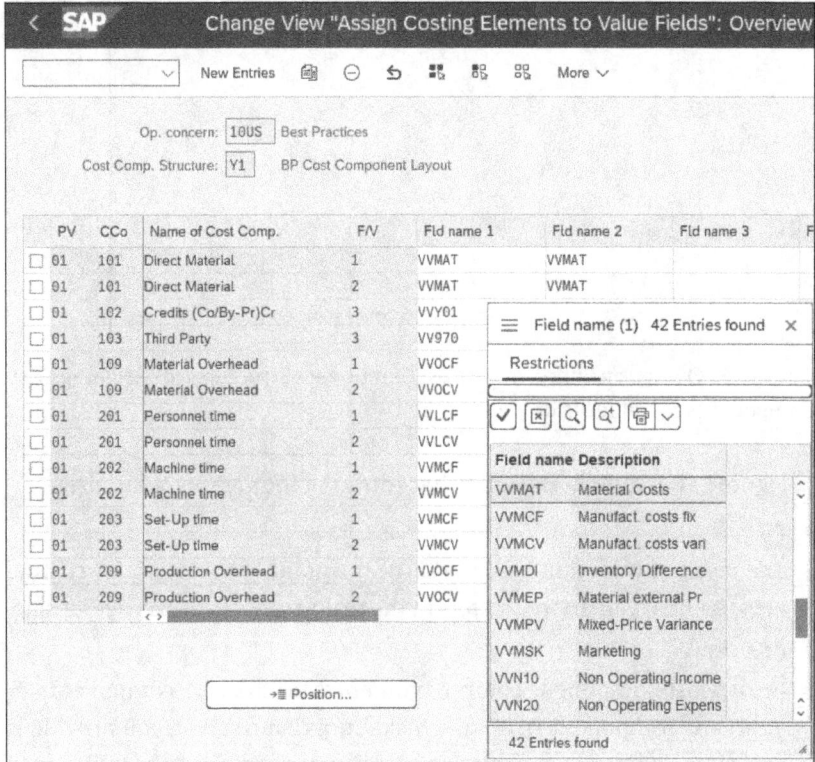

Figure 5.52 Assign Cost Components to Value Fields

If you're using various different costing variants, you have to maintain different value fields for each costing variant in the **Fld name 2** and **Fld name 3** columns. This is necessary because the value fields would be overwritten otherwise.

5.8 Cost of Goods Sold Transfer and Split

Now, let's create a sales order and bill it to see if the COGS gets split in the value fields according to the cost components. In Figure 5.53, a sales order is displayed for **Material FG-161**. The sales condition **VPRS** for the **Internal price** is in the sales order with a value of **77.85 USD** per **EA**.

Create a sales order

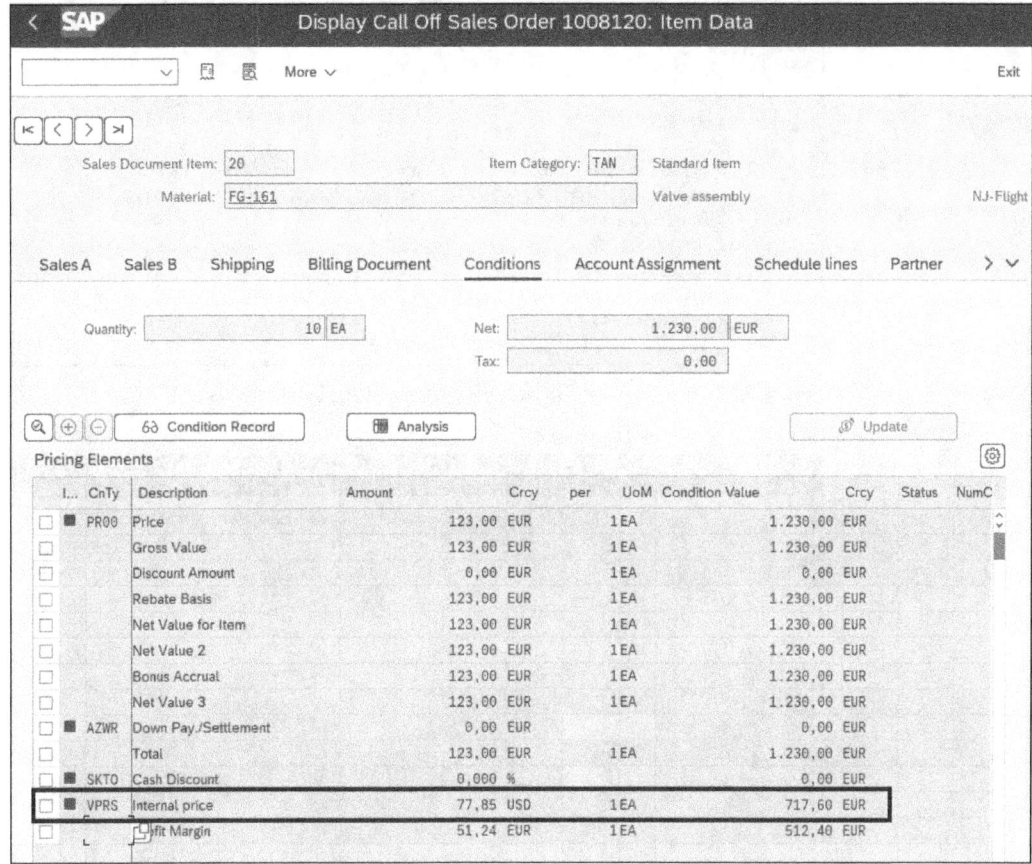

Figure 5.53 Review the Internal Price in the Sales Order

Let's check the value of the goods issue. In Figure 5.54, you see the relationship browser with all relevant accounting documents. Double-click on **Accounting document 4900013996** to display the goods issue posting.

Check goods issue posting

Figure 5.55 shows the accounting document of the goods issue. The value of the posting is **77.85 USD**, which equals the value in the sales order in Figure 5.53.

Checking accounting documents

5 Costing-Based Profitability Analysis

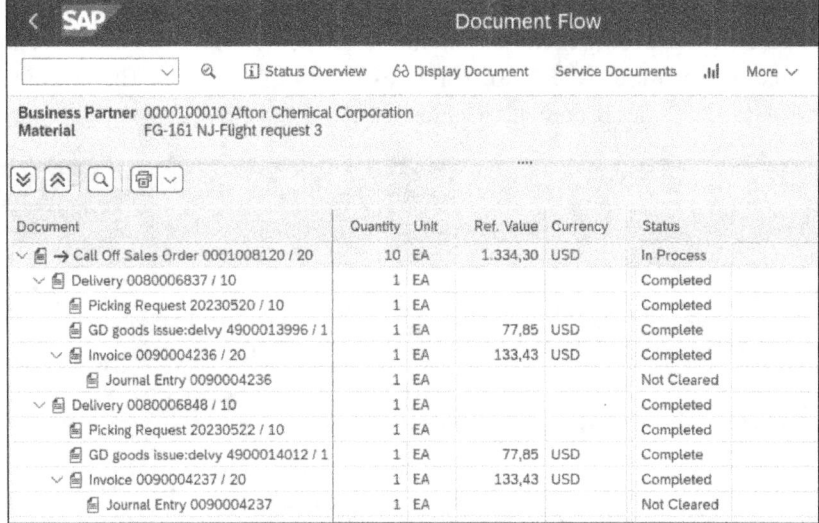

Figure 5.54 Relationship Browser for Accounting Documents

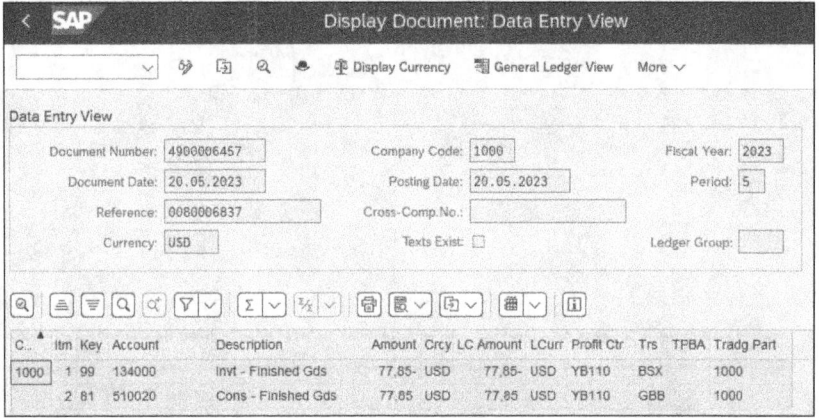

Figure 5.55 Display Goods Issue Accounting Document

Checking the material master Next, you can check the price in the material master in Figure 5.56, which is **77.85 USD** per **EA**. This matches the internal price in the sales order and the value in the goods issue.

Execute cost comparison In Figure 5.58, you see the value fields of the profitability document. The value for the goods issue matches the amount in the sales order, material master, and product costing. In other words, the configuration settings are correct, and the COGS in costing-based profitability analysis can be reconciled with financial accounting.

Checking the material cost estimate In Figure 5.57, you can look at the material cost estimate split in the cost components of the cost component structure. The value for 1 EA is 77.85 USD. So far, all values are matching. Finally, let's look at the profitability document.

5.8 Cost of Goods Sold Transfer and Split

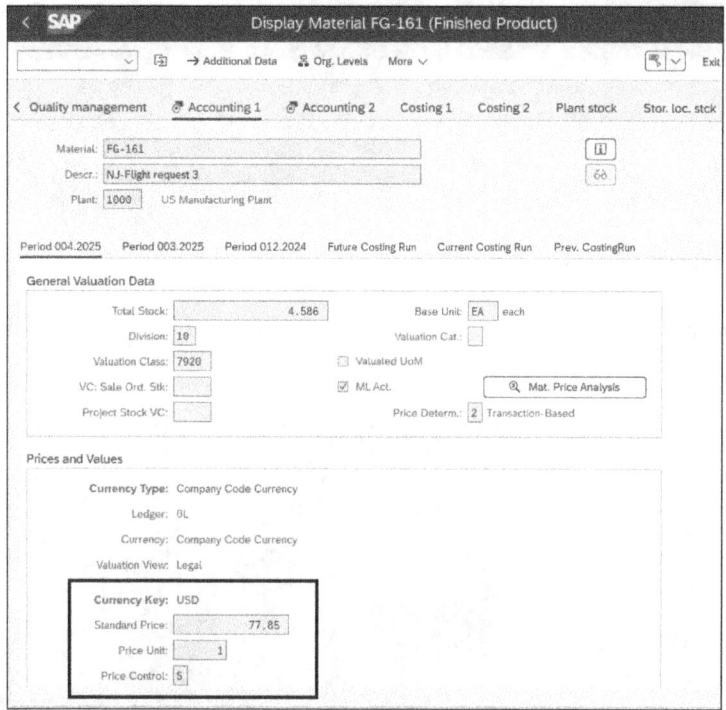

Figure 5.56 Review the Standard Price in the Material Master

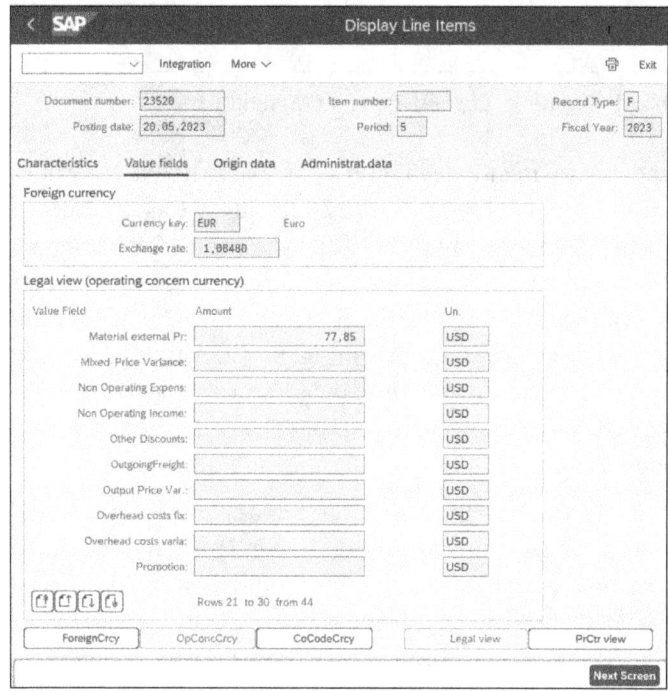

Figure 5.57 Review Value Fields in the Profitability Document

5 Costing-Based Profitability Analysis

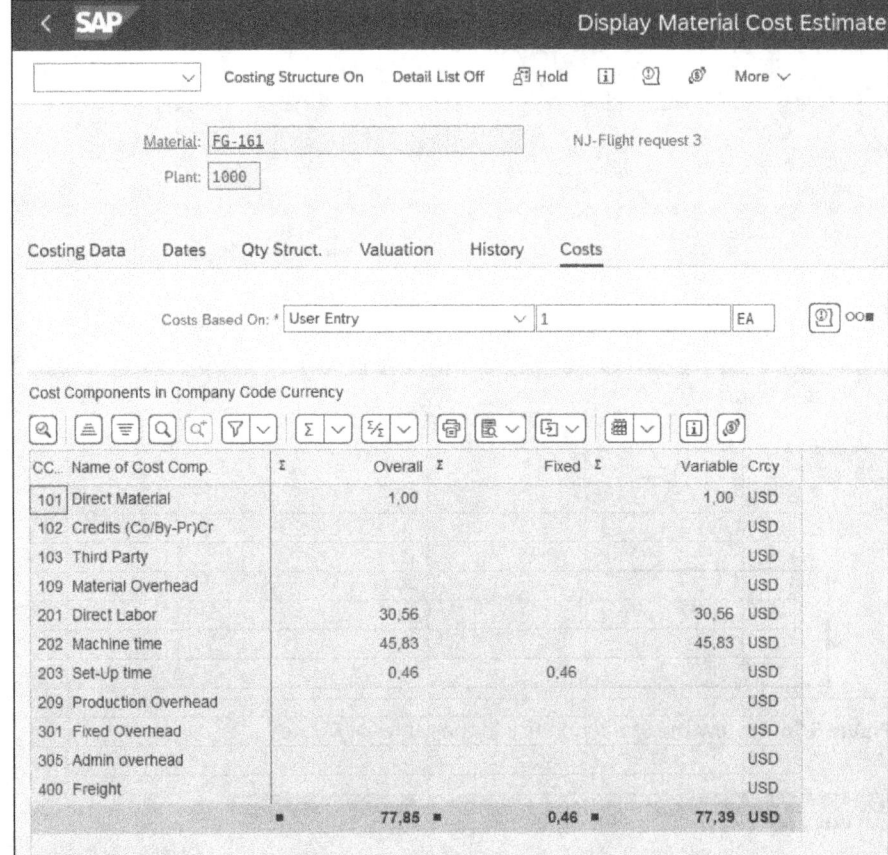

Figure 5.58 Display the Material Cost Estimate for Material

Re-derive values in the profitability document

Earlier in this section, we mentioned that it's possible to derive the values in a costing-based profitability segment of a billing invoice again after the document is posted. To do so, you need to add the point of valuation **02** (periodic revaluation of actual data) to the valuation strategy shown earlier in Figure 5.45. You also need to add the point of valuation **02** to the assignment of the costing key shown previously in Figure 5.50. This functionality comes in handy if you had an error in the configuration of the transfer and the splitting of COGS. You can revaluate your billing invoices with Transaction KE27 (Periodic Valuation).

In this section, you learned how to transfer COGS and split COGS in costing-based profitability analysis. You learned how to configure the transfer to still be able to reconcile your financial accounting with costing-based profitability analysis. If there were any errors in your COGS derivation, we shared with you a transaction to subsequently transfer your profitability documents.

5.9 Variance Calculation

In Section 5.7, we quickly touched on variances in the production process and that you need to settle them to costing-based profitability analysis. In this section, we want to explain how variances occur and what configuration settings are required to settle them correctly.

In Figure 5.59 in the next subsection, you can see on the right side an example profitability report in costing-based profitability analysis. We'll explain how costs from production will find their way to costing-based profitability analysis.

We explained the upper part of the example profitability report already; you know how sales, sales rebates, and COGS find their way into costing-based profitability analysis, as explained in Section 5.6 and Section 5.7. So, you understand the processes and value flows in costing-based profitability analysis to get to gross margin. After the gross margin, you see production variances. Production variances are determined on production orders at month end, as well as on production cost centers. Let's look at the value flow of the variances.

Gross margin in costing-based profitability analysis

5.9.1 Cost Center Types

You can see in Figure 5.59 a number of different cost center types. They all receive their costs from different functionalities such as financial accounting, asset accounting, human resources, accounts payable, and so on. Now let's take a more detailed look into the types of cost centers and what kind of costs they are collecting:

Different cost center types

- **Allocation Cost Center**
  On an allocation cost center, costs are collected that can't be assigned directly to a functional cost center, for example, energy and rent. You receive one bill for the whole company amount, but it's too complicated to split that bill when you enter it in financial accounting and assign it to the different cost centers. Instead, you collect all costs on one cost center and can allocate them at month end with a cost center assessment to other cost centers based on various criteria, for example, the size of the site.

- **Service Cost Center**
  There are usually several different service cost centers in a company that has a production aspect, but even nonproducing companies have service cost centers. Those are cost centers for maintenance, IT, HR, production support, and so on. All of those cost centers provide services to other functions. Service cost centers either get allocated to a cost center they

are providing services for (similar to the allocation cost centers), for example, by number of employees assigned to a cost center, or they allocate the activities by number of hours they provided service for another cost center.

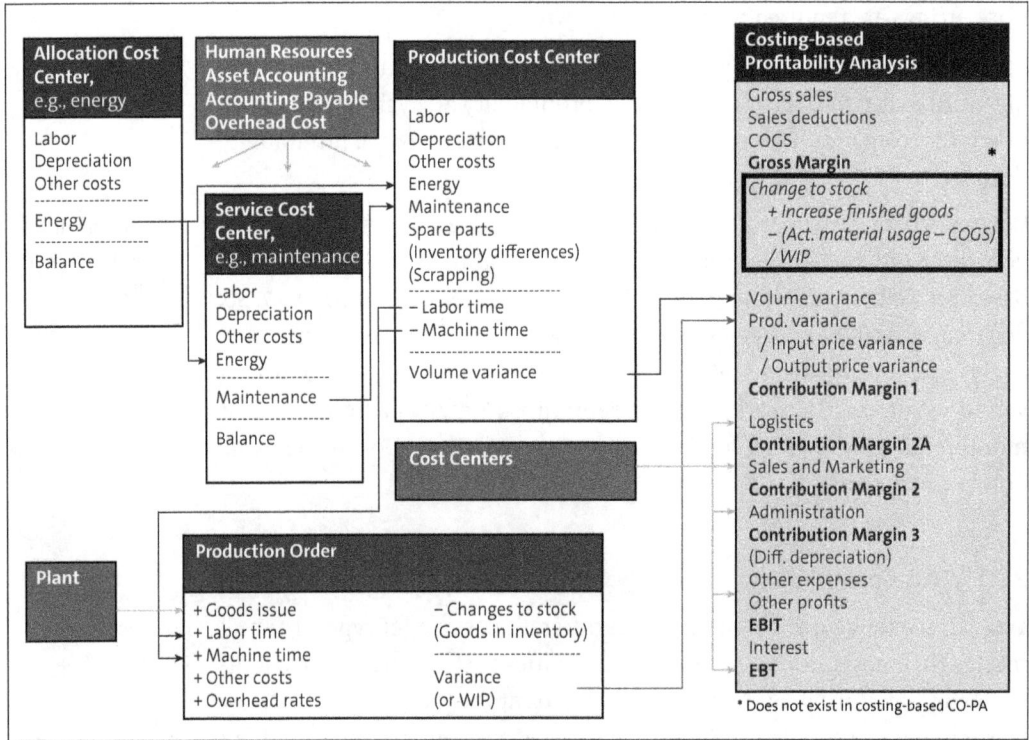

Figure 5.59 Value Flow for Variances

- **Production Cost Center**

    On a production cost center, you collect all costs associated with production, such as labor cost, depreciation of machines, energy, and so on. The objective is to assign the total production costs to the goods manufactured so that at month end, the balance of the cost center equals to zero. This requires a lot of detailed planning and rarely happens as there are usually variances. The costs on the production cost centers are used to determine the price for a personnel hour and machine hour, as well as any other prices that are required for the production.

Price determination on production cost centers   In the planning process, the expected total production costs are determined for each cost center. On the basis of the total costs, output quantity, and personnel hours, you can determine the price. Here, you assume that the cost center is credited to 100%. You can determine your prices manually, but the

SAP system also provides an automatic price calculation option. These prices are then used in the material cost estimate of the goods manufactured to cost the personnel hours and machine hours that are required to produce the product. The routing defines how many machine hours or personnel hours are used to manufacture the product. The COGM for the finished goods is determined from the material costs, the costs for working hours/machine hours, and potential overhead rates. The COGM is the standard price of a product and is updated in the material master. These costs are the basis for the inventory valuation and therefore are used for any valuation of material movements in the system. Any balance at month end on a production cost center has to be allocated to costing-based profitability analysis with a cost center assessment cycle. You'll learn how to create cost center assessments in Section 5.10.

- **Overhead Cost Center**
  Overhead cost centers collect costs that can't be allocated to the production process or aren't of service for any other cost center but are still required in a company to operate, such as sales, marketing, finance, and so on. Those cost centers get allocated directly to costing-based profitability analysis at month end.

- **Production Order**
  When you create a production order, the system resolves the BOM and routing to determine the target costs of the order. With goods issued on the order and confirmations of hours for the production order, costs are debited to the order. If the manufacturing process for the order has been completed, the receipt of the manufactured product is posted to the warehouse/inventory. The goods movement for the delivery of the finished goods to the warehouse is valuated using the standard price (or the price according to the price control in the material master), which reflects the price determined in the material cost estimate.

After the production is complete and the order shows a balance, this is referred to as a variance. The total of the actual costs doesn't correspond to the calculated costs. There can be several reasons for this, for example, more or fewer material components were consumed, more or fewer working hours were necessary, or price fluctuations occurred for material components. These variances are determined at month end and are settled to costing-based profitability analysis. The settlement of the production order also creates a financial posting. The change of stock is corrected by the variance. We'll explain this in more detail with an example later in this section.

Variance on production orders

Figure 5.60 provides a simplified overview of the production value flow.

5   Costing-Based Profitability Analysis

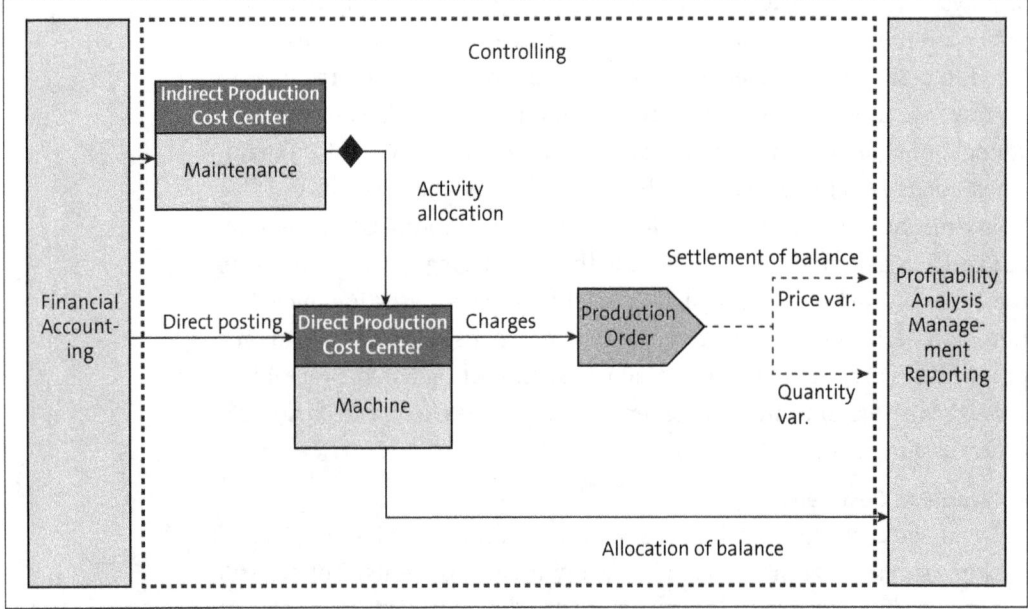

Figure 5.60 Value Flow in Production

Work in process
: At the end of the month, WIP is determined on the basis of the production orders/process orders that have neither status **TECO** (**Technically Completed**) nor status **DLV** (**Delivered**). When the orders are settled, the system generates a posting that doesn't affect the net income and capitalizes the actual costs of the order as WIP in the balance sheet.

Variance calculation
: For orders that have already been delivered or completed, the system determines variances and the type of variance. For this purpose, the SAP system uses various default variance categories, which group variances into those on the output side and those on the input side. When the orders are settled, the system posts a price difference in financial accounting and corrects the change in the stock account. This posting doesn't affect the net income. In traditional SAP systems, the system uses the account that is defined in the materials management in SAP S/4HANA account determination of account grouping PRD-PRF to post variances in financial accounting.

Splitting price differences
: In costing-based profitability analysis, you can settle the variances and post the values on separate value fields by variance category.

5.9.2   Variance Calculation and Settlement of Process Order

In Chapter 4, we discussed how to configure and execute the variance calculation. Let's discuss how we set up the the settlement process for costing-based profitability analysis.

5.9 Variance Calculation

To settle variances, you have to create a PA transfer structure. To create or display a PA transfer structure, go to Transaction KEI1 or use the configuration path **Controlling** • **Profitability Analysis** • **Flows of Actual Values** • **Settlement of Production Variances** • **Define PA Transfer Structure for Variance Settlement**.

Creating PA transfer structures: Transaction KEI1

In Figure 5.61, you see an overview of the PA transfer structure that exists in the system. Select the PA transfer structure **A1 – Settle Production Variances** by selecting the checkbox at the beginning of the line. Navigate to the **Dialog Structure**, and go to the **Assignment lines** folder.

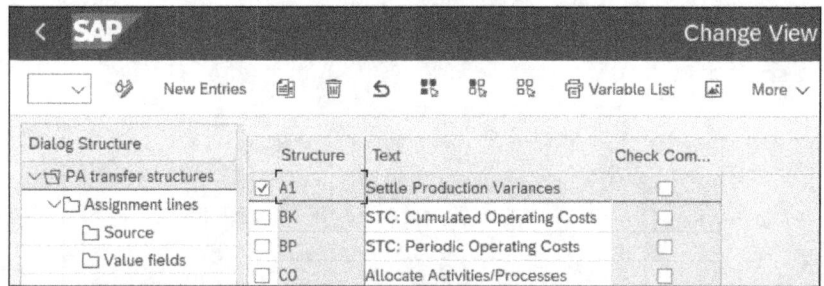

Figure 5.61 Overview of PA Transfer Structures

In Figure 5.62, you see the assignment lines of the PA transfer structure. To settle each variance category to a different value field, you need at least one assignment line per variance category. Select the checkbox at the beginning of **Assignment 10** for **Input price variance**, and navigate to the **Source** folder in the **Dialog Structure**.

Display the assignment lines of PA transfer structures

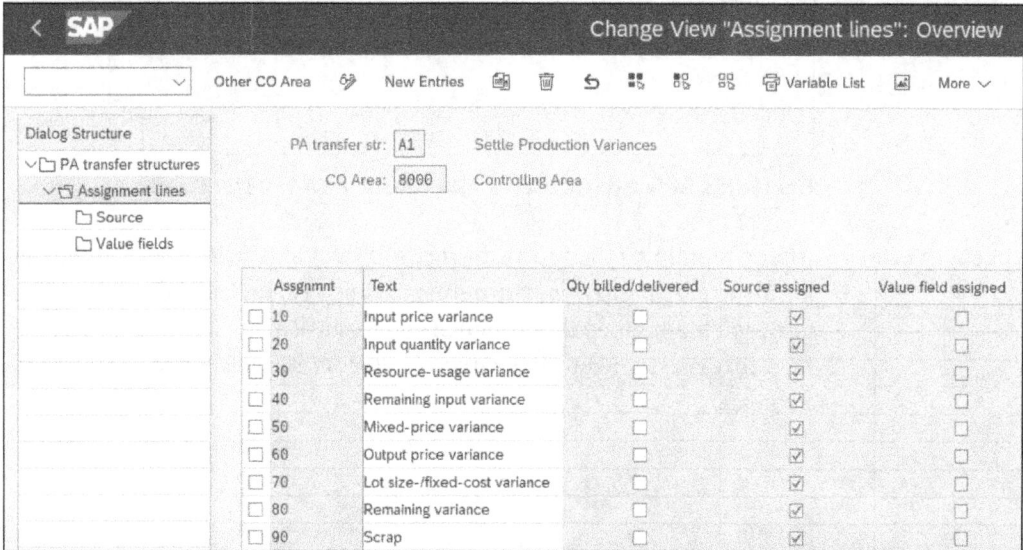

Figure 5.62 Assignment Lines in the PA Transfer Structure

5 Costing-Based Profitability Analysis

Display the PA transfer structure source

In the source, you define which cost elements will be settled with the selected assignment line. In Figure 5.63, select the cost element interval from **0** to **Z** in the **Cost Element** area. In the **Source** area, select **Variances on production orders** and the variance category that you want to settle with the assignment line. For this example, choose **PRIV** (**Input Price Variance**).

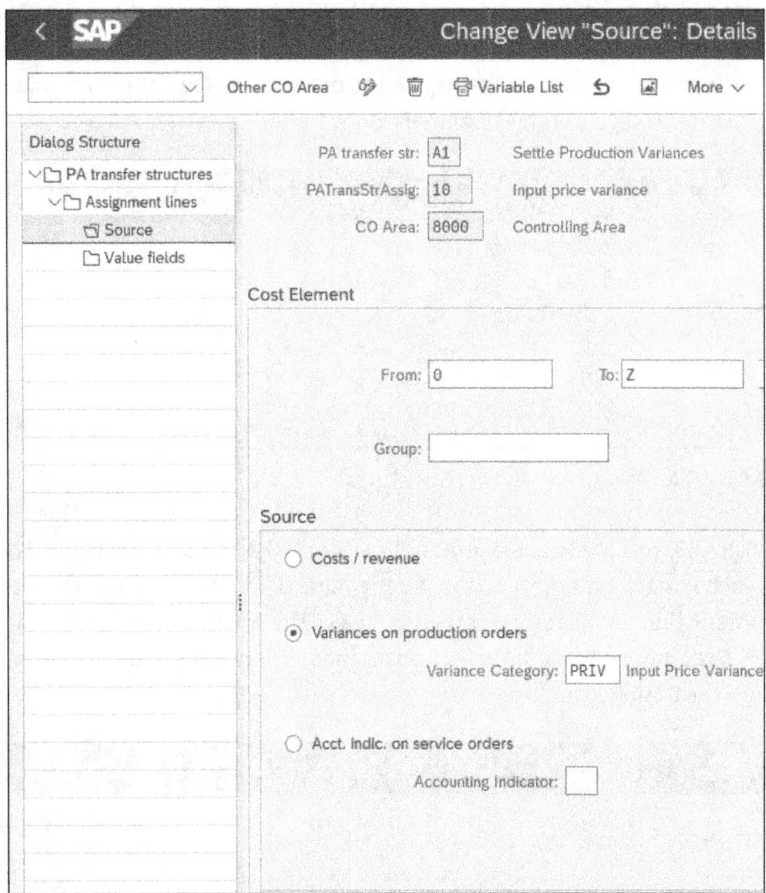

Figure 5.63 Define the Source of Assignment in PA Transfer Structure

Display PA transfer structure value fields

After maintaining the source, navigate to the **Value fields** folder in the **Dialog Structure**. In this area, you define to which value field the variance category you maintained in the **Source** will be settled. In Figure 5.64, the value field that is assigned to the variance category scrap is **VV110 – Input Price Variance**.

5.9 Variance Calculation

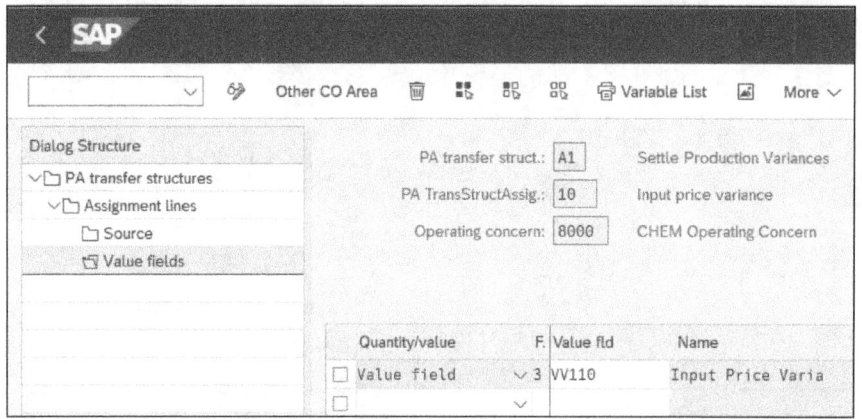

Figure 5.64 Value Fields Assigned to Source in the PA Transfer Structure

After reviewing the PA transfer structure, you need to make sure it's assigned to a settlement profile. Go to configuration path **Controlling • Profitability Analysis • Flows of Actual Values • Settlement of Production Variances • PA Transfer Structure • Assign Settlement Profiles** to review the settlement profile. In Figure 5.65, select the checkbox at the beginning of line **Z800 – Process Order**, and look at the details with 🔍 (Details).

Maintaining settlement profiles

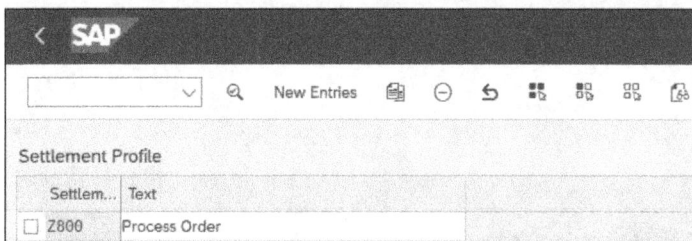

Figure 5.65 Review Settlement Profile

In Figure 5.66, you see the details of **Settlement Profile Z800 – Process Order**. In the **Default Values** area, check that the previously reviewed PA transfer structure, **A1** for **Settle Production Variances** is assigned. In the **Indicators** area, select the **Variances to Costing-Based PA** checkbox; then, variances are settled in costing-based profitability analysis when the orders are settled. If you don't select this, the system doesn't create a profitability analysis document for costing-based profitability analysis. Now, make sure that in the **Valid Receivers** area, the receivers **Material** and **Profit. Segment** (profitability segment) are allowed receivers.

Display settlement profile details

5 Costing-Based Profitability Analysis

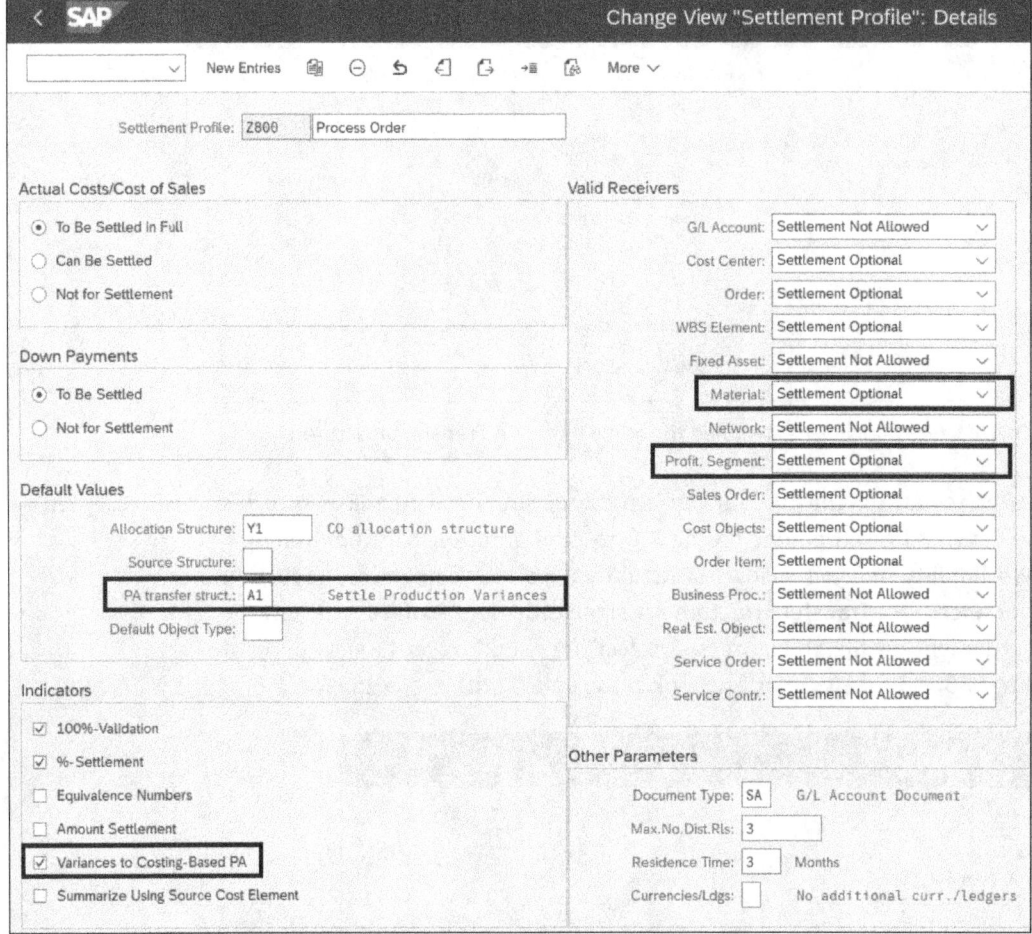

Figure 5.66 Settlement Profile Details

To run the settlement, go to the Run Actual Settlement app or the Run Settlement app. In addition, you can settle orders via collective processing with the Settle Orders app.

Orders are settled on a monthly basis. The order settlement doesn't differentiate from the margin analysis, which is why we won't further explain it in this chapter.

5.10 Cost Center Assessment

To transfer cost center balances to costing-based profitability analysis, you have to create cost center assessment cycles. The cost center assessment cycles are executed on a monthly basis. The creation of cost center assessment cycles is very similar to the margin analysis. You create the cost center assessments with the Manage Allocations app. For further information, review the cost center assessment for margin analysis in Chapter 4.

Before you create a cost center assessment cycle, you should put some thought into the cost center assessment process flow by considering the following parameters:

- **Sender**
  Which cost centers or cost center groups will have a balance at month end that needs to be transferred to costing-based profitability analysis?

- **Cost element**
  Which cost elements do you want to transfer? We recommend working with cost center groups as they are easier to adjust when a new G/L account is created. To have an overview of all your assessment cycles and their senders as well as receivers, it's recommended to create an overview document that lists all the details. Otherwise, you might miss a new cost center or cost element in your month-end cost center assessment and spend unnecessary time and effort finding the error and correcting it.

- **Receiver**
  After defining the sender cost center, you have to specify which value field will receive the costs in profitability analysis. Additionally, you have to specify the characteristics and characteristic values of the receiver in costing-based profitability analysis. The number of receivers is determined by the combination of all values (10 customers and 10 material groups lead to 100 profitability segments).

- **Assessment cost element**
  To transfer costs to costing-based profitability analysis with a cost center assessment, you have to create a secondary cost element with cost element category 42 for assessments.

5.11 Direct Account Assignment

Direct account assignment allows you to assign values directly to the profitability document without any order settlements or cost center assessments. Direct account assignment is often used to post scrapping costs, for example. If you assign scrapping costs to cost centers, the information on the material is lost during the allocation of the costs to costing-based profitability analysis. However, if you assign scrapping costs directly to costing-based profitability analysis, you can transfer more detailed information such as the material and all from the material-derived characteristics to costing-based profitability analysis.

5 Costing-Based Profitability Analysis

Maintaining direct account assignments: Transaction KEI2
Call Transaction KEI2 or follow the configuration path **Controlling · Profitability Analysis · Flows of Actual Values · Direct Posting from FI/MM · Maintain PA Transfer Structure for Direct Account Assignments** to create a direct account assignment for costing-based profitability analysis.

Review PA transfer structures
In Figure 5.67, you see an overview of all PA transfer structures. Select the **FI** checkbox in the **Structure** column. This PA transfer structure is created by default from SAP. Navigate to the **Assignment lines** folder in the **Dialog Structure**.

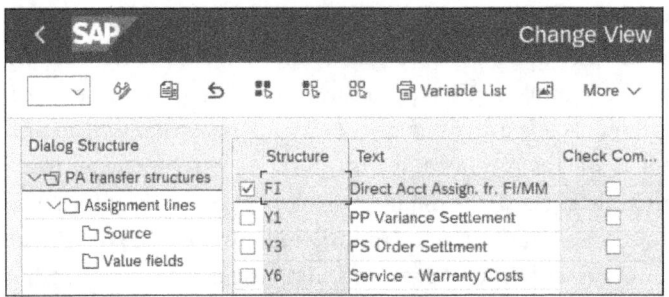

Figure 5.67 Overview of PA Transfer Structures

Review assignments
In Figure 5.68, you see an overview of already created assignments in the PA transfer structure **FI**. You can create a new assignment with ☐ (New) or F5.

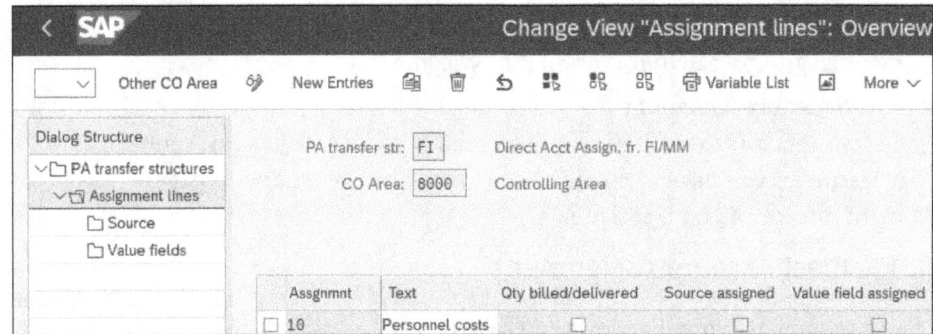

Figure 5.68 Overview of Assignment Lines in the PA Transfer Structure

Create new assignments
Figure 5.69 shows different assignments. Let's take a closer look at **100** for **Scrap**. Next, navigate to the **Source** folder in the **Dialog Structure**.

Define the source
In Figure 5.70, select **Cost Element Group SCRAP_CE** as the source. All postings in this cost element should be transferred to the value field you'll maintain in the next step. Here, navigate to the **Value fields** folder in the **Dialog Structure**.

5.11 Direct Account Assignment

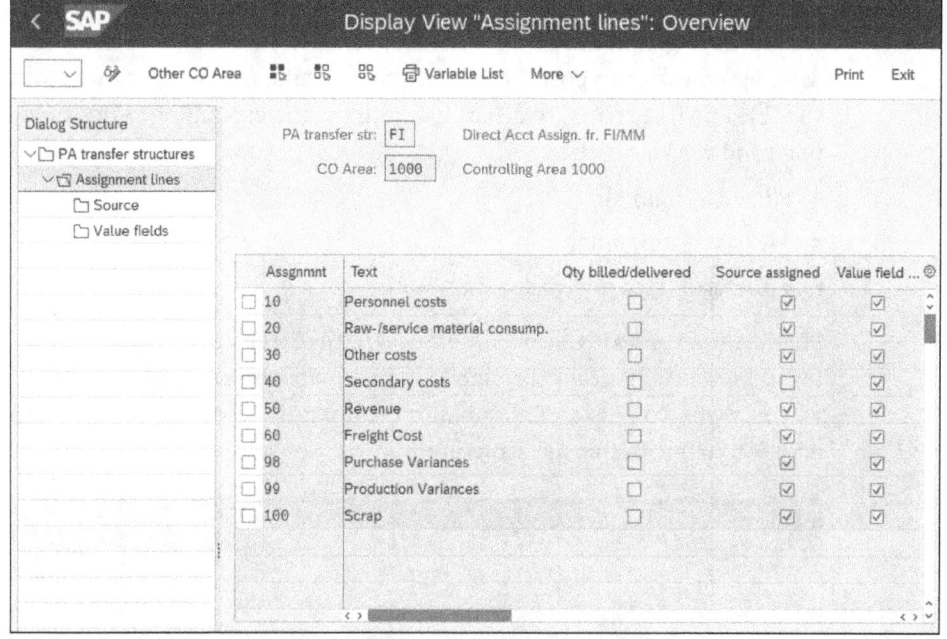

Figure 5.69 Create Assignment in the PA Transfer Structure

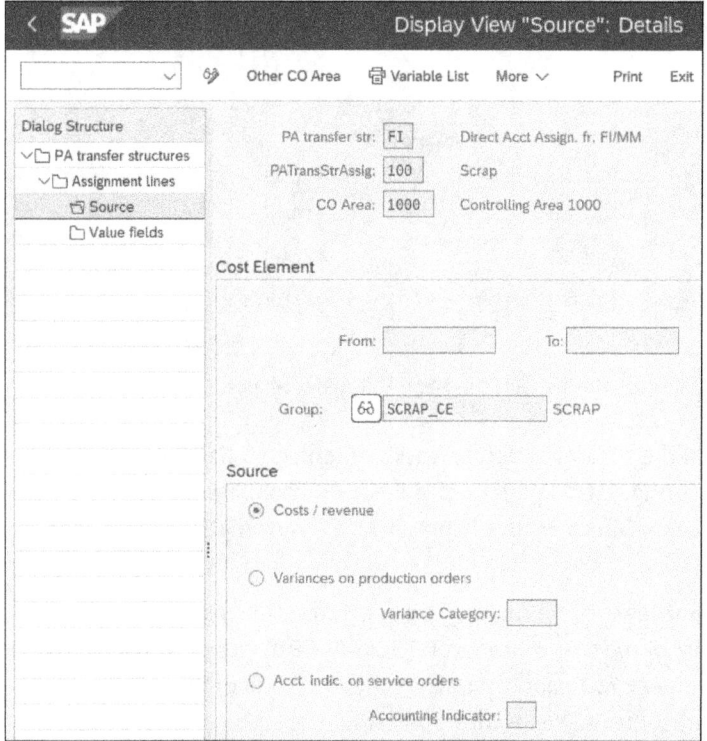

Figure 5.70 Maintain the Source in Value Field Maintenance

5   Costing-Based Profitability Analysis

Assigning value fields

With **New Entries** or F5, you can assign a value field in Figure 5.71. In this example, assign the **VVSCR** value field for **Scrap**. In column **F**, you can define whether you want to post the total costs to a value field or split them into variable and fixed costs. To do so, maintain the corresponding indicator for fixed and variable costs:

- **1** (fixed amounts)
- **2** (variable amounts)
- **3** (totals of fixed and variable amounts)

This example posts inventory differences to the **Scrap** value field. This posting is done automatically through materials management in SAP S/4HANA, which is why you have to maintain the automatic account assignment in addition to the PA transfer structure.

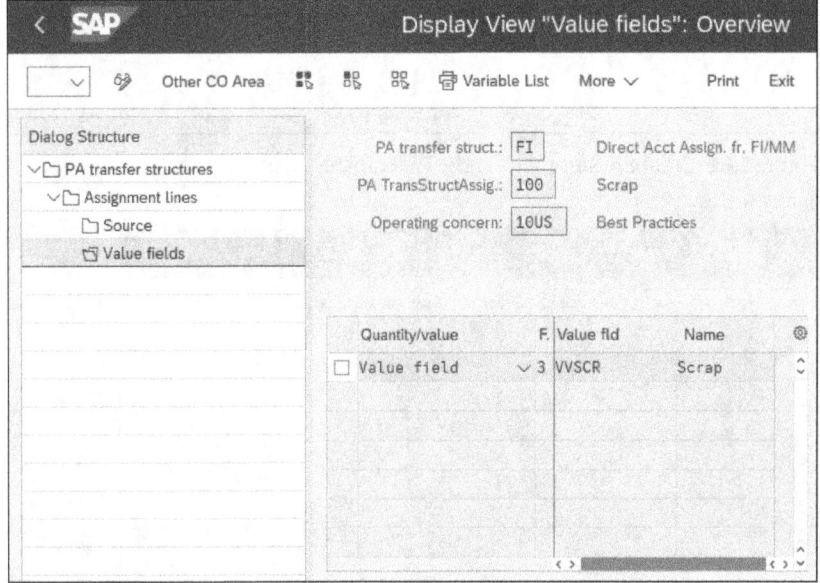

Figure 5.71 Maintain Value Fields in the PA Transfer Structure

Maintaining default account assignments: Transaction OKB9

To maintain the automatic account assignment, go to Transaction OKB9 or follow the configuration path **Controlling • Profitability Analysis • Flows of Actual Values • Direct Posting from FI/MM • Automatic Account Assignment**.

With **New Entries** or F5, you can create an entry for cost element **520060**, which is part of the cost center group **SCRAP_CE** in company code **1000**, as shown in Figure 5.72. In addition, select the checkbox in the **PrfS** (profitability segment) column to allow direct posting to the profitability segment.

280

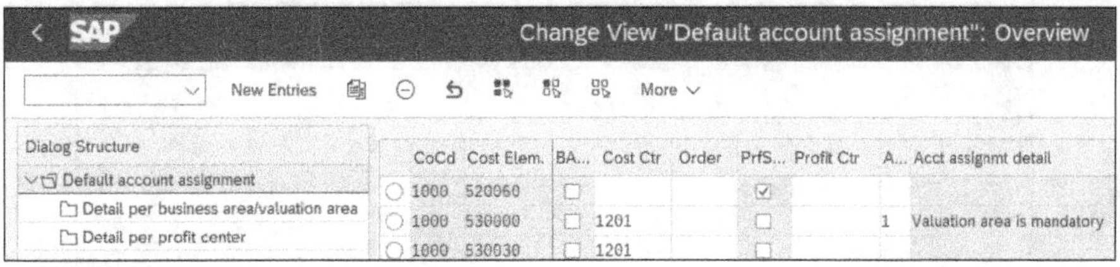

5.11 Direct Account Assignment

Figure 5.72 Create an Automatic Account Assignment

Let's post a journal entry to SAP S/4HANA and look at the document. In Figure 5.73, you can display the financial accounting documents. You see that there has been a profitability document created for costing-based profitability analysis.

Review material document

The profitability segment of costing-based profitability analysis opens showing in Figure 5.74 that the system derived the characteristics of the journal entry which was posted, such as **Product** and **Plant**.

Review characteristics in the profitability segment

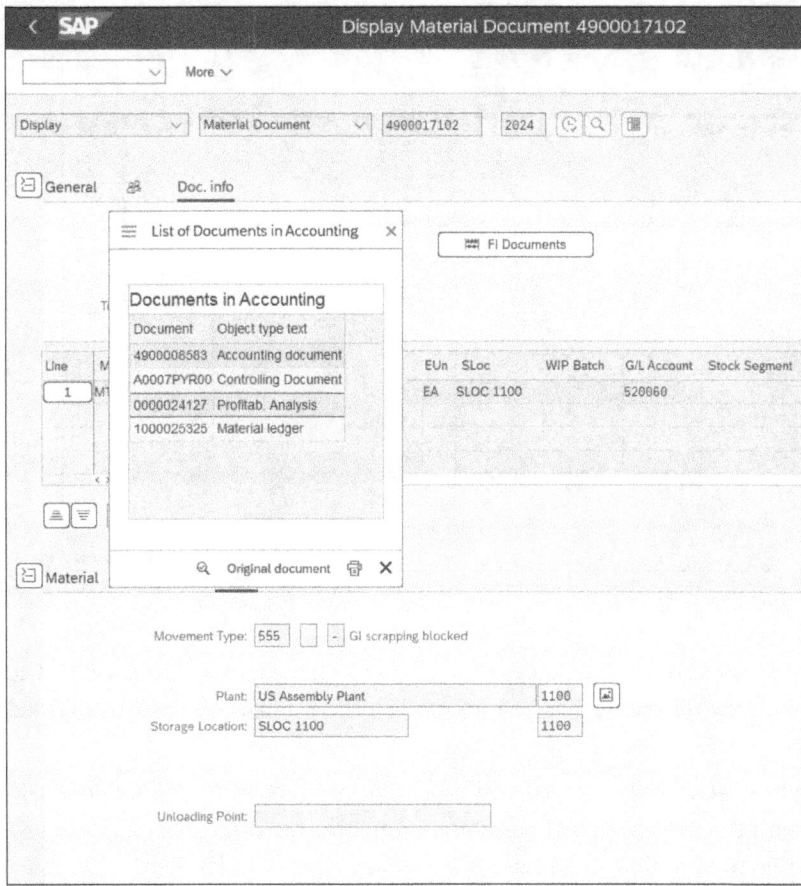

Figure 5.73 Display Details of the Journal Entry

281

5   Costing-Based Profitability Analysis

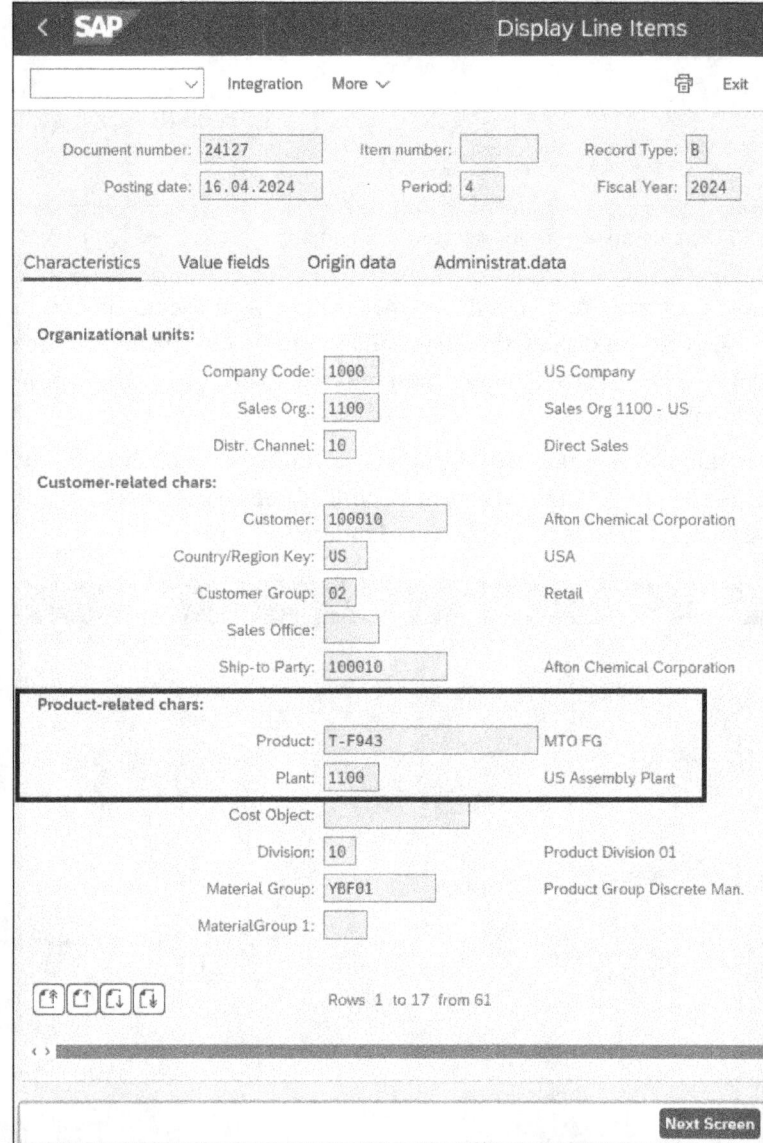

Figure 5.74 Display Characteristics of the Profitability Segment

Review value fields in the profitability document

Let's look at the **Value fields** in the tab of the same name in Figure 5.75. The inventory difference has been posted to the **Scrap** value field, as defined earlier in Figure 5.71.

In this section, we showed you how to post directly to a profitability segment. This is extremely helpful for automated postings that originate in a different module such as materials management in SAP S/4HANA. It allows you to derive more characteristics than when the cost would be posted on a cost center or an internal order first.

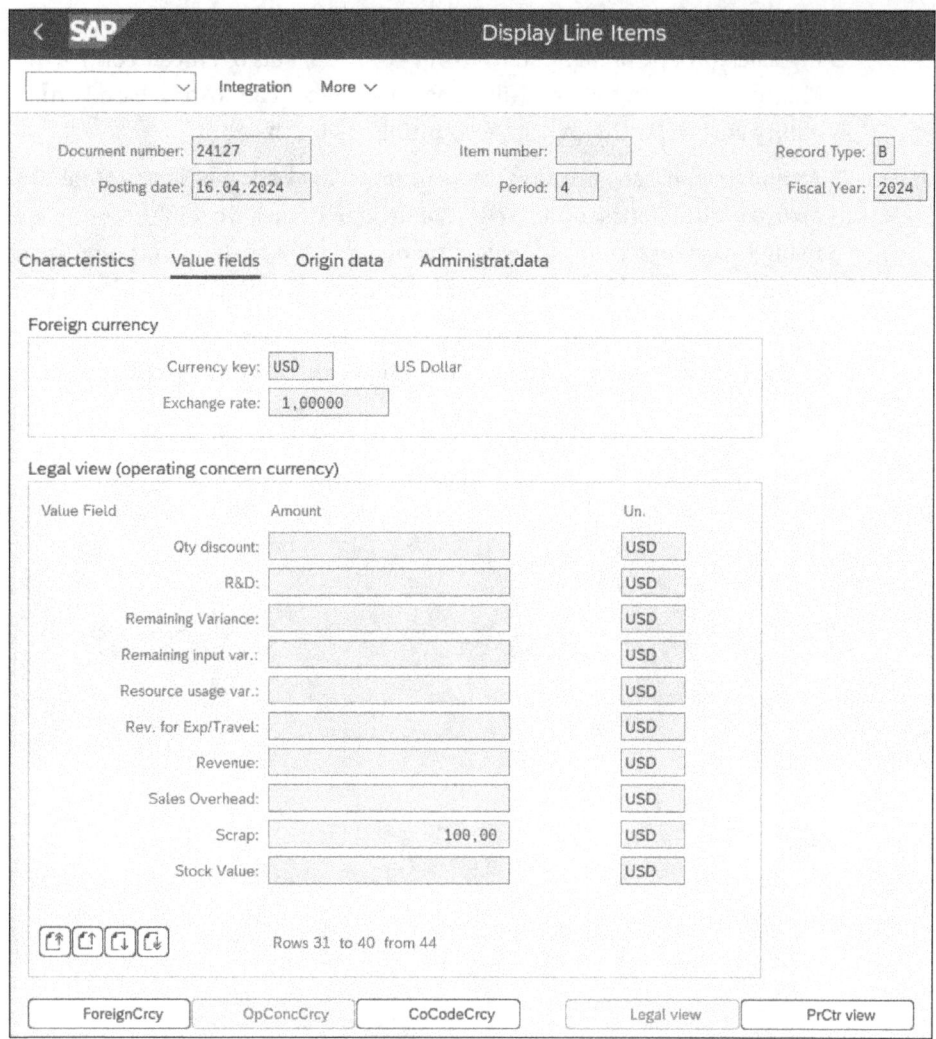

Figure 5.75 Display Value Fields of the Profitability Document

5.12 Summary

This chapter explained how value fields for amounts and quantities are created. It showed you how to get an overview of all value fields that are available in the system and which details you need to maintain in the creation of value fields. Only after the assignment to the operating concern are value fields available for use in costing-based profitability analysis.

We showed you how costing-based profitability analysis receives values. We explained the value flow, which is important to understand to make the right choices in configuration for costing-based profitability analysis.

You saw how to configure costing-based profitability analysis from incoming sales orders, billing data transfer to COGS, and overhead controlling. The configuration settings will allow you to reconcile costing-based profitability analysis with financial accounting at month end.

Although you can also create documents in costing-based profitability analysis with Transaction KE21N (Create Line Items), we don't recommend using this transaction as the documents are only posted in costing-based profitability analysis, so no reconciliation with financial accounting is possible.

In the next chapter, you'll learn how to execute planning in margin analysis.

Chapter 6
Planning

This chapter provides useful information on the future of planning in SAP S/4HANA Finance. It describes the changes for planning in margin analysis and explains how these planning strategy changes affect your business processes.

The planning functionalities in margin analysis and costing-based profitability analysis allow you to plan sales, cost of goods sold (COGS), and overhead costs on different levels and dimensions. You can plan your contribution margin with any characteristic combination, under the assumption that the characteristic is created and assigned to the operating concern. This allows you a lot of flexibility in your planning process.

The planning in margin analysis and costing-based profitability analysis is integrated, meaning you can create your plan data in margin analysis and costing-based profitability analysis, but you can also receive data from other modules such as internal order settlements, cost center assessments, or standard cost for materials from different material cost estimates.

Integrated planning

You can also create your planning in different versions. Every version reflects a different scenario with different assumptions. The planning in margin analysis and costing-based profitability analysis isn't time specific. You can plan for multiple years in the future, and, in costing-based profitability analysis, you can even plan based on the calendar week level. A lot of planning tools also allow you to roll over the planning from the previous year as a starting point. Along with these, many more tools are available, such as top-down planning, that help you create your planning scenarios. We'll introduce these tools in this chapter.

Planning with different versions

With SAP S/4HANA Finance, the planning strategy in the SAP system changes. This particularly affects margin analysis as well as the components for financial accounting and controlling in the SAP system. The go-to tool for planning in financial accounting and controlling is SAP Analytics Cloud. Planning in costing-based profitability analysis will still take place in the SAP S/4HANA core functionality. As there are no further developments to costing-based profitability analysis expected, there have been no changes to the planning functionalities either. We won't describe planning functionalities in costing-based profitability analysis in this chapter.

SAP S/4HANA planning strategy

285

6 Planning

Sections in this chapter

In this chapter, we'll discuss the changes in the planning strategy in SAP S/4HANA Finance and its impacts on margin analysis as well as the future road map. Then, we'll show you how to create plan data in margin analysis.

By the end of this chapter, you should understand how to create plan data in margin analysis. You'll also have gained an overview of all available planning tools to support your planning process.

6.1 Planning in Margin Analysis

Universal Journal for plan data: table ACDOCP

Planning in SAP S/4HANA is changing dramatically from the classic planning transactions that we know from SAP ERP. SAP S/4HANA introduced table ACDOCP, which is the planning equivalent to Universal Journal table ACDOCA for the actuals. Table ACDOCP is a step toward providing full integration between planning and actuals in SAP S/4HANA. With the SAP Fiori app Import Financial Plan Data (F1711), you can import plan data into table ACDOCP. However, this data isn't automatically available for all controlling transactions, such as plan allocations, activity planning, and so on, which is why users were still using classic plan transactions from SAP ERP. All controlling planning functionalities are available in SAP Analytics Cloud and replicate into table ACDOCP for further processing in SAP S/4HANA.

SAP Analytics Cloud

The strategic road map for planning in SAP S/4HANA identifies SAP Analytics Cloud as its planning tool. All planning functionality that we know from SAP ERP is retiring. The classic planning transactions, such as Transaction KP06 for cost center planning, are already deactivated in SAP S/4HANA. You can reactivate them (see SAP Note: 2474069 – Reactivate G/L Planning) because, to date, not all functionalities are accounted for in SAP Analytics Cloud, but with every release, more functionalities are being added to SAP Analytics Cloud to provide at least the same functionality as in SAP ERP if not more.

Note that the classic planning functionalities can only be reactivated in SAP S/4HANA—not in SAP S/4HANA Cloud. It's not recommended to reactivate the classic planning functionalities (used in SAP ERP) if you're implementing SAP S/4HANA but to implement your planning functionalities in SAP Analytics Cloud instead.

Architectural landscape

So how does SAP Analytics Cloud fit into the architectural landscape of SAP S/4HANA? In Figure 6.1Figure 6.1, you can see that SAP Analytics Cloud sits on top of SAP S/4HANA with live data connection with SAP Analytics Cloud to give you access to transactional data in real time. SAP Analytics Cloud also allows you to import data from other SAP sources (e.g., SAP Ariba and SAP SuccessFactors), as well as non-SAP sources. As you remember, the actual data of the margin analysis is saved in the Universal Journal (table ACDOCA), and the planning data is saved in the planning Universal Journal (table ACDOCP) together with all other financial data. There are no separate planning tables for margin analysis, which is why you don't see any additional planning tables in Figure 6.1.

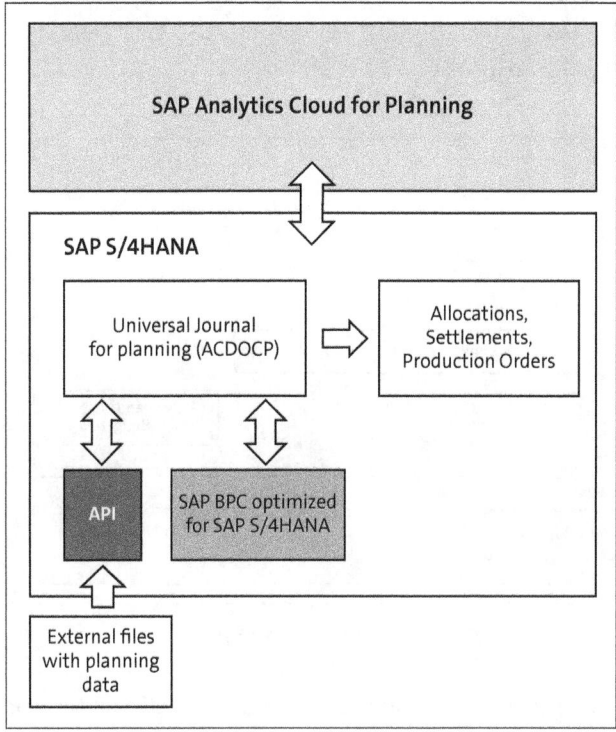

Figure 6.1 Architecture of SAP Analytics Cloud and SAP S/4HANA

The update of plans and forecast occurs in near real time and offers a full integration of various planning models. A lot of planning input templates are delivered with SAP Analytics Cloud that help you enter your planning data at different levels of detail. SAP Analytics Cloud comes with different predictive algorithms and planning capabilities that allow you to analyze your data and adjust your planning with predictive insight on the fly. So what planning functionalities are actually available in SAP Analytics Cloud? In Figure 6.2, you can see a business process overview of financial planning. SAP Integrated Business Planning (SAP IBP) will continue to be the central planning tool for supply chain planning. SAP Analytics Cloud and SAP IBP are integrated and allow both the receiving and sending of data for revenue and production cost planning.

Functionalities of SAP Analytics Cloud

> **Note**
> To learn more about how to integrate SAP Analytics Cloud with SAP IBP, check out SAP Note: 2721602 – Connection to IBP with SAP Analytics Cloud.

6 Planning

Controlling planning functionalities

As mentioned earlier, all planning functionalities in controlling have been added to SAP Analytics Cloud, so there's no more reason to continue planning with the classic planning functionalities that we know from SAP ERP. Although not related to planning, creating budgets for cost centers and activating an availability control on cost centers are also relatively new functionalities with SAP S/4HANA.

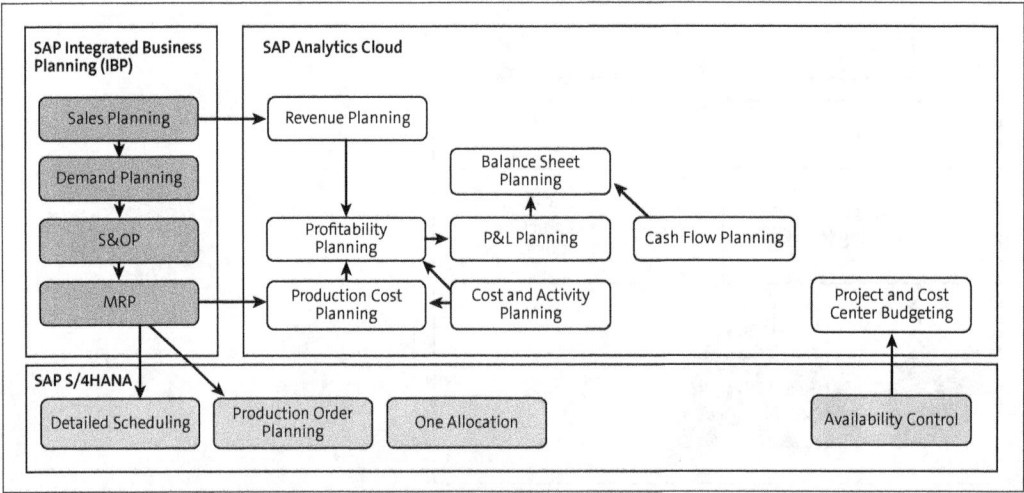

Figure 6.2 Business Process Overview of Financial Planning

Benefits of SAP Analytics Cloud

In the next sections, we'll show you how the planning functionalities for margin analysis look in SAP Analytics Cloud. Before we move on, here's a summary of the benefits of planning in SAP Analytics Cloud:

- **Faster decision-making**
  Integration with the Universal Journal and other SAP systems allows you to have immediate insight and react quickly to predictions/adjusted forecasts.
- **Simplified planning**
  The planning templates are intuitive and include analytic functionality that allows you to finish your operating planning in a shorter time so that you can focus on your strategic planning.
- **Instant insight**
  SAP Analytics Cloud updates your forecast in real time.
- **Predictive forecasting**
  SAP Analytics Cloud uses historical data to forecast likely future business scenarios.

- **Increased collaboration**
  SAP Analytics Cloud has a calendar functionality and allows you to create tasks and communicate with each other via notifications.
- **Visualization**
  Preconfigured dashboards give you an immediate overview of the planning and allow you to identify possible errors or gaps.

We'll show you these benefits in detail in the next sections.

6.2 Sales Planning

The revenue planning functionality in SAP Analytics Cloud is fully integrated with SAP IBP and allows data to be sent and received between the two systems. SAP Analytics Cloud delivers preconfigured sales and revenue plan input templates for sales prices, sales quantity planning, and revenue and deduction calculation. The master data and transaction data from SAP S/4HANA can be loaded in SAP Analytics Cloud to create plan data to reference actual data. Sales planning can be executed on a high level to a very granular level, in which every characteristic that you assigned to your operating concern is readily available for the creation of your plan data.

Revenue planning

Let's look at some input forms within SAP Analytics Cloud. There are several input forms for the different planning areas. Figure 6.3 shows the input form for the sales **Price Planning** based on the base unit of measure. You could include more characteristics in the planning layout if you want to create your planning data on a more detailed level. You can also copy prices or increase prices by a certain percentage to do simulations.

Plan data input forms

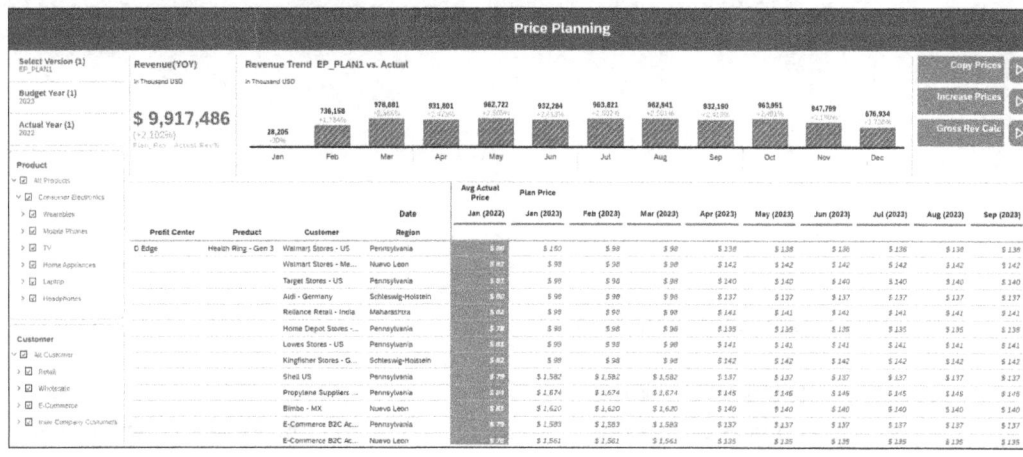

Figure 6.3 Sales Price Planning

6 Planning

Quantity planning and analysis

After the sales **Price Planning**, you can go to the **Quantity Planning and Analysis** template, as shown in Figure 6.4, to plan your sales quantities.

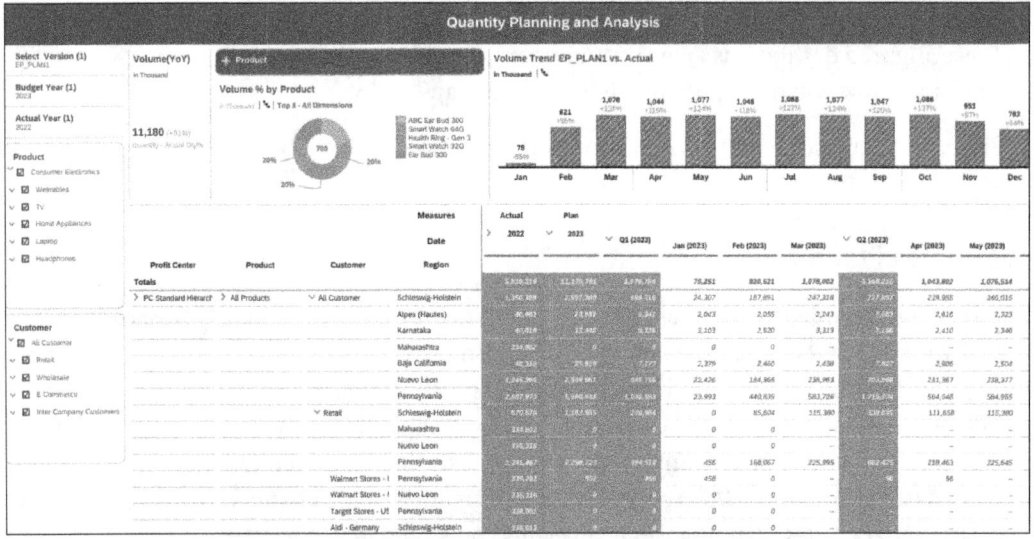

Figure 6.4 Sales Quantity Planning

Discount planning

Then, you can move on to the **Discount Planning** template. In Figure 6.5, you can plan absolute or percentage sales deductions. You can also create formulas to add more detailed logic to determine the sales deductions.

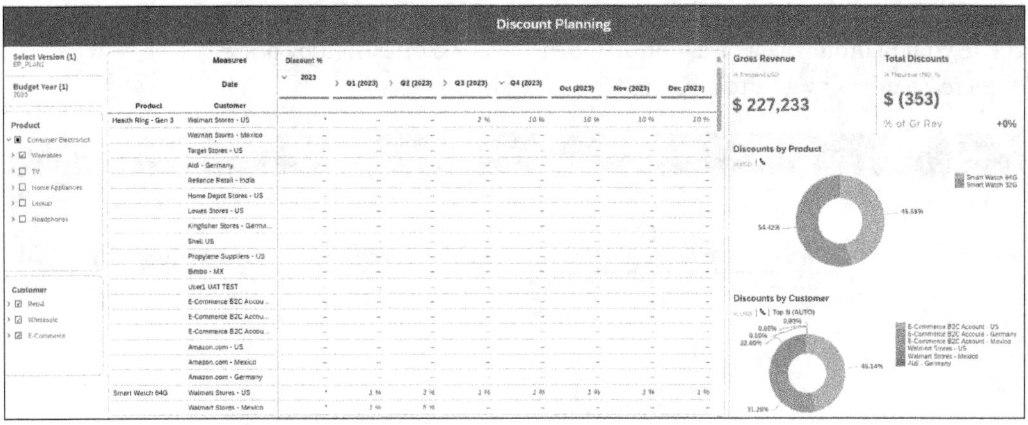

Figure 6.5 Discount Planning

P&L planning

Next, let's check the result of the planning for sales price, sales quantity, and sales deductions in the **Marginal Contribution Report** shown in Figure 6.6. From the inputs in the previous planning templates, the **Net Sales** is being calculated.

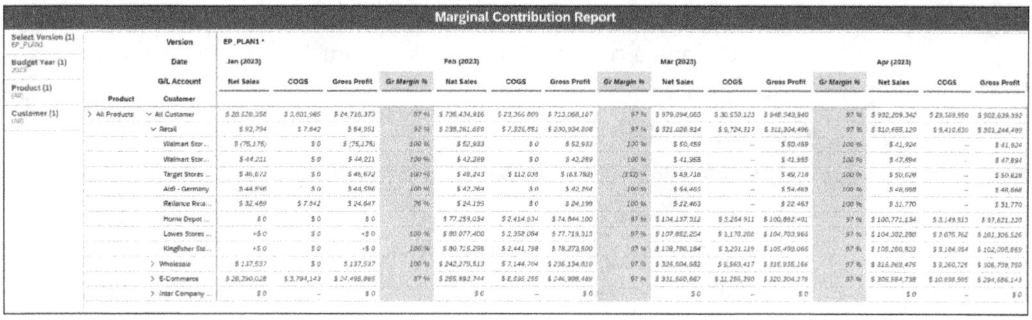

6.2 Sales Planning

Figure 6.6 P&L Planning

As in the classic planning, SAP Analytics Cloud also has planning aids:

- **Prepare Actuals**
  This planning aid is used to calculate the P&L. You've seen previously that we made entries in different planning templates such as sales price, sales quantities, and so on. With the **Prepare Actuals** functionality, those planning templates are used to calculate and create a plan P&L/Marginal Contribution.

- **Copy Actuals to Plan**
  With this planning aid, you can copy actuals in a planning version to use them as a basis to create plan data.

Planning aids in SAP Analytic Cloud

Many more planning aids are available as well. You can also do the following with your planning data in SAP Analytics Cloud:

- **Data action (including an advanced formula)**
  You can apply from easy to very complex formulas in every input planning sheet.

- **Data lock**
  You can lock data so that it's no longer changeable.

- **Smart copy**
  You can copy data sets or only single values.

- **Version management**
  You can work with different versions and therefore save different simulations of your plan data.

- **Undo/redo steps and undo/redo history**
  You can easily undo/redo steps even in the history if the data has already been saved.

- **Input task**
  You can create a task and assign it to a coworker who can submit the task after it's resolved. You'll get a notification if the task has been submitted.

Planning functionality in SAP Analytic Cloud

291

- **Data point comment**
  You can add a comment at any data point level that explains the planning data.
- **Advanced filter**
  You can create a filter with multiple conditions.
- **Member on the fly**
  You can create master data that isn't available in SAP S/4HANA, for example, to display new products or new customers for the year you're planning.
- **Value driver trees**
  This functionality is delivered out of the box and allows you to create what-if scenarios in your planning models.
- **Predictive**
  After you upload historical data into SAP Analytics Cloud, the machine learning capabilities will look for trends and include those in their predictive analysis.
- **Forecast/rolling forecast**
  You can create forecasts based on actuals that update themselves according to trends. In the rolling forecast, you limit the data range of the system to look back/look ahead to analyze data and include it in the forecast.
- **Advanced planning**
  Advanced planning comes with three functionalities that help you modify your plan data:
  - **Spreading**: Assign costs that you planned on a higher hierarchy level to lower levels. The costs are spread equally to the lower level if done automatically, but you can also spread your data manually.
  - **Distributing**: Move costs within the same hierarchy level to other characteristics which just a few clicks.

    Assigning: Add or replace values in planning. The system assigns an amount based on criteria and logic you defined.

As you can see, there are a lot of different possibilities to play with your planning data and make the planning process more efficient. Let's look at the reporting of the plan data that is delivered with SAP Analytics Cloud. In Figure 6.7, you see the **Sales Performance** report, which gives you an overview by key figure and allows you to see other plan versions and historical data. That way, you get a quick overview of your plan and also see important key figures at a glance. The report also allows you to execute what-if analysis.

6.3 Cost of Goods Sold Planning

Figure 6.7 Sales Plan Overview

6.3 Cost of Goods Sold Planning

SAP Analytics Cloud also enables COGS planning. As shown in Figure 6.2 earlier, product cost planning interfaces with supply chain planning. In Figure 6.8, you see the COGS planning; however, there's also another scenario in SAP Analytics Cloud that allows you to import cost estimates, evaluate the sales quantities with the cost estimates, and split the cost according to cost components. You have the same possibilities if not more to determine your plan COGS in SAP Analytics Cloud and integrate them in your overall planning scenario.

COGS planning

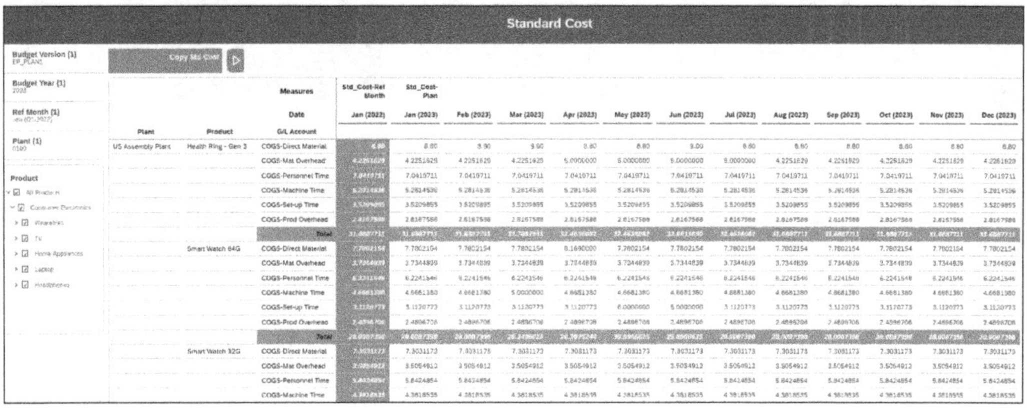

Figure 6.8 COGS Planning

6.4 Overhead Cost Planning

Cost center planning

Overhead cost planning also takes place in SAP Analytics Cloud. You can plan your cost centers on a cost center and cost element basis. Usually for margin analysis planning, you'll create a cost center allocation in the SAP Fiori app Manage Allocations, as explained in Chapter 5. You have access to this data and the plan allocations in SAP Analytics Cloud as well and can analyze the impact of the allocations on your plan data.

6.5 Summary

This chapter described the planning in margin analysis. The future tool for planning on the road map has been announced to be SAP Analytics Cloud.

Compared with the classic planning transactions, the planning in SAP Analytics Cloud is much more intuitive, and the calendar functionalities allow a streamlined, integrated process. Available features allow interaction with colleagues and status management for planning tasks that facilitate a smooth planning process. SAP Analytics Cloud offers a great user experience.

Chapter 7
Reporting

This chapter describes the reporting options in profitability analysis with SAP S/4HANA Finance and provides an insight into SAP Fiori apps for margin analysis.

SAP S/4HANA Finance provides various reporting options. The reports are more intuitive and allow for analyzing large data volumes from a high level to a very granular level of detail. Similar to planning, the changes in reporting mainly affect margin analysis, as this is the only form of profitability analysis that is getting any further developments in SAP S/4HANA. The reporting functions in costing-based profitability analysis basically remain the same; there are no innovations expected. The major enhancement for costing-based profitability analysis is that the SAP HANA database improves the performance considerably and avoids long runtimes. As there are no standard reports for costing-based profitability analysis, reports still have to be created via the Report Painter, which is available as an SAP Fiori app.

Reporting functionalities in SAP S/4HANA

7.1 Profitability Analysis Reporting Basics

The reporting in SAP S/4HANA for profitability analysis targets the following customer complaints:

Pain points of profitability analysis reporting

- It's hard to find the right information in reporting.
- A high level of manual effort is required to detect the root cause of issues, for example, dropped margins.
- No analytics-driven processes are used in reporting.
- The system doesn't provide adaptability of the analytics structure to changes in reality, such as changes to a profit center structure.

SAP S/4HANA embedded analytics changed the game for reporting mainly thanks to the Universal Journal as a single source of truth, which is often also compared to a large pivot table. The Universal Journal is the central repository for all information regarding finance and financial planning and analysis (FP&A), which includes margin analysis.

7 Reporting

Today, a lot reporting can be done easily in SAP S/4HANA, which is taking away the need for combining financial data with characteristics from profitability analysis, the reconciliation of financial accounting and profitability analysis as well as long runtimes for reports. SAP S/4HANA embedded analytics is the combination of SAP S/4HANA data with intelligent analytics, such as machine learning and visualization capabilities.

SAP S/4HANA embedded analytics

Actionable insights are embedded in business processes to shift the time invested in number crunching to investing time on acting to achieve accelerated and more accurate decisions. With SAP S/4HANA embedded analytics, actionable insights are embedded in the business process. The source of the data is based on the Universal Journal, which delivers real-time insights and a flexible drilldown to the lowest level of information. SAP Fiori apps show all relevant information for reporting in one screen and allows graphical views paired with an intuitive navigation. We'll show you some of those SAP Fiori apps in this chapter. The goal is to have a close integration of the analytics capabilities within the operational transactions to make it easy to act on the right information without having to switch between screens.

Reporting functionalities in SAP S/4HANA

SAP S/4HANA divides its reporting functionalities into three pillars:

- **Simple data models:**
  - Real-time data availability
  - Easy drill down
  - Accelerated performance
  - Single source of truth
  - simple data structures
- **Intuitive user experience (UX):**
  - Role-based users
  - Availability on any device
  - Combined screens with transactional and analytical information
  - All information provided in one screen
  - One unified UX
  - SAP S/4HANA embedded analytics
  - Cockpit with exception-based worklists
- **Intelligent technologies:**
  - Integration of predictive insights and machine learning
  - Usage of machine learning to detect anomalies
  - Process automation
  - Simulation capabilities
  - Next-generation business processes

These three pillars reinforce the SAP's vision regarding the following analytics capabilities with the help of machine learning:

- Actionable insights
- Automated root-cause detection
- Unified hierarchy maintenance to adapt reporting to real-time events

Another important development in reporting is the usage of SAP Analytics Cloud. We already talked about and showed some of the functionalities of SAP Analytics Cloud in Chapter 6. SAP Analytics Cloud is the strategic platform for planning and analytics in SAP. SAP Analytics Cloud can access SAP S/4HANA data in real time. Regarding analytics capabilities, SAP Analytics Cloud comes with predelivered content for reporting and delivers dashboards that can be easily created to satisfy your individual reporting needs. An SAP Analytics Cloud dashboard consumes a core data services (CDS) view sitting, for example, on top of the Universal Journal. With the transient analytical query technique, the data is made available in SAP Analytics Cloud without duplicating the data in the SAP Analytics Cloud environment. It's very beneficial to combine the reporting capabilities in SAP S/4HANA with SAP Analytics Cloud. **SAP Analytics Cloud**

SAP Analytics Cloud also allows you to create stories. A *story* is the visualization of data. The basis of creating a story is to create a model. The models are the foundation of the story and provide the data model for analysis. The story allows you to visualize your data and helps you discover insights hidden within your data. There are a lot of predefined stories or story templates in the form of a chart, geo map, table, 8-image layout, shape, or text. The stories can be launched from SAP Fiori launchpad in the form of an SAP Fiori app. The SAP Fiori app is automatically generated when the story is created. **Stories**

The integrated financial architecture of SAP S/4HANA and SAP Analytics Cloud allows you to analyze, simulate, plan, predict, discover, and collaborate not just on the legal entity level but also on the consolidated level so you can concentrate on acting on data instead of reconciling data. **Integrated financial architecture**

Besides SAP Fiori and SAP Analytics Cloud, there are other reporting tools such as SAP Datasphere, which is a comprehensive data service built on SAP Business Technology Platform (SAP BTP) that enables every data professional to deliver seamless and scalable access to mission-critical business data.

In the next sections, we'll introduce you to the reporting functionalities in SAP Fiori to show the breadth of choices in this area as well as the reporting capabilities.

7 Reporting

7.2 Predictive Accounting

Predictive accounting

In Chapter 4 we discussed predictive accounting, which is available with the use of margin analysis. Predictive accounting also comes with several SAP Fiori apps to support reporting functionalities. Predictive journal entries don't have an entry in table BKPF (Accounting Document Header) but existing SAP Fiori apps and reports have been adjusted to be able to display predictive journal entries. Any CDS-based financial accounting reports can immediately make use of predictive journal entries and therefore turn any report into a report with predictive insights.

Predictive journal entries

You can distinguish between predictive journal entries and "real" journal entries in reporting by displaying the ledger. predictive journal entries are only created in the extension ledger, which sits on top of the leading ledger. It's important for the audit trail to display the source ledger when analyzing predictive insights.

7.2.1 Displaying Predictive Documents

Display predictive accounting documents

To display predictive accounting documents, you can go directly to the Universal Journal, but in a productive system, you hardly have any access to view database tables. You can use any journal entry display app to view predictive accounting documents. Figure 7.1 shows the **Display Journal Entries In T-Account View** tile for the app we'll use to display predictive journal entries.

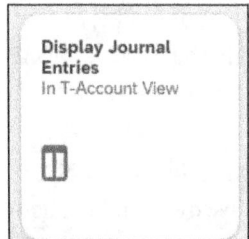

Figure 7.1 Display Journal Entries In T-Account View App

In the selection screen of the SAP Fiori app in Figure 7.2 (upper half), enter the document number of a predictive **Journal Entry** ("PA00001PE0/800") as well as the **Fiscal Year** ("2023"), the **Ledger** ("E1"), and the **Company Code** ("8000"). Click **Go** to display the predictive journal entry.

Predictive journal entry layout

In the lower half of the same screen, you can now see the predictive journal entry displayed. You can change the layout to show additional characteristics and display any characteristic that is assigned to the margin analysis.

7.2 Predictive Accounting

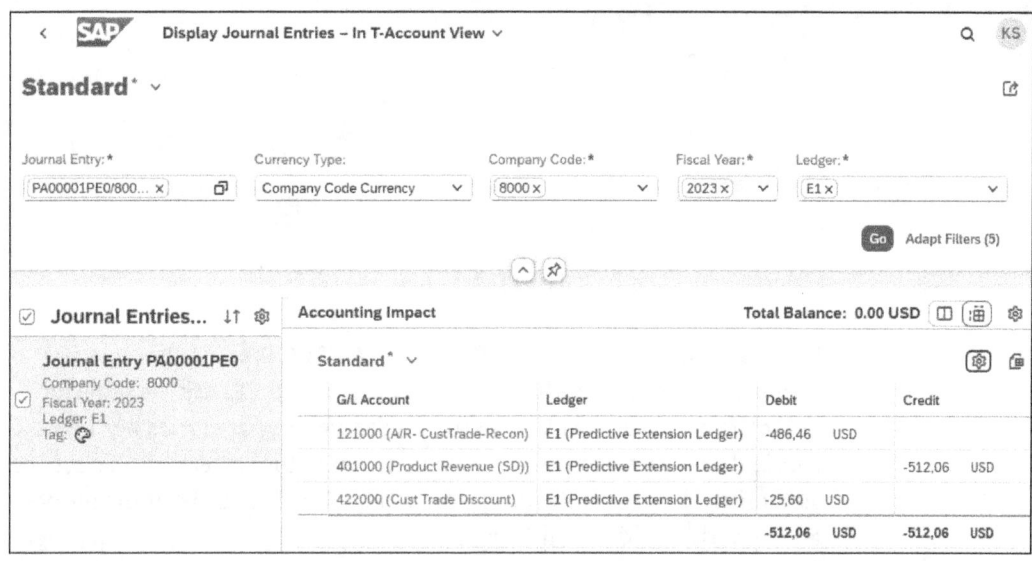

Figure 7.2 Display Journal Entries App: Selection Screen

You can also display the Journal Entry (In T-Account View) app by flipping the switch right above the journal entry to **Default:** (left side).

7.2.2 Displaying the Profit and Loss and Balance Sheet Report

Predictive journal entries can be displayed as part of a regular profit and loss (P&L) and balance sheet report. You need to ensure that the general ledger (G/L) accounts used to post the predictive accounting documents are assigned to the financial statement version to be able to display them in the P&L and balance sheet report.

Displaying the Financial Statement Version

With the Manage Global Hierarchies app (see its tile shown in Figure 7.3), you can display the financial statement version that is assigned to your chart of accounts and make sure all the G/L accounts used in your predictive journal entries are assigned to a financial statement item. All financial statement versions are created like any other hierarchies (e.g., cost center, profit center) with the Manage Global Hierarchies app. Financial statement versions that are created via the SAP GUI aren't displayed but are available for change in the Manage Global Hierarchies app, so it's important to maintain your financial statement versions here. A relatively new functionality, Excel Upload, helps you reduce the workload when creating a financial statement version from scratch.

Display financial statement version

299

7 Reporting

Figure 7.3 Manage Global Hierarchies App Tile

In Figure 7.4, check that the G/L account **401000 Product Revenue (SD)** is assigned correctly to the financial statement version. Check the same for all other G/L accounts involved in the predictive journal entries both on the P&L and balance sheet sides. On the right side of the screen, you can see the assignment to a semantic tag. In Section 7.2.4, we'll talk about the purpose and importance of semantic tags.

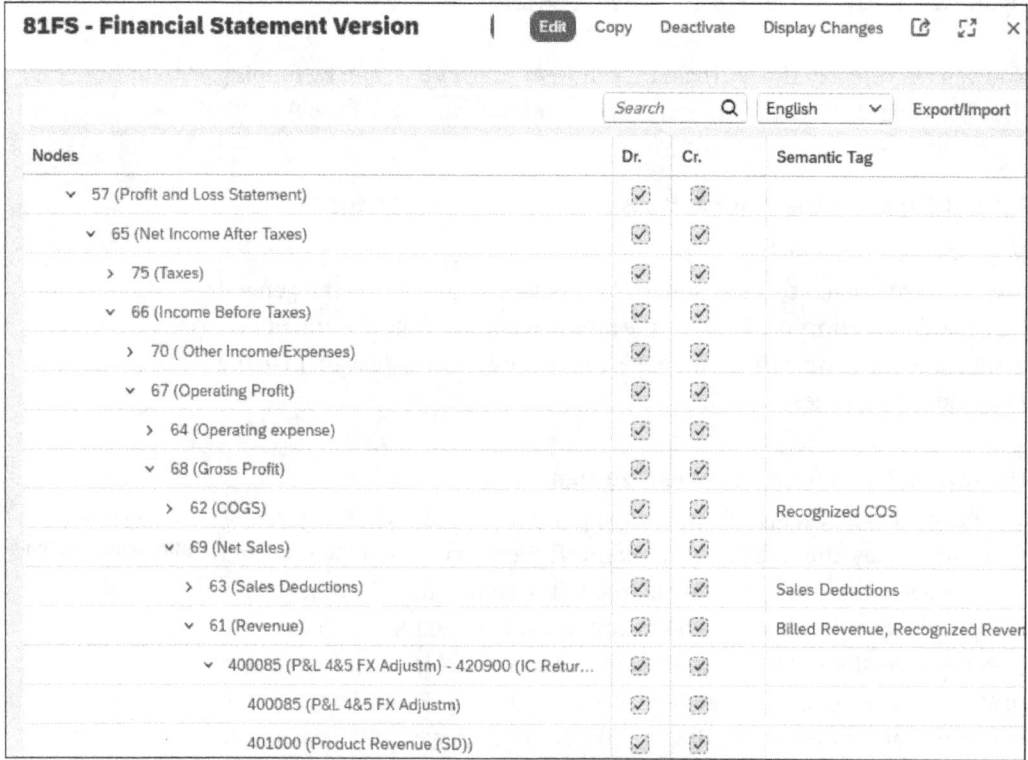

Figure 7.4 Review the Financial Statement Version

It's recommended to keep the authorizations restricted to a small group as the financial statement versions are very important for the external reporting of the P&L and balance sheet.

Displaying Profit and Loss and Balance Sheet Report

After you've ensured all G/L accounts are assigned to the financial statement version, you can display your P&L and balance sheet report. To best compare the difference between the extension ledger with the predictive journal entries and the leading ledger without the predictive journal entries, execute the report for the ledger comparison with the Ledger Comparison app (see its tile in Figure 7.5).

Display the P&L and balance sheet report

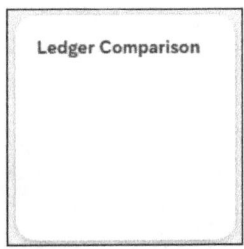

Figure 7.5 Ledger Comparison App Tile

The selection screen of the ledger comparison report of the financial statement is quite large. In **General Selections** shown in Figure 7.6, maintain the **Currency Type** "10" for the company code currency, **Company Code** "8000", and **Account Number** "401000".

Ledger comparison report

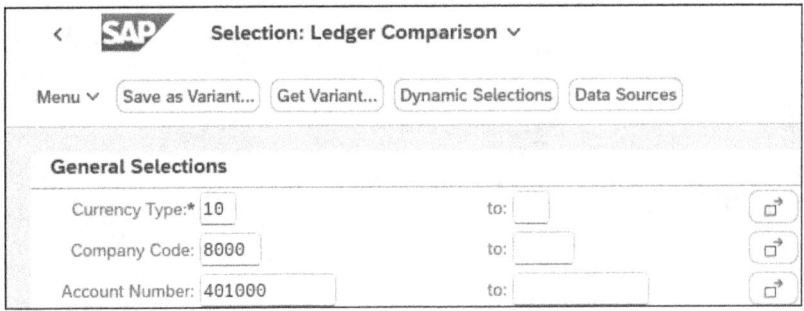

Figure 7.6 Maintain General Selections in the Ledger Comparison Report

In the **Report selections** area in Figure 7.7, maintain the financial statement version "0010" as well as **Ledger** "OL" (leading ledger) and the **Comparison Ledger** "E1" (predictive accounting ledger). In addition, maintain the time frame to compare data in. In this example, this is **From period** "1" and **To period** "12" in **Fiscal year** "2023". Execute the financial statement ledger comparison report with **Execute** or F4.

Maintain the selection screen

7 Reporting

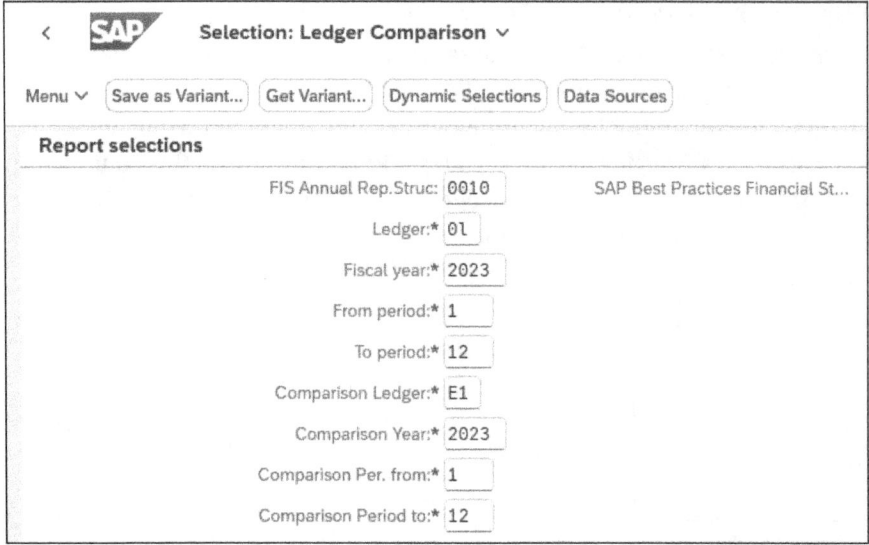

Figure 7.7 Maintain Report Selections in the Ledger Comparison Report

Analyze the ledger comparison report — In Figure 7.8, you can see the **Execute Ledger Comparison: Overview** report for account "40100".

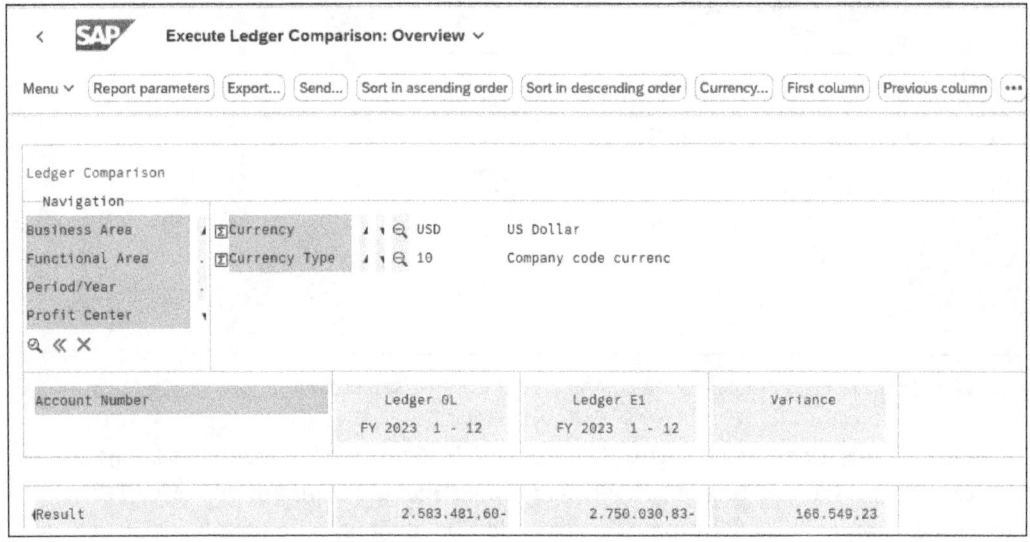

Figure 7.8 Review the Ledger Comparison Report for G/L Account 40100

The first column, **Account Number**, shows the selection criteria of the report's selected G/L accounts. The second column, **Ledger 0L**, shows the values for the leading ledger in period **1 - 12** of **FY 2023**. The third column, **Ledger E1**, shows the values of leading ledger **0L** plus the predictive journal entries in period **1 - 12** of **FY 2023**.

For the G/L account 40100, you can see a difference of 166.549.23 USD between the leading ledger and the extension ledger. The difference can be explained with predictive journal entries—there are sales orders with expected revenues that increased the revenue.

Reviewing Incoming Sales Orders

In Figure 7.9, you can see one of the sales orders that is responsible for the difference in the leading ledger and the extension ledger. The sales order has been delivered, but the order is not invoiced, which is why 9.520,00 USD appears in the predictive ledger but not in the leading ledger. In addition, no accounting document has been created yet. Any changes to the sales order quantity or prices will be reflected in a cancellation of the predictive journal entry and reposting with the corrected value.

Review incoming sales orders

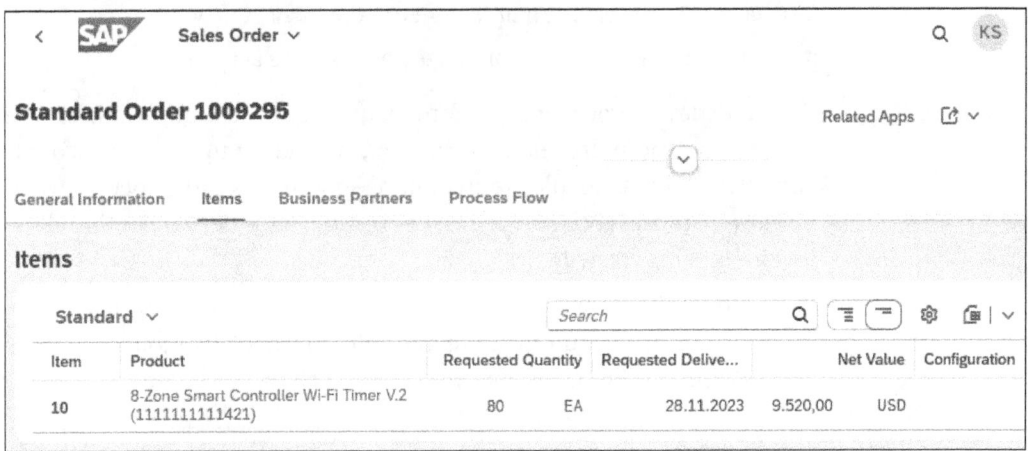

Figure 7.9 Sales Order Overview

7.2.3 Monitoring Predictive Accounting

Predictive accounting mainly happens in the background. Because you don't see any message or calculation when you save the sales order, nor a predictive accounting document in the document flow, it's hard to spot errors. To make sure all documents have been transferred correctly to predictive accounting, you can use the Monitor Predictive Accounting (Predictive Quality) app (Figure 7.10). On the app tile, a key performance indicator (KPI) shows the success rate of creating predictive journal entries from incoming sales orders. In this example, the number is rather low at 9%. Double-click on the tile to see more details.

Monitor predictive accounting

7 Reporting

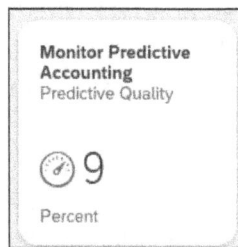

Figure 7.10 Monitor Predictive Accounting (Predictive Quality) App Tile

Error categories The Monitor Predictive Accounting app shows you a summary of the errors in the three following categories:

- Number of sales order errors in relation to total number of sales orders
- Number of sales order errors by line item showing where the error occurred in the creation of the predictive journal entry
- Number of sales order errors sorted by message age

You can start reprocessing the erroneous sales orders directly from the screen. Most errors are related to master data and configuration and occur later in the process. Predictive accounting spots errors early, not only when a follow-up process occurs, such as the creation of the delivery or the billing invoice is being created.

To minimize the errors in the incoming sales order processing, it's recommended to run the automatic reprocessing of predictive journal entries every 15 minutes in the productive system, as shown in Figure 7.11.

Figure 7.11 Automatic Processing of Predictive Journal Entries

If you're on SAP S/4HANA Cloud, the job is scheduled automatically after predictive accounting is activated. With Transaction SA38, you can schedule the job for automatic reprocessing of erroneous sales orders. In Figure 7.12, you can see report **FINS_PRED_AIF_REPROCESS** that has to be scheduled for the automatic reprocessing of predictive journal entries.

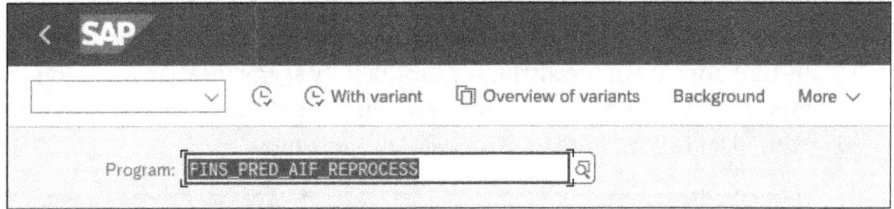

Figure 7.12 Schedule Report for Automatic Reprocessing of Sales Orders

7.2.4 Incoming Sales Order Report

The Incoming Sales Order Report app allows you to analyze incoming sales orders, cost of goods sold (COGS), and margin. You need to assign the G/L accounts in the financial statement version to semantic tags to see the margin in the incoming sales order report because the margin is calculated within the financial statement version and read from semantic tags.

Incoming sales order report

Definition of Semantic Tags

You may be wondering what a semantic tag is and how to assign a semantic tag to an account in the financial statement version. Many clients complained about the maintenance of different structures such as financial statement versions, cost element groups, and so on for consistent reporting. It's time intensive to keep all structures up-to-date, to not miss any accounts in reporting, and then analyze the report to find out which structure or group is inconsistent. To remedy this, semantic tags were introduced.

Semantic tags

Semantic tags define KPIs by classifying accounts, for example, as billed revenue. SAP predelivers 60 semantic tags for use in SAP standard KPI reports such as cash flow, contribution margin, or project profitability. Semantic tags can be used universally in financial accounting and FP&A, and they allow drilldown to the lowest level, which is the G/L account. You can configure semantic tags and link them with G/L accounts and nodes in the financial statement version. If you change any G/L accounts in the financial statement version, the semantic tag will be automatically updated and reflect the change without any manual intervention.

The objective of semantic tags is to remain stable; you shouldn't create different semantic tags by country or legal entity. The semantic tags display

Objective of semantic tags

7   Reporting

the KPIs you're running your business by and the measures you want to judge your performance by.

The idea is to deliver dashboards and reporting out of the box, rather than to deliver a tool set such as the Report Painter or Report Writer that was used in SAP ERP for reporting. Out-of-the-box reporting with semantic tags will reduce your implementation time and cost.

Attributed profitability segments   Semantic tags give you information along the way especially paired with the functionality of the attributed profitability segments that allow you to immediately derive margin analysis components. Closing tasks will be reduced and allow more time for decision-making.

> **Note**
>
> In SAP Note 2538634 – Semantic Tagging for FSV and CDS Query Extension, you'll find more information about semantic tags.

Configuration of Semantic Tags

Configuring semantic tags   Semantic tag groups are available to help you structure and organize the semantic tags. You can display and create semantic tag groups by following configuration path **Financial Accounting • General Ledger Accounting • Master Data • G/L Accounts • Semantic Tags for Balance/Profit and Loss Structures • Define Semantic Tag Group**.

In Figure 7.13, you can see the semantic tag groups that exist within standard SAP. You can create new semantic tag groups to structure your semantic tags with **New Entries** or [F5].

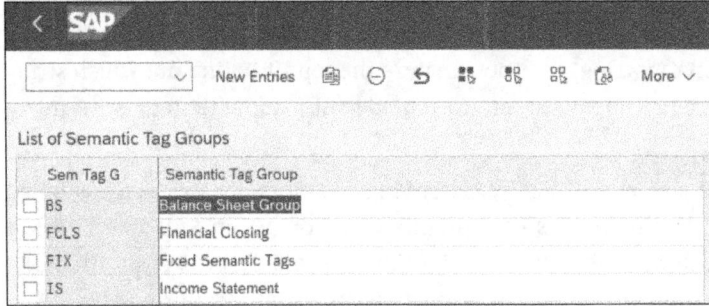

Figure 7.13 Review Semantic Tag Groups

Now that you've reviewed the semantic tag groups, you can review the semantic tags by following configuration path **Financial Accounting • General Ledger Accounting • Master Data • G/L Accounts • Semantic Tags for Balance/Profit and Loss Structures • Define Semantic Tags for Balance Sheet-/Profit and Loss Structures**.

Overview of semantic tags

In Figure 7.14, you see an overview of all the semantic tags in the SAP S/4HANA system.

Change View "List of Semantic Tags": Overview

New Entries More

List of Semantic Tags

Sem. Tag	Semantic Tag Long Name	Semantic Tag Name	Sem Tag G	Par SemTag
3RDPTYEXP	Third Party Expense	Third Party Expense	IS	
ACCPAY	Accounts Payable	Accounts Payable	BS	
ACCPAY_OTH	Accounts Other Payables	Other Payables	BS	
ACCREC	Accounts Receivable	Accounts Receivable	BS	
ACCREC_OTH	Accounts Other Receivable	Other Receivable	BS	
ACR_COST	Accrued Cost	Accrued Cost	BS	
ACR_REV	Accrued Revenue	Accrued Revenue	BS	
ACT_COST	Actual Cost	Actual Cost	IS	
ADJ_COS	COS Adjustments	COS Adjustments	IS	
ADJ_REV	Revenue Adjustments	Revenue Adjustments	IS	
ADMCST	Administration Cost	Admin Cost	IS	
ALLOC	Allocations	Allocations	IS	
AMORINASST	Amortization of Intangible Assets	Amtzn Intgnbl Assets	IS	
ASSET	Assets	Assets	FIX	
BILL_REV	Billed Revenue	Billed Revenue	IS	
BNFTEXPN	Employee Benefits Expense	Benefits Expense	IS	
BNSCMSNEXP	Bonus & Commission Expense	Bonus & Commsn Expn	IS	
CHGFARET	Gain/Loss from Retirement of Fixed Assets	Gain/Lss Rtrmt FA	IS	
CHGLTINV	Long-term Investments	Long-term Invmts	BS	
COGS	Cost of Goods Sold	Cost of Goods Sold	IS	
COGS_3PAR	Third Party	Third Party	IS	ACT_COST
COGS_DMAT	Direct Material	Direct Material	IS	
COGS_MATI	Machine Time	Machine Time	IS	ACT_COST
COGS_OMAT	Overhead Cost Material	OH Cost Material	IS	ACT_COST
COGS_OPRO	Overhead Cost Production	OH Cost Production	IS	ACT_COST
COGS_PERT	Personnel Time	Personnel Time	IS	ACT_COST
COGS_SUTI	Setup Time	Setup Time	IS	ACT_COST

Figure 7.14 Overview of Semantic Tags

The different columns show the following content:

- **Sem. Tag**
  The semantic tag technical name.
- **Semantic Tag Long Name**
  The long name of the semantic tag.
- **Semantic Tag Name**
  The name of the semantic tag.
- **Sem Tag G** (semantic tag group)
  The assignment to a semantic tag group from Figure 7.13 shown earlier.

- **Par SemTag** (parent semantic tag)
  The parent semantic tag is used for budget availability control.

  The budget availability control checks all costs assigned to the semantic tag **ACT_COST**. This is a semantic tag that is predefined by SAP. If you execute any changes to this semantic tag, the budget availability control won't work correctly anymore. You can report on a more detailed account level in the project budget report, but you have to assign **ACT_COST** as a parent semantic tag to those more detailed semantic tags to guarantee correct functionality.

With **New Entries** or [F5], you can create additional semantic tags.

Semantic tags in reporting

To use semantic tags in reporting, they need to be assigned to the financial statement version. To assign semantic tags to a financial statement version, follow configuration path **Financial Accounting • General Ledger Accounting • Master Data • G/L Accounts • Semantic Tags for Balance/Profit and Loss Structures • Assign Semantic Tags to Balance Sheet-/Profit and Loss Structures**.

Assign semantic tags

In Figure 7.14, you can assign additional semantic tags via **New Entries** or [F5]. For the calculation of the margin in the incoming sales order report, you need to assign the semantic tags in Table 7.1 to a financial statement item.

Semantic Tag	Semantic Tag Name
RECO_COS	Recognized cost of sales
RECO_REV	Recognized revenues
REC_MARGIN	Recognized margin
SALES_DED	Sales deductions
BILL_REV	Billed revenue

Table 7.1 Relevant Semantic Tags for Incoming Sales Order Report

With **New Entries** or [F5], you can assign semantic tags to the following:

- Financial statement items
- G/L accounts
- Functional areas

In the example in Figure 7.15, assign semantic tags to a financial statement item by choosing **Assign Semantic Tag to FS Item**, and then press [Enter].

Assign semantic tags RECO_REV

In the next screen shown in Figure 7.16, assign the **Semantic Tag RECO_REV** to **FS Item 63** in **FS Version 1720**. It's important that you execute the incoming sales order report with this financial statement version or no data will be displayed in the report. Save your entries with **Save** or [Ctrl]+[S].

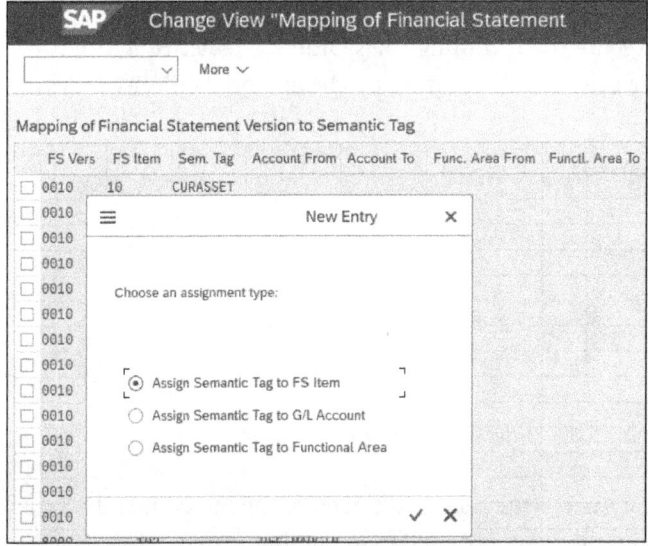

Figure 7.15 Map a Semantic Tag to the Financial Statement Version

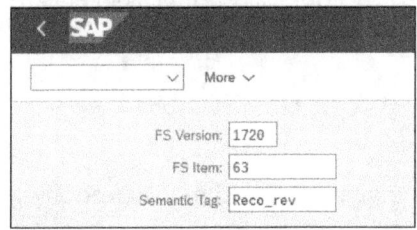

Figure 7.16 Assign Semantic Tag

After you've saved your assignment, you can see the details of the newly created entry in Figure 7.17. You can see that if you chose to assign the semantic tag to a G/L account or functional area, you would be able to assign an interval to the semantic tag.

Details of semantic tags assignment

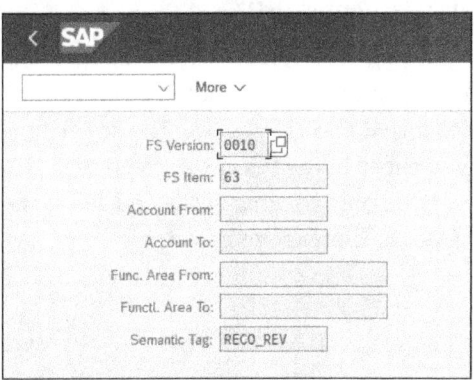

Figure 7.17 Details of the Added Entry

Execute the Incoming Sales Order Report

Execute the report

Now, you can execute the Incoming Sales Orders (Predictive Accounting) app for incoming sales order reporting by clicking on its tile, as shown in Figure 7.18.

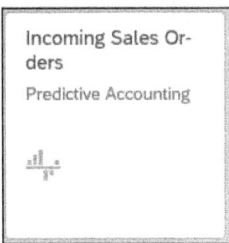

Figure 7.18 Incoming Sales Orders (Predictive Accounting)

Incoming sales order report selections

In Figure 7.19, you can see the selection screen of the incoming sales order report. In this example, you only need to maintain the mandatory fields:

- **Sales Order Selection**
  Either display **All Incoming Orders** (chosen for this example) or **Remaining Orders**.

- **Prediction Ledger**
  Choose **Prediction Ledger E1**.

- **Sales Organization**
  Choose all the **Sales Organizations** that are assigned to the Controlling area.

- **Semantic Tags**
  Choose the **Semantic Tags** (**Revenue**, **Margin**, and **COS** [cost of sales]) as well as the Figure 7.16**Start Date**. Choose from various periods, such as a date range, quarters, or a specific month. In this example, choose **20.02.2023 to 20.02.2024 e** to display all dates until today's date.

To execute the report, confirm the selections with **OK**.

In Figure 7.20, you can see the **Incoming Sales Order** report. The screen is split in two parts. In the upper part, you can see a graphical display of the **Recognized Revenue** and the **Recognized COS** (i.e., COGS). The key figures are based on the semantic tags you chose in the selection screen in Figure 7.19. You can adjust the report very flexibly to your needs and select the chart type you want to see or change the dimensions of the chart.

7.2 Predictive Accounting

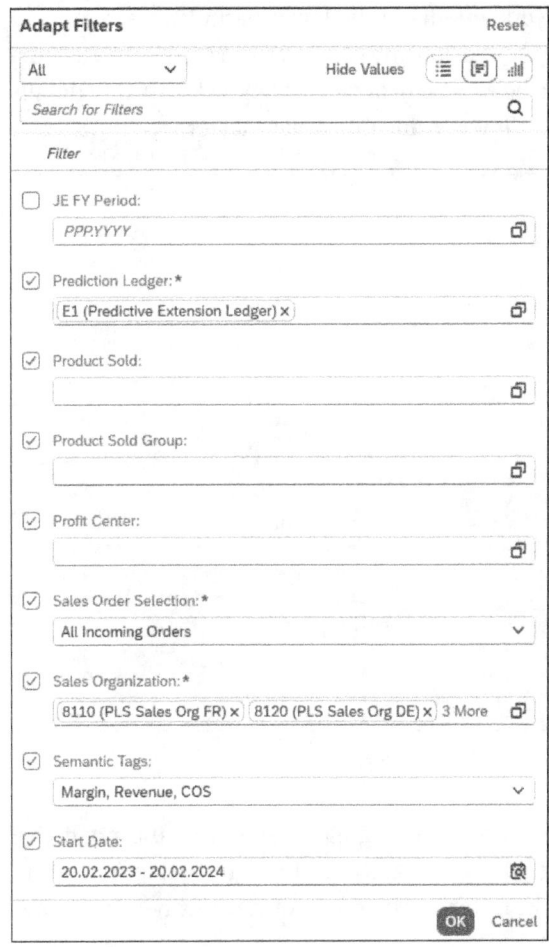

Figure 7.19 Selection Screen for Incoming Sales Orders

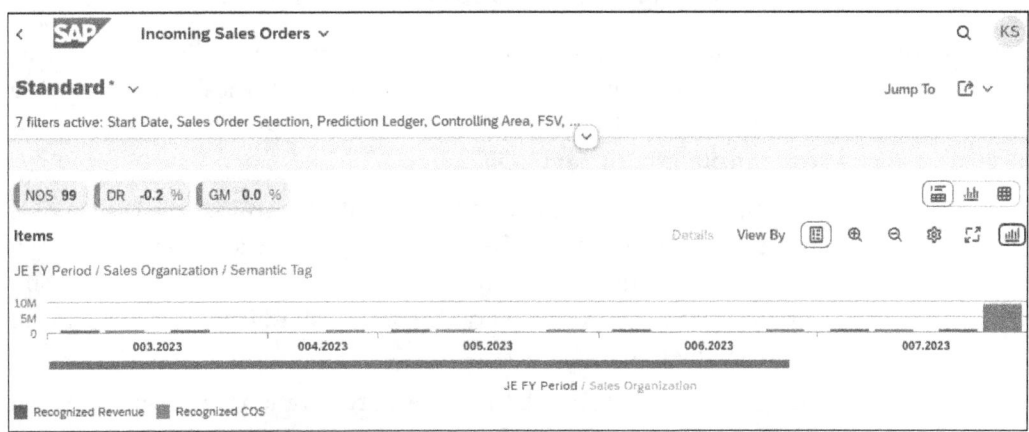

Figure 7.20 Incoming Sales Order Graphical Display

In Figure 7.21, you see the documents in the lower part of the screen. You can adjust the screen flexibly and add any column from the Universal Journal. The line items can also be exported, or you can click into further details such as **Journal Entry**, **Billing Invoice**, and **Customer Master**.

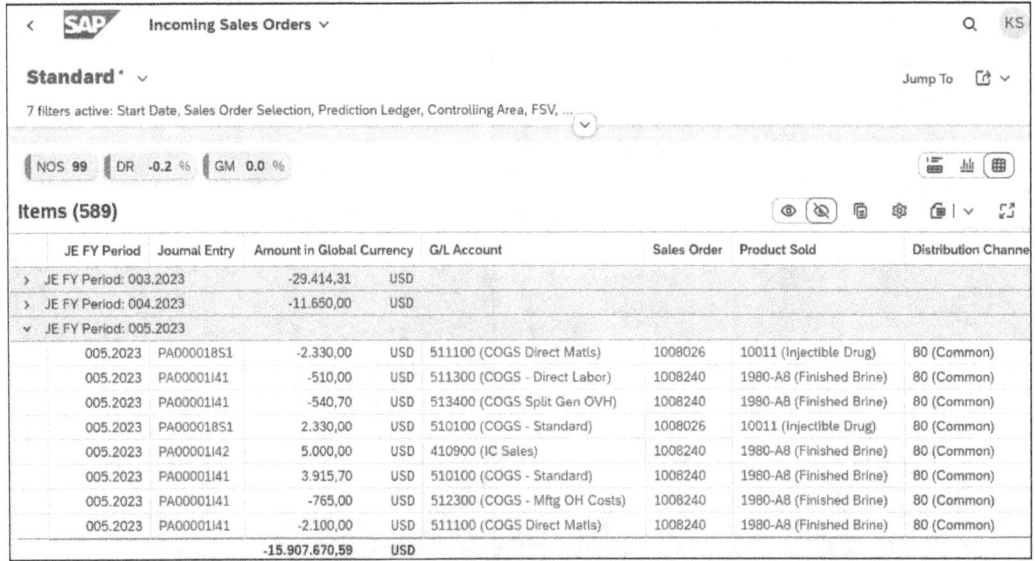

Figure 7.21 Incoming Sales Order Data Analysis

The **Incoming Sales Order** report gives a great overview of and predictive insight into the incoming sales and margin for the foreseeable future and allows the company to make faster and better-informed business decisions.

7.3 Showing Line Items for Margin Analysis

Line-item reports

The line-item report for margin analysis allows you to display a list of all documents that have been posted in actual or plan. The reports resemble a document list and allow you to show any characteristic of the operating concern or the document itself.

You can review line items posted in margin analysis with the Display Line Items app, as shown in Figure 7.22. The app is only applicable for margin analysis, not costing-based profitability analysis. For costing-based profitability analysis, the KE24 app can be activated, which works the same as Transaction KE24 in SAP GUI.

As we learned in Chapter 1, the highest organizational element of margin analysis is the operating concern. When you look at the selection criteria of the Display Line Items report in Figure 7.22, there is no field to choose the

7.3 Showing Line Items for Margin Analysis

operating concern. Before executing the transaction, you need to set the operating concern in your user settings.

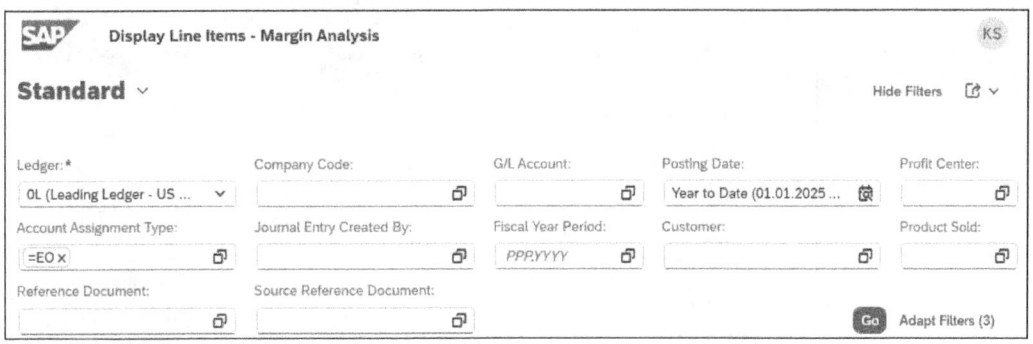

Figure 7.22 Selection Criteria for the Line-Item Report in Margin Analysis

To change your user settings, go back to the home screen of the SAP Fiori launchpad, and click on your initials in the upper-right corner of the screen, as shown in Figure 7.23.

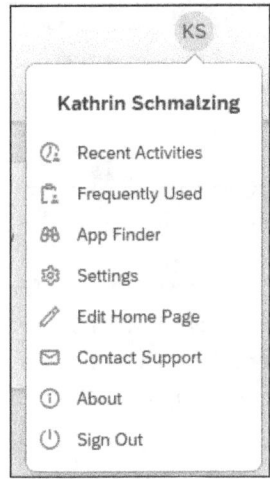

Figure 7.23 Access User Settings

Navigate to **Settings** and then to **Default Values**, and scroll down until you find the **Controlling** section, in which you'll find the **Operating concern**. As shown in Figure 7.24, enter the **Operating concern** "Z100", and press Enter, before saving the settings and returning to the app, as shown in Figure 7.25.

7　Reporting

Figure 7.24 Save the Operating Concern in User Settings

Adapt filters of line-item reports

Now, we can execute the line-item report for margin analysis. The app for this has some standard selection criteria you can adjust by clicking on Adapt Filters. In Figure 7.25, you can see all the characteristics of margin analysis; actually, you can see almost all fields of the Universal Journal (table ACDOCA).

Analyze line-item report

After adapting your filters and confirming your selection with **OK**, you can narrow down your selection for the report by entering values in the selection criteria. With Go, you execute the report and see all relevant line items based on your selection criteria. In Figure 7.26, you can see all journal entries that we selected in the report with all the characteristics that we chose in the settings of the report.

7.3 Showing Line Items for Margin Analysis

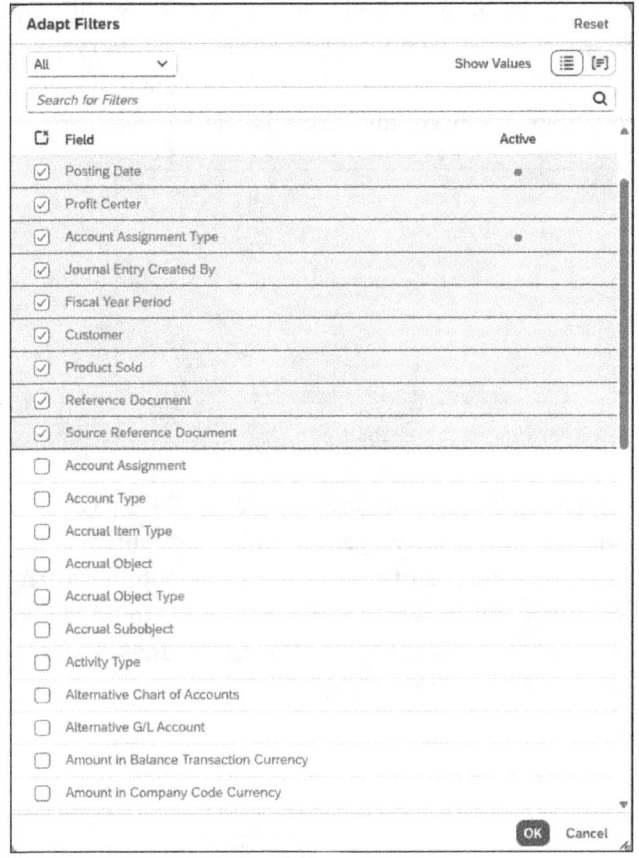

Figure 7.25 Adapt Filters of the Line-Item Report

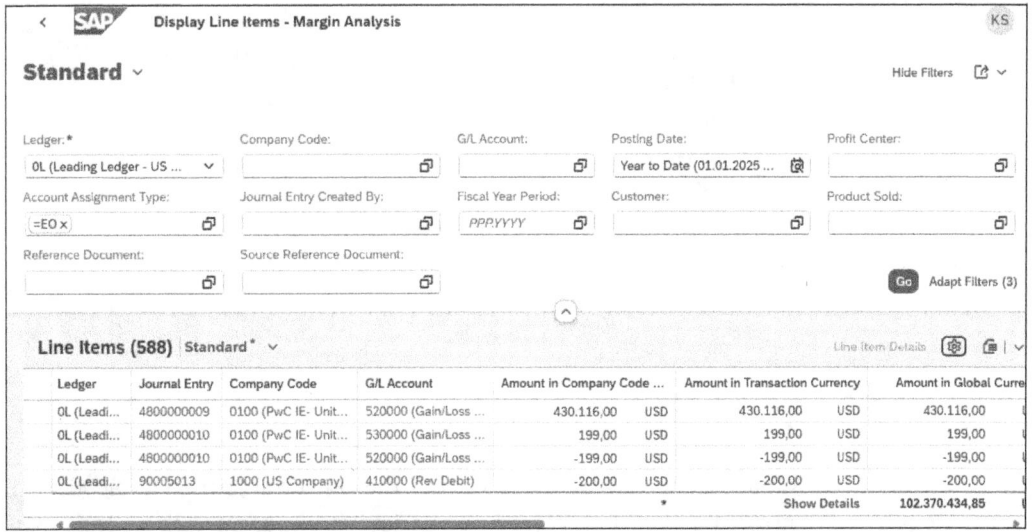

Figure 7.26 Analyze the Line-Item Report

7   Reporting

Profitability segment details

You can hover over a journal entry to display the original document or to jump into other reports to do a more thorough analysis. In Figure 7.27, you can see that we can jump into two apps from here: **Manage Journal Entries** and **Manage Journal Entries – New version (recommended)**.

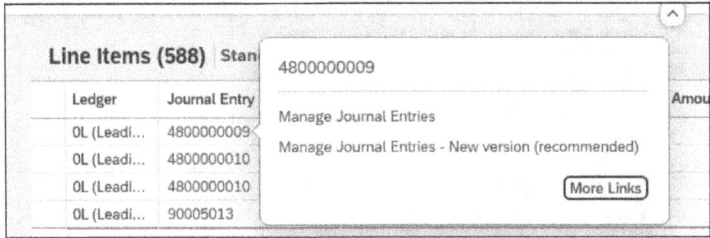

Figure 7.27 View Additional Reporting Capabilities

Add additional Links

You also have the opportunity to jump into other apps. With a click on More Links, you can display all apps that you can link to the journal entry. In Figure 7.28, you can check the box to add them to the popup shown earlier in Figure 7.27 when you hover over the document. Keep in mind that you need appropriate authorization to link additional apps and then jump into those apps from the Display Line Items app.

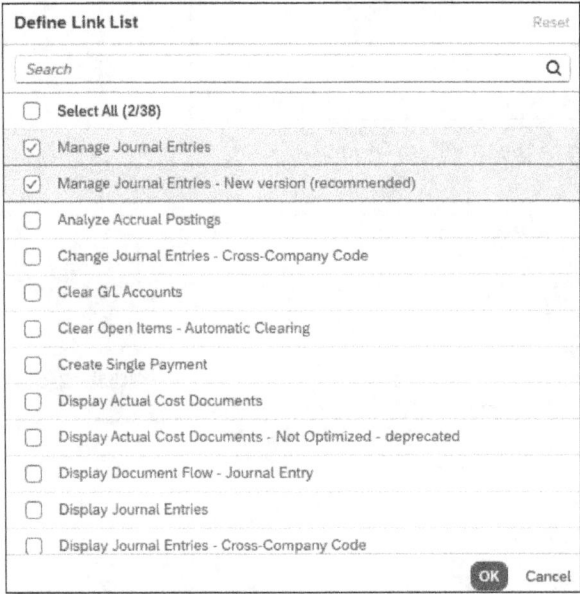

Figure 7.28 Choose More Links

As margin analysis is integrated into the Universal Journal (table ACDOCA), you can use almost every journal entry app to display margin analysis characteristics, which increases the flexibility of reporting significantly.

7.4 Profitability Analysis Reporting

In this section, we'll show you some SAP Fiori app reports for margin analysis. For costing-based profitability analysis, all reports have to be created manually from scratch. Transaction KE30 can be used for the same, and this transaction is also available as a SAP Fiori.

7.4.1 Gross Margin

The Gross Margin (Presumed/Actual) app (see its tile in Figure 7.29) is used to analyze margin analysis data. With this app, you can get insights about predictive data with forward-looking KPIs that use predictive data, such as incoming sales orders. You can also add plan data in this report to compare your actuals with plan.

Figure 7.29 Gross Margin (Presumed/Actual) App

In the selection screen in Figure 7.30, maintain the **Ledger** ("Predictive Extension Ledger E1"), the **Controlling Area** ("1000"), and the **FSV** (financial statement version; "0010") as mandatory selection criteria. When you scroll down in the selection screen, you have many more characteristics to limit your selection. Execute the report with **OK**.

The report displays and analyzes side by side predicted, actual, and planned revenues; cost of sales; margins; and sales deductions based on incoming sales orders. You can filter the report based on time periods and other various margin analysis characteristics. In Figure 7.31, you can see in the upper part of the screen some graphs that show the following:

- **Presumed Margin by Customer**
- **Presumed Margin by Product Group**
- **Presumed Margin by Profit Center**
- **Presumed Margin by Fiscal Year** (when you scroll to the right)

7 Reporting

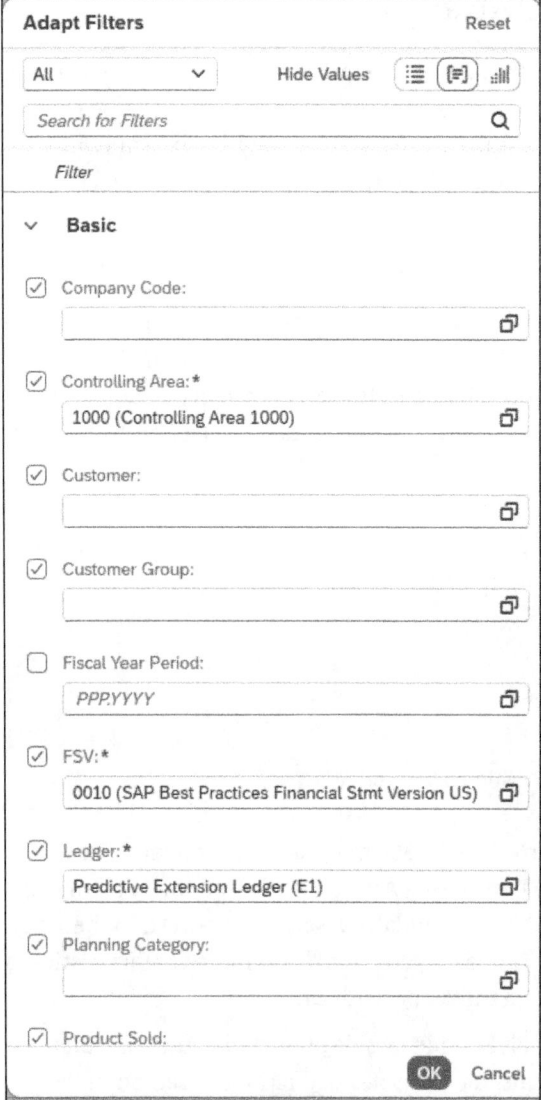

Figure 7.30 Selection Criteria for the Gross Margin App

The lower part of the screen shows the line items that make up the presumed margin, revenue, and so on. Double-click the line item to display the original document information, such as the journal entries, sales orders, customers, and products. You can analyze your data directly from the list report in the lowest level of detail.

7.4 Profitability Analysis Reporting

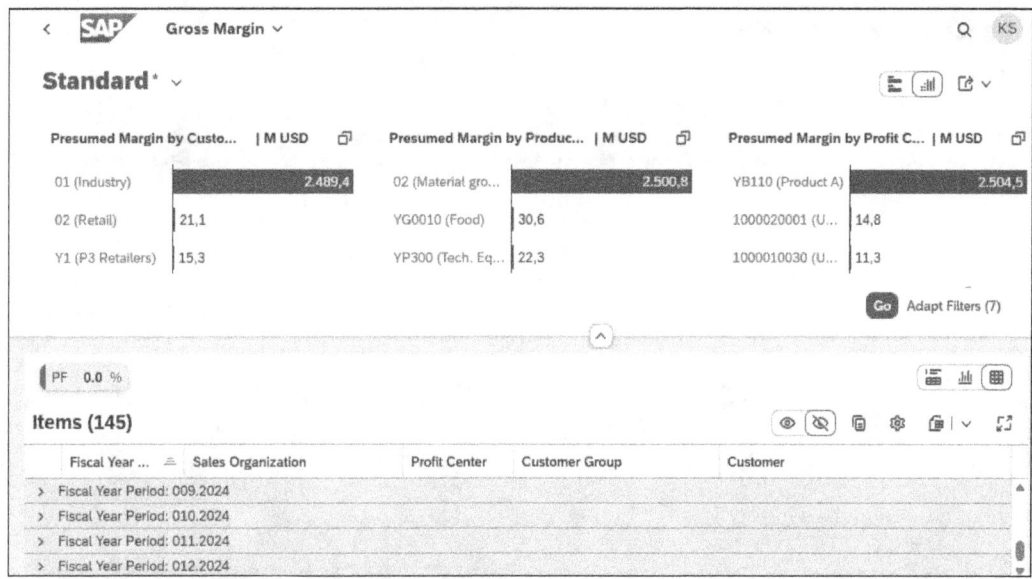

Figure 7.31 Margin Analysis Report for Gross Margin App

7.4.2 Sales Accounting Overview

The Sales Accounting Overview app (see its tile in Figure 7.32) provides an overview of the major sales KPIs. You can customize which apps will be displayed in the overview.

Sales Accounting Overview app

Figure 7.32 Sales Accounting Overview App

The app allows you to analyze profit and contribution margins by different criteria, and you can jump in from the smaller tiles into reports based on the data you want to analyze. With the Sales Accounting Overview app, you can see at one glance which KPIs require more attention and then deep-dive from there into a more detailed analysis. Figure 7.33 shows an example of how the Sales Accounting Overview app can look.

319

7 Reporting

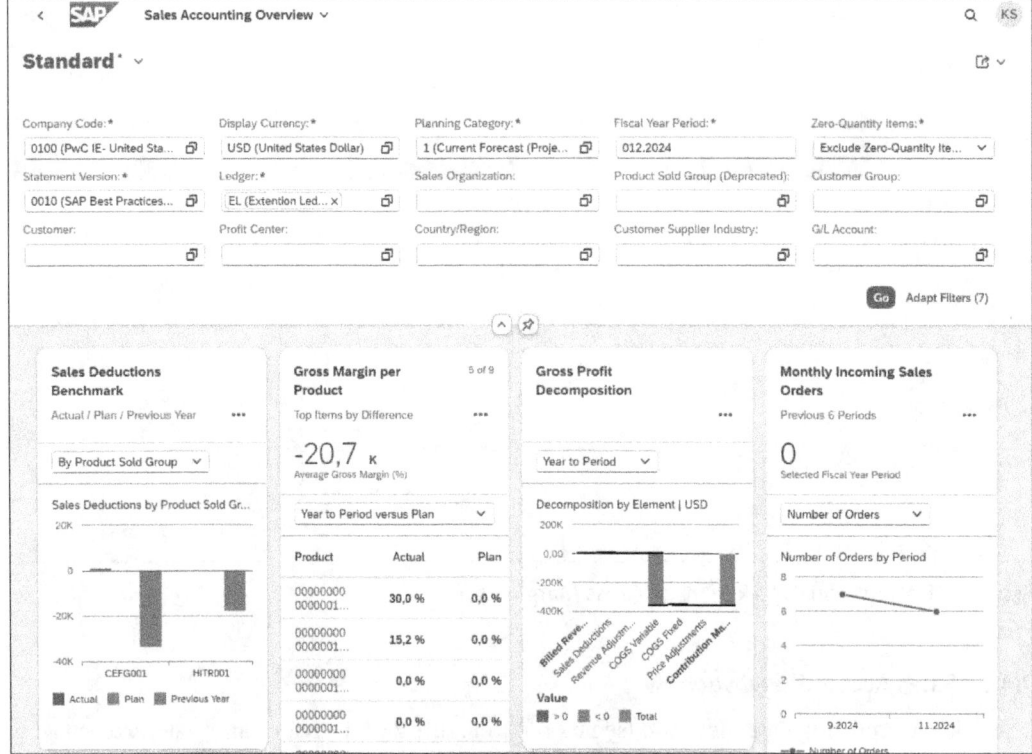

Figure 7.33 Example of Sales Accounting Overview App

7.4.3 Product Profitability with Production Variances

Product Profitability with Production Variances app

The Product Profitability with Production Variances app , which is only available in SAP S/4HANA public cloud, allows you to analyze your contribution margin at the product level. You can analyze your products according to the fixed costs, variable costs, and standard cost component split. You have the characteristics of the product as well as the margin analysis characteristics available to analyze your margin. The report allows you to analyze the quantities sold and the margin. ou can analyze your products according to the fixed costs, variable costs, and standard cost component split. You have the characteristics of the product as well as the margin analysis characteristics available to analyze your margin. The report allows you to analyze the quantities sold and the margin.

Product profitability semantic tags

A prerequisite for the line items shown in the report is that they are assigned to a profitability segment and that the characteristic for the product is filled. You can display the values in the product profitability report both in global currency and company code currency. The report doesn't show you the revenue or costs by account but instead by semantic tag. A broad range of semantic tags are available in the SAP standard, but you can also create your own KPIs with semantic tags.

7.4.4 Profit and Loss Statements

Margin analysis is completely integrated into the Universal Journal (table ACDOCA), which is why margin analysis characteristics can be displayed in nearly all finance reports. In Figure 7.34, let's have a look at the P&L (Actuals) app to confirm the display of margin analysis characteristics.

Figure 7.34 P&L (Actuals) App

The selection criteria doesn't have the operating concern, as both the operating concern and the controlling area have to be maintained in the user settings to be able to execute some of the reports that are relevant to controlling and margin analysis. The selection criteria also doesn't include any margin analysis characteristics. Let's execute the P&L app for **Ledger 0L** and **Fiscal Year of Ledger 2025** by clicking on Go, as shown in Figure 7.35.

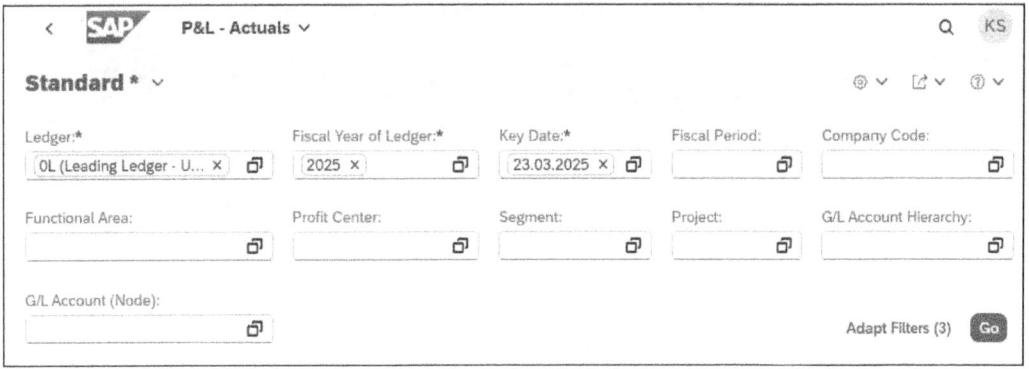

Figure 7.35 Maintain Selection Criteria for P&L

On the right-hand screen in Figure 7.36, you can see the different values available in columns in the report, which are the values in the different currencies and the quantities.

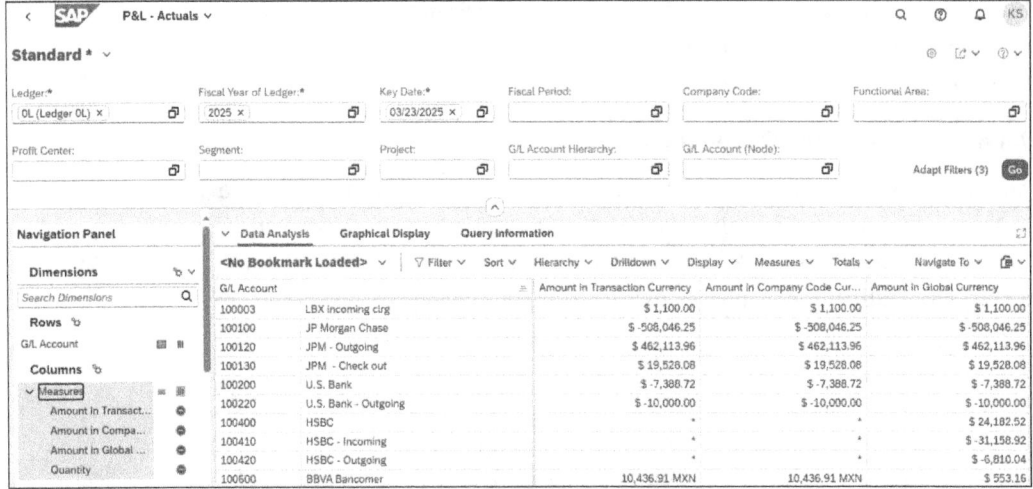

Figure 7.36 Display Values in Report Columns

By scrolling down, you can see all additional characteristics that are available in the report. You can pull them via drag and drop in either the rows or the columns of the report. In the available fields, you have all the margin analysis characteristics at your disposal. Figure 7.37 shows the **Customer** in the report, along with the COGS per customer.

Figure 7.37 Review COGS per Customer

The app further allows you to sort, filter, and pull in hierarchies for any characteristics, as shown in Figure 7.38.

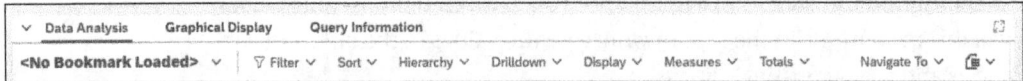

Figure 7.38 Functionality Enabling Different Views by Characteristics

7.4 Profitability Analysis Reporting

As you can see, any finance app can support margin analysis and therefore significantly enhance your reporting capabilities.

7.4.5 Market Segments Actual

The P&L reporting app also displays balance sheet accounts. The Analyze Market Segments (Actuals) app (tile shown in Figure 7.39) only displays P&L accounts and offers more flexibility than the P&L app.

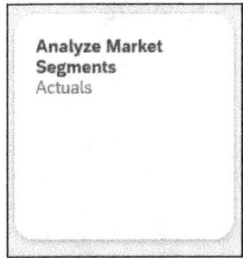

Figure 7.39 Analyze Market Segments

The Analyze Market Segments app doesn't have any selection criteria but shows you on the left side of the screen all the different characteristics from margin analysis and finance for your analysis (Figure 7.40). You can drag and drop those characteristics in the **COLUMNS** and **ROWS** areas to display them in the reporting.

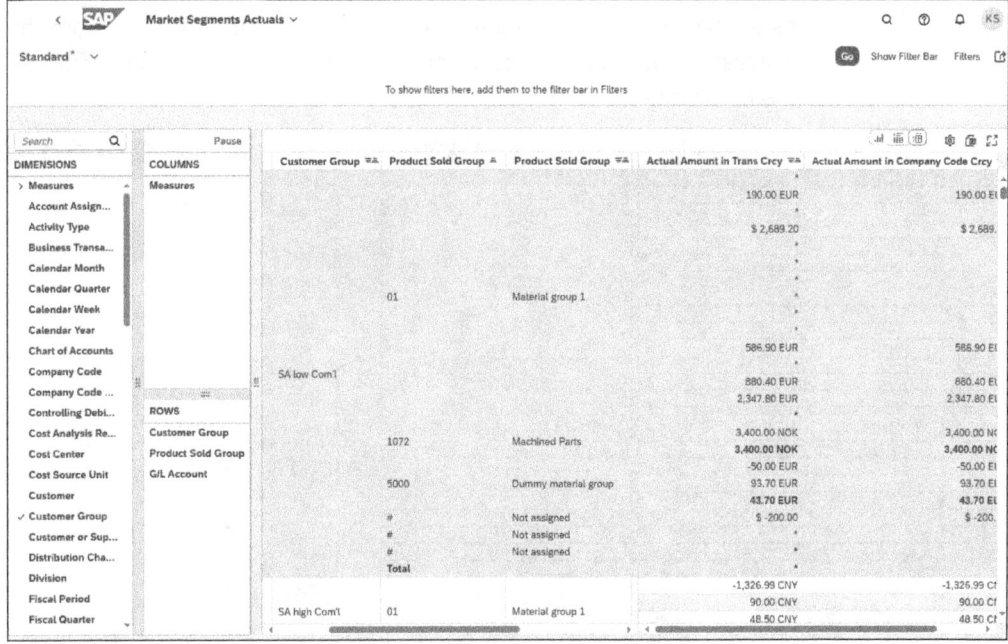

Figure 7.40 Review the Analyze Market Segments App

323

7 Reporting

7.4.6 Report Writer

Report Writer for costing-based profitability analysis

For costing-based profitability analysis, there are no new SAP Fiori apps being developed. The reporting is based on the line-item reports or any report that you create for yourself with the Report Writer. For margin analysis, it's recommended to work solely with CDS views as well as existing SAP Fiori apps and not to leverage the Report Writer tools anymore.

7.5 Other Reporting Possibilities with Development

Reports with development

Apart from the standard SAP Fiori apps and reports, you can create your own reports for profitability analysis. For the creation of your own reports, you need development skills. Although we won't discuss how to develop reports in this section, we'll make you familiar with the different possibilities that SAP provides to create your own reports.

7.5.1 ABAP Reports

Programs and reports in SAP are written in ABAP, as it's the core programming language of SAP ERP and also SAP S/4HANA, although most of the code has been rewritten and optimized for SAP S/4HANA.

ABAP code

In Figure 7.41, you see an example of ABAP code in the ABAP Debugger. This example shows that we were about to create a journal entry in Transaction FB50 when we activated the Debugger after entering the first line item to review what the program is going to do next. You can make out from the left screen some technical characteristics such as **HKONT**, which is the G/L account in table BSEG.

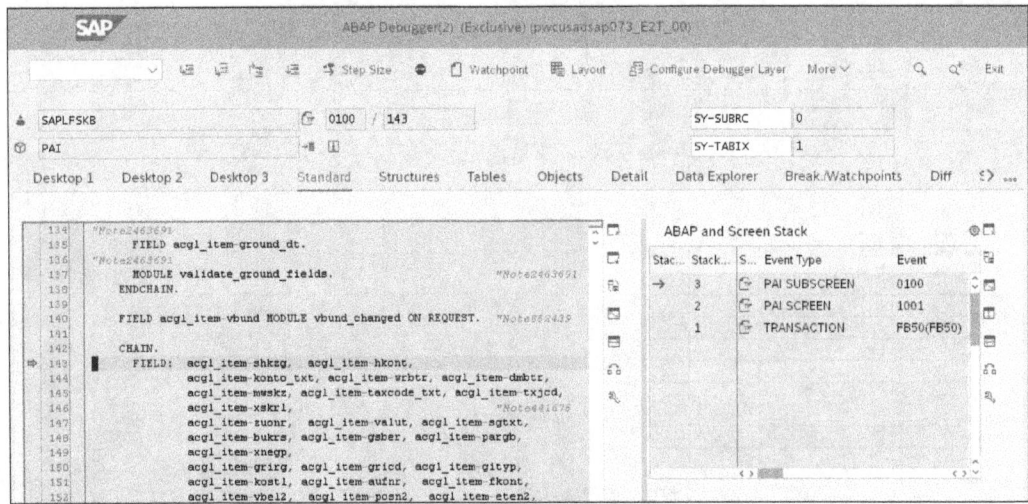

Figure 7.41 Code Review in the ABAP Debugger

7.5 Other Reporting Possibilities with Development

You can extend your functionalities with your own logic using ABAP code in programming user exits and Business Add-Ins (BAdIs). It's not possible to use this kind of programming in SAP Business Technology Platform essentials, but only in SAP S/4HANA Cloud Private Edition and on-premise versions.

User exits and BAdIs

When you migrate your data from SAP ERP to SAP S/4HANA, you need to analyze your existing custom ABAP code to see if you can migrate it to SAP S/4HANA or if it's even necessary to migrate because a lot of new functionality has been introduced with SAP S/4HANA.

In SAP S/4HANA 2023, SAP introduced the ABAP platform with a lot of new functionalities and tools. With the ABAP platform, SAP introduced the ABAP RESTful programming model for SAP Fiori. It includes essential extensions of the ABAP language, development tools, and frameworks that enable the efficient end-to-end development of SAP HANA–optimized SAP Fiori apps.

ABAP platform

ABAP allows you to create any custom report combining data from SAP database tables and displaying or even changing this data in the format you desire. A report that we've created quite often is a product group report that shows both several characteristics from margin analysis but also working capital per product group.

Create custom reports

7.5.2 Core Data Services Views

A CDS view allows you to effectively and comfortably connect to and access data. Technically, a CDS view is a new data modeling infrastructure based on SQL. SAP delivers standard CDS views, but you can also create your own CDS views.

CDS views

Figure 7.42 shows how you can use the stacked model to create reports based on CDS views. In the bottom layer, you find the database tables, which are copied 1:1 in the basic view. In the basic view, you can apply filters to increase performance or rename some standard table fields.

Report based on CDS view

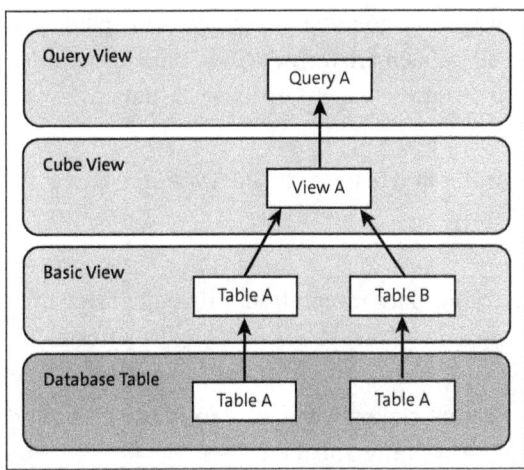

Figure 7.42 Stacked Model for CDS Views

325

In the cube view, different tables are combined via table joints. You might remember this technique from creating a query. You can also apply calculations in the cube view.

In the next layer, the query view, you can apply filters, and this is the view that is exposed to the app. The CDS views can be copied and transported and therefore are easily reusable as well as quite easy to build.

CDS views and ABAP CDS views can also be used with ABAP and have many more functionalities in this combination.

CDS views for margin analysis SAP offers a lot of standard CDS views that can be copied and reused for your custom-specific reports. For margin analysis, there is the CDS view called Market Segments – Plan/Actuals (C_MARKETSEGMENTPLANACTQ2501). This CDS view can be used to get information about actual and plan values per customer group, product sold group, and G/L account.

7.5.3 SAP Datasphere

Profitability Reporting is often created in a data warehouse, to be able to consume large data quantities with the ability to slice and dice them on the lowest level of characteristics. SAP Datasphere allows you to create profitability reporting based on client requirements with the capability to include data from multiple SAP or non-SAP systems.

SAP Datasphere is a cloud-based data management and analytics platform designed to unify, integrate, and analyze data across both SAP and non-SAP sources. SAP Datasphere is available for both on-premise and the cloud. It serves as a central data hub to empower organizations to connect, model, govern, and share their data assets efficiently and securely.

SAP Datasphere is the successor of SAP Data Warehouse Cloud, and it's expanding its foundational capabilities with enhanced features for data integration, semantic modeling, and governance. While both platforms support data management, SAP Datasphere introduces broader functionality with a strong focus on virtualization and enterprise-wide data transparency.

The following lists the key features and benefits of SAP Datasphere to highlight the main features:

- **Data experience**
  SAP Datasphere brings together data integration, cataloging, semantic modeling, data warehousing, and virtualization in a single platform.
- **Data integration**
  SAP Datasphere connects seamlessly to diverse data sources—cloud and on-premise, SAP and non-SAP systems, and data lakes.

- **Data modeling**
  Users can transform and replicate data using intuitive graphical tools or advanced SQL editors, with support for both low-code and no-code development.
- **Data catalog**
  A unified catalog enables users to discover, classify, and prepare data assets for reuse across the organization.
- **Data virtualization**
  SAP Datasphere provides real-time access to data across sources without the need for physical movement, ensuring a consistent and efficient data view.
- **Data governance**
  Tools are included for managing security, privacy, and compliance, which enables the publication of trusted, high-quality analytics-ready data.
- **SAP integration**
  SAP Datasphere natively integrates with SAP solutions such as SAP Analytics Cloud for streamlined reporting and insights.
- **End-to-end data lineage**
  SAP Datasphere visualizes data flow from source to consumption, enhancing transparency, traceability, and trust in data.

7.5.4 SAP Business Technology Platform

SAP Business Technology Platform (SAP BTP) offers a powerful and flexible foundation to support your entire reporting and analytics landscape. It connects easily to both internal and external data sources, ensuring your reports and dashboards are always built on the most up-to-date information. With built-in tools for transforming and cleaning data, SAP BTP helps ensure the data you work with is accurate and ready for analysis.

At the heart of SAP BTP is SAP Analytics Cloud, which allows you to create interactive reports, dashboards, and analytics applications. Whether you need day-to-day operational insights or more advanced predictive analytics, it gives you the flexibility to meet a wide range of business needs. When your reporting needs are more unique or specialized, SAP BTP also gives you the ability to build custom applications. These solutions can be tailored to your exact requirements, offering the specific insights your team needs to make better decisions.

In summary, SAP BTP brings together the tools and capabilities business users need to manage data, build reports, and unlock insights—all in one integrated platform

7.6 Summary

This chapter made you familiar with the reporting capabilities for margin analysis and shared that no new developments for reporting of costing-based profitability analysis have been made nor can be expected. All reporting for costing-based profitability analysis is following the transaction in SAP GUI, which also can be activated as a SAP Fiori app.

The most innovative reporting can be found in the SAP Fiori apps, which is also where we can expect more development and more innovations. Reports will become more intuitive and integrated and will have a more attractive visualization. We can also expect more simulation capabilities and the inclusion of predictive insights in the reporting of the contribution margin.

Appendices

A Changes to the Data Model 331
B Additional Sources of Information 333
C The Author 335

Appendix A
Changes to the Data Model

With the introduction of table ACDOCA, that is, the Universal Journal, the data model in the SAP system changed, as described in Chapter 1.

The *index tables* listed in Table A.1 are no longer actively in use and only available as compatibility views in the system.

Table	Description
BSIS	Accounting: Secondary Index for G/L Accounts
BSAS	Accounting: Secondary Index for G/L Accounts (Cleared Items)
BSID	Accounting: Secondary Index for Customers
BSAD	Accounting: Secondary Index for Customers (Cleared Items)
BSIK	Accounting: Secondary Index for Vendors
BSAK	Accounting: Secondary Index for Vendors (Cleared Items)
FAGLBSIS	Accounting: Secondary Index for G/L Accounts
FAGLBSAS	Accounting: Secondary Index for G/L Accounts (Cleared Items)

Table A.1 Obsolete Index Tables

The *totals tables* listed in Table A.2 are no longer actively in use and only available as compatibility views in the system.

Table	Description
GLT0	G/L Account Master Record Transaction Figures
GLT3	Summary Data Preparations for Consolidation
FAGLFLEXT	General Ledger: Totals
KNC1	Customer Master (Transaction Figures)
LFC1	Vendor Master (Transaction Figures)
KNC3	Customer Master (Special G/L Transaction Figures)

Table A.2 Obsolete Totals Tables

A Changes to the Data Model

Table	Description
LFC3	Vendor Master (Special G/L Transaction Figures)
COSS	CO Object: Cost Totals for Internal Postings
COSP	CO Object: Cost Totals for External Postings

Table A.2 Obsolete Totals Tables (Cont.)

The tables listed in Table A.3 are no longer actively in use in order to avoid data redundancy. They are only available as compatibility views in the system.

Table	Description
COEP	CO Object: Line Items (by Period)
COBK	CO Object: Document Header
ANEP	Asset Line Items
ANEA	Asset Line Items for Proportional Values
ANLP	Asset Periodic Values
CKMI1	Index for Accounting Documents for Material
BSIM	Secondary Index, Documents for Material
MLHD	Material Ledger Document: Header
MLIT	Material Ledger Document: Items
MLCR	Material Ledger Document: Currencies and Values
MLCRF	Material Ledger Document: Field Groups (Currencies)
MLCD	Material Ledger: Summarization Records (from Documents)

Table A.3 More Obsolete Tables

Appendix B
Additional Sources of Information

- Simplification List SAP S/4HANA 2023
  http://s-prs.co/v608901
- SAP S/4HANA Release Highlights 2023 for Private Cloud
  http://s-prs.co/v608902
- SAP S/4HANA Release Highlights 2023 for On-Premise
  http://s-prs.co/v608903
- Highlights of SAP S/4HANA Public Cloud
  http://s-prs.co/v608904
- What's New in SAP S/4HANA Cloud Public Edition 2408 – Product Release Highlights
  http://s-prs.co/v608905
- SAP Business AI: Release Highlights
  http://s-prs.co/v608905
- SAP Fiori Apps Reference Library
  http://s-prs.co/v608907
- Margin Analysis Help Portal
  http://s-prs.co/511711
- Margin Analysis blog
  http://s-prs.co/v608908
- SAP Support Portal
  http://s-prs.co/511709
- SAP Learning Hub
  http://s-prs.co/511710
- Universal Parallel Accounting
  http://s-prs.co/v608909
- Universal Parallel Accounting in SAP S/4HANA
  http://s-prs.co/v608910
- Universal Parallel Accounting Scope Information
  https://me.sap.com/notes/3191636/E
- Activation of Universal Parallel Accounting
  https://me.sap.com/notes/3577390/E

Appendix C
The Author

Kathrin Schmalzing is a partner at one of the "big four" consulting firms. Before she transitioned to consulting, Kathrin first worked in the automotive industry in the controlling department, giving her a clear view of the challenges faced by end users and the daily tasks in controlling. Kathrin has been working with SAP solutions since 2002 and has been involved in multiple end-to-end implementations of SAP S/4HANA solutions at an international level. Her current focus is leading global SAP S/4HANA transformations.

Index

A

ABAP code 325
ABAP RESTful Application Programming
   Model 325
Account-based profitability analysis ... 19
Account determination 148
ACDOCP 286
Actual assessment cycle 196
Actual value flow 209
   margin analysis 127
Administration cost center 188
Allocation cost center 160, 269
Allocation structure 178
   create 178
Analysis of characteristic derivation ... 118
Archiving predictive journal entries . 136
Assessment cost element ... 189, 193, 277
Assign controlling area 52
Automatic account assignment 280
Automatic account determination
   sales and distribution 143
Automatic planning 255

B

Billed quantity 223
Billing 252
Billing data transfer 141
Billing invoice 33
Billing quantity in SKU 224
Billing type 253
Bottom-up predictions 128
Budget availability control 308

C

Characteristic 24, 30, 104
   activate 88
   change 74
   create 73
   custom 23
   detail screen 75
   fixed 23, 71
   from SAP table 73, 76
   type 73
   value 95
   without value maintenance 73, 83
   with own value maintenance 73, 80
   with reference to existing values 73, 86
Characteristic derivation ... 64, 92–93, 95,
   98, 103, 109, 112
   clear 92, 109
   derivation rule 92
   enhancement 92, 112
   move 92
   sequence 113
   table lookup 92
Classic planning functionalities 286
COGS split 35, 151
Company code currency 54, 56
Contribution margin accounting 31
Controlling area 41, 52, 62
Controlling area currency 54, 56
   determine 54
Controlling planning functionalities 288
Core Data Services (CDS) 325
Cost center 252
   types 188
Cost center assessment 188, 246–247,
   252, 277
   receiver 188
Cost component 249
   layout 151
   structure 263
Cost element 188, 277
Costing-based profitability analysis 19,
   30, 39, 60, 211, 230, 246
   basic settings 209
   tables 36
Costing key 255
   assign 259
Costing variant 251, 253, 256, 264
Cost item 263
Cost of goods manufactured 246
Cost of goods sold 151, 240, 246
Cost of goods sold planning 293
Cost of sales accounting 230, 246
Cross-company code cost accounting . 42
Cumulative, confirmed quantity 224
Currency 45
   in the controlling area 56
Currency category 64
Currency type 55
Customer service 146

337

Index

D

Data model 36, 331
Default account assignment 247
Default variance category 162, 272
Define the receiver category 180
Derivation
 analysis 116
 rule 96
Derive costs of goods manufactured 246
Deviations of the input or output side 247
Direct account assignment .. 33, 198, 251, 277
Document currency 54

E

Early close 140
Embedded analytics 295
Exchange rate type 64
Exclusive access to cost estimate 258
Extension ledger 58, 128
External predictive posting 140

F

Field status group 199
Fiscal year variant 41, 46
Fixed amount 264
Foreign currency 64
Formula 31

G

Global company currency 54
Goods issue for delivery 246
Goods issue posting 246
Gross weight 223
Group currency 54

I

Incoming sales order 33, 129, 232, 235
 quantity 223
 report 305, 310
 with date of entry 225
Index tables, obsolete 331
Input form 289
Input price variance 164
Input quantity variance 165

Integration of cost elements in G/L accounts 27
Interfaces to profitability analysis 32
Internal order 178, 182
Internal view on data 20

L

Ledger comparison report 301
Line-item report 312
Local currency 54
Logistics invoice receipt 33
Lot size variance 165

M

Maintain characteristic values 91
Maintain settlement profile 181
Maintain versions 63
Manual planning 255
Manufacturing costs 246
Margin analysis 19
 integration in Universal Journal 28
 reconciliation 127
 value flow 127
Master data
 costing-based profitability analysis 26
 maintenance 250
Material cost estimate 256
Mixed-price variance 165

N

Net weight 224
Number range interval 61–62
Number range maintenance 59
Number status 60

O

Operating concern .. 41, 51–52, 63, 65–66, 74, 79, 104
 assign controlling area 41–42
 define currency 45
 generate retroactively 68
 organizational structure 41
 transport 66, 69
Operating concern currency 46, 64
Order quantity 224, 235
Order settlement 178, 247
Order type 182
Output price variance 165

Index

Overhead cost center 188
Overhead costing 34
Overhead cost planning 294

P

P&L planning 290
Parallel local currency 54
Parallel valuation 46
PA transfer structure 275, 280
   create 273
Period accounting 246
Periodic valuation of actual data 255
Planning 285
   aids 291
   strategy, SAP S/4HANA 286
Postprocessing of billing documents 148
Prediction ledger 58
Predictive accounting 59, 298
   documents 298
   for purchasing 138
   for recurring entries 140
   functionalities 128
   sales order item category 129
Predictive data 127
   costing-based profitability analysis 219
Predictive event 140
Predictive journal entries
   committments 139
   migration 137
Predictive journal entry 130
   document number 132
   posting date 131
Price calculation, automatic 161, 271
Price determination
   sales order 141
Price difference 247, 252
Process order/production order 252
Product costing 34
Production, variance 35
Production cost center 161, 270
Profitability
   segment 145
Profitability analysis
   activate 52
   basic settings 41
   comparison of account-based/costing-based 34
   costing-based 30, 60, 211
   in SAP ERP 20
   postings 21
   structure 30
   types 21
Profitability segment 21
Profit center valuation 46

Q

Quantity 235
Quantity field 35, 235
   transfer 223
Quantity value field 216

R

Realignment 95
Real-time valuation of actual data 255
Receiver 189, 277
Receiver tracing factor 194
Reconciliation process 250
Reconciliation with financial accounting 210, 246
Record type 235
Remaining input variance 165
Remaining variance 164–165
Reporting 295
Reporting of the plan data 292
Report writer 324
Required delivery quantity 224
Resource-usage variance 165
Revenue account determination 146, 240
Revenue planning 289

S

Sales and distribution 223
Sales cost center 188
Sales order
   account assignment 221
   pricing 220
   pricing conditions 220
Sales planning 289
Sales price planning 289
Sales quantitiy planning 290
Sales quantity 235
SAP Analytics Cloud 286, 297
SAP ERP 146
SAP Note 1077293 42
SAP Note 160892 51
SAP S/4HANA
   conditions 234
   sales and distribution 223

Index

SAP tables for characteristics 77–78
Scrap 163
Scrap variance 164
SD account determination 145
Segment 193
Semantic tags 305
Sender 188, 277
Sender cost center 194
Service cost center 188
Service cost centers 160, 269
Settle internal orders 178
Settlement cost element 178
Settlement of internal orders 246
Settlement profile 182, 275
Source 274
Special matter 253
Splitting profile 152
Splitting the costs of goods sold 152
Statistical condition 143, 252
System requirements 14

T

Table
    ACDOCA 54, 186
Table lookup 92
Target cost version, additional 166
Time of valuation 251
Top-down prediction 128
Total production costs 161, 270
Totals tables, obsolete 331
Transaction
    CO88 174
    KANK 166
    KE24 226
    KE27 255
    KE3I 66
    KE4I 234, 246
    KE4M 223, 235
    KE4R 263
    KE4U 253
    KE4W 240
    KEAO 42, 68, 88, 218
    KEA5 .... 73, 212, 216
    KEAT 243
    KEDR 93, 96, 98
    KEI1 273
    KEI2 278
    KEKE 52
    KEKF 225
    KEN1 59
    KEN2 61

    KEPM 255
    KES1 91
    OKB9 280
    OKKP 54
    OKV1 162
    OKV6 165
    OKVG 164
    OKVW 163
    VKOA 145
Transfer of incoming sales orders 225
Transfer price 46
Transport an object class 67
Transport a rule entry 96
Type of variance 162, 272

U

Unit of measure 35
Universal Journal ... 36–38, 45–46, 54, 186
    for plan data 127
User exit 252

V

Valuation 256
    using material cost estimates 254
    using profitability analysis conditions 254
    using transfer prices 254
    using user-defined valuation routines . 254
    with various costing variants 253
Valuation strategy 253, 256
Value field 26, 30, 211, 234, 239, 256
    for amounts 212
    for quantities 212
    list 214
Value flow 32, 151, 248
    analysis 245
    in production 271
Variable
    amount 264
Variance 162, 271
Variance calculation ... 151, 162–163, 165, 172, 246, 252, 272
Variance category 169, 252
Variance document 166
Variance key 162–163, 172
Variances on the input side 162, 164, 172, 272
Variances on the output side 162, 165, 272

Variance variant 165
Version 62
　lock 64
Volume 224

W

Work in process ... 150, 162, 247, 252, 272

- Set up financial accounting and controlling processes in SAP S/4HANA
- Configure your system with step-by-step instructions
- Prepare for testing, go-live, and production support

Stoil Jotev

Configuring SAP S/4HANA Finance

Starting a new SAP S/4HANA Finance implementation? Get it right the first time! From setting up an organizational structure to defining master data, this comprehensive guide to configuring SAP S/4HANA Finance walks you through each key task. Follow step-by-step instructions organized by functional area: general ledger, accounts payable and receivable, margin analysis, group reporting, and more. Customize SAP S/4HANA to meet your FI/CO needs!

744 pages, 3rd edition, pub. 12/2024
E-Book: $84.99 | **Print:** $89.95 | **Bundle:** $99.99

www.sap-press.com/5920

- Perform your daily accounting tasks in SAP S/4HANA Finance
- Follow step-by-step instructions to run general ledger accounting, accounts payable and receivable, asset accounting, and more
- Explore customer project scenarios and use event-based revenue recognition

Jonas Tritschler, Stefan Walz, Reinhard Rupp, Nertila Mucka

Financial Accounting with SAP S/4HANA: Business User Guide

Running accounting in SAP S/4HANA Finance? This is your hands-on guide! Learn how financial accounting works with the Universal Journal and understand key organizational structures. Get step-by-step instructions to perform your finance tasks for the general ledger, accounts payable, accounts receivable, fixed assets, and customer projects and billing. With details on both classic SAP GUI transactions and modern SAP Fiori apps, as well as coverage of new event-based revenue recognition, you'll find everything you need in these pages!

571 pages, 2nd edition, pub. 08/2023
E-Book: $74.99 | **Print:** $79.95 | **Bundle:** $89.99

www.sap-press.com/5698

Interested in reading more?

Please visit our website for all new book
and e-book releases from SAP PRESS.

www.sap-press.com